工业机器人技术应用

· 中文版 ·

主　编：王同庆、商丹丹
副主编：周　京、韩志国、马绪鹏、喻　秀

天津出版传媒集团
天津科学技术出版社

图书在版编目(CIP)数据

工业机器人技术应用:汉英对照/王同庆,商丹丹主编. —— 天津:天津科学技术出版社,2023.9
 ISBN 978-7-5742-0641-0

Ⅰ.①工… Ⅱ.①王…②商… Ⅲ.①工业机器人-汉、英 Ⅳ.①TP242.2

中国版本图书馆 CIP 数据核字(2022)第 201198 号

工业机器人技术应用:汉英对照
GONGYE JIQIREN JISHU YINGYONG:HANYING DUIZHAO

责任编辑:刘 磊

出　　版:	天津出版传媒集团
	天津科学技术出版社

地　　址:天津市西康路 35 号
邮　　编:300051
电　　话:(022)23332695
网　　址:www.tjkjcbs.com.cn
发　　行:新华书店经销
印　　刷:北京盛通印刷股份有限公司

开本 710×1000　1/16　印张 27.25　字数 300 000
2023 年 9 月第 1 版第 1 次印刷
定价:68.00 元

编委会

主　　编　王同庆　商丹丹
副 主 编　周　京　韩志国　马绪鹏　喻　秀
参编人员　王　称　孟　波　林兆娟　王丹阳

前 言

自从 1959 美国第一台工业机器人诞生以来，机器人产业就显示出了强大的生命力。现在以美、日、德为代表的工业发达国家工业机器人已经广泛地应用于制造业诸多领域。工业机器人已经成为先进制造业中不可替代的重要装备，工业机器人的应用也成为衡量一个国家制造水平和科技水平高低的重要标志。工业机器人不仅可以提高生产效率和产品质量，还可以代替人工进行高强度、在恶劣环境和危险环境下作业，与其他的机器、设备相比，机器人具有高度的适应能力，可实现生产过程的全自动化。

随着劳动年龄人口的逐渐减少，国内制造业(规模以上单位)职工平均工资水平也不断提升，制造业企业用工成本处于快速提升阶段。劳动力成本的上升直接影响了制造业企业的健康发展和利润水平，自动化程度较低的劳动密集型生产企业人力成本日益增加，制造业利用廉价劳动力竞争的模式亟待改变。以自动化设备代替人工的需求迫切，在此背景下中国的工业机器人市场进入了快速崛起阶段。

"机器换人"过程中工业机器人对企业的积极影响显著。纵观发达国家工业化进程，自动化设备作为提高生产效率的关键手段，与工业制造技术相结合，在传统装备制造业生产方式的革命性变革进程中起到至关重要的作用。与人工相比，自动化制造设备具有工作效率高、制造精度高等特点，随着企业人工成本的不断上升，智能制造设备在帮助企业优化生产、提高产品质量的同时，还帮助企业降低了运营成本，提升了利润空间。

当前，智能制造是先进制造技术最典型的制造模式之一，是新世纪制造业的发展方向和新一轮工业革命的核心内容，同时也是中国制造走向世界的突破口和主攻方向。发展智能制造，将信息化和工业化深度融合，打造经济发展新动能，对推动产业提质增效、转型升级，实现制造强国战略目标具有决定性的意义。而工业机器人在智能制造系统中扮演着不可或缺的重要角色。

本书根据工业机器人实际生产岗位技能需要，系统地介绍了工业机器人的基本概念、结构组成、安装与调试、基本操作、示教编程、通信设置、离线编程和项目应用等内容,涉及搬运机器人、码垛机器人、焊接机器人和激光切割机器人等主要应用。本书内容新颖、易教易学,注重学生知识全面性的培养。通过学习本书,学生可对工业机器人有一个总体认识和全面了解。

  本书的培养目标贴近工业机器人实际生产工况要求,结合职业院校主流智能制造虚拟制造仿真实训室、智能制造切削单元装调维修实训室、智能制造搬运单元实训室、智能制造综合应用实训室等先进的教学实训设备,开发了配套课程资源和项目案例。为使中国智能制造技术职业教育更好地满足国际需求,特结合我国职业技能等级标准,遵循 EPIP 教学模式,采用中英文双语对照方式出版本书。

  本书适合作为高职高专工业机器人技术、智能制造技术、机电一体化技术和电气自动化技术等相关专业的教材,也可以作为工程技术人员的职业技能培训、自学的参考资料。

目 录

第 1 章　工业机器人概述 ································· 1
 1.1 机器人简介 ····································· 1
 1.2 工业机器人的定义 ····························· 1
 1.3 机器人发展历史 ································ 2
 1.4 机器人发展方向 ································ 6

第 2 章　工业机器人基本组成及技术参数 ············· 9
 2.1 工业机器人基本组成 ··························· 9
 2.2 基本组成部分 ·································· 13
 2.3 工业机器人主要技术参数 ····················· 15
 2.4 工业机器人的分类及典型应用 ················ 19

第 3 章　工业机器人安装与调试 ······················ 25
 3.1 机器人拆包装与固定 ·························· 26
 3.2 机器人调试流程 ······························· 30

第 4 章　工业机器人的基本操作 ······················ 35
 4.1 认识示教器 ···································· 35
 4.2 配置必要的操作环境 ·························· 38
 4.3 单轴运动的手动操纵 ·························· 48
 4.4 机器人的零点校准 ····························· 59
 4.5 机器人的重新启动 ····························· 64
 4.6 机器人的紧急停止与恢复 ····················· 65

第 5 章　RobotStudio 软件应用 ······················· 67
 5.1 RobotStudio 软件介绍 ························· 67
 5.2 RobotStudio 安装 ······························ 69
 5.3 RobotStudio 软件界面 ························· 73

  5.4 创建虚拟工作站 ································ 76
第6章 ABB 机器人的 I/O 通讯 ···························· 83
  6.1 ABB 机器人 I/O 通讯的种类 ···················· 84
  6.2 DSQC651 介绍 ·································· 85
  6.3 DSQC651 配置 ·································· 88
  6.4 Profibus 通讯 ···································· 101
  6.5 示教器可编程按键设定 ························ 107
第7章 ABB 机器人编程基础 ···························· 111
  7.1 机器人系统术语 ································ 111
  7.2 ABB 机器人的程序结构 ······················ 115
  7.3 ABB 机器人程序数据 ························ 117
  7.4 ABB 机器人常用指令 ························ 136
第8章 ABB 机器人程序编写 ···························· 155
  8.1 示教编程 ·· 156
  8.2 离线编程 ·· 171
第9章 机器人保养与常见故障维修 ···················· 187
  9.1 ABB 机器人保养 ······························ 187
  9.2 机器人常见故障维修 ························ 192
  9.3 机器人安全操作规范 ························ 198
第10章 工业机器人实践应用项目 ······················ 201
  10.1 ABB 机器人基本操作 ······················ 201
  10.2 机器人零点转数计数器更新 ·············· 202
  10.3 I/O 信号的配置、监控及快捷键定义 ······ 203
  10.4 工具坐标系的设定与简单示教编程 ······ 204
  10.5 工件坐标系的设定与二维离线编程 ······ 205
  10.6 码垛程序设计与示教 ························ 206

第 1 章 工业机器人概述

1.1 机器人简介

什么是机器人？机器人的英文为 Robot，是一种能够半自主或全自主工作的智能机器。Robot 一词最早出现在 1920 年捷克斯洛伐克作家卡雷尔·凯佩克（Karel Capek）的科幻小说《罗萨姆万能机器人公司》中，是根据捷克语 Robota（苦役、苦工、奴隶）和波兰语 Robotnik（工人）两个词创造出来的，即代表进行苦力劳动，完成繁重的工作。在该剧中，机器人按照其主人的命令默默地工作，没有感觉和感情，以呆板的方式从事繁重的劳动。该剧预告了机器人的发展对人类社会的悲剧性影响，引起了人们的广泛关注，被当成了"机器人"一词的起源。但直到 20 世纪中叶，"机器人"才作为专业术语被加以引用。

随着现代科技的不断发展，机器人这一概念逐步演变成现实。在现代工业的发展过程中，机器人逐渐融合了机械、电子、运动、动力、控制、传感检测、计算技术等多门学科，成为现代科技发展中极为重要的组成部分。

1.2 工业机器人的定义

目前，虽然机器人面世已有几十年历史，但仍然没有一个统一的定义。其原因之一就是机器人还在不断的发展，新的机型、新的功能不断涌现。国际上对机器人的定义有以下几种。

（1）美国机器人协会（The Robot Institute of America，RIA，1979 成立）：一种用于移动各种材料、零件、工具或专用装置的，通过程序动作来执行种种任务的，并具有编程能力的多功能操作机。

(2)日本工业机器人协会(Japan Industrial Robot Association,JIRA)：工业机器人是一种带有存储器件和末端操作器的通用机械，它能够通过自动化的动作替代人类劳动。

(3)中国工业机器人研究学者：机器人是一种自动化的机器，所不同的是这种机器具备一些与人或生物相似的能力，如感知能力、规划能力、动作能力和协同能力，是一种具有高度灵活性的自动化机器。

(4)国际标准化组织(International Organization for Standardization, ISO)：工业机器人是一种仿生的、具有自动控制能力的、可重复编程的多功能、多自由度的操作机械。

定义虽不同，但有一定的共性，即：工业机器人是由仿生机械结构、电机、减速机和控制系统组成的，用于从事工业生产，能够自动执行工作指令的机械装置。它可以接受人类指挥，也可以按照预先编排的程序运行，现代工业机器人还可以根据人工智能技术制定的原则和纲领行动。

一般情况下，工业机器人应该具有以下四个特征。

A.特定的机械结构。

B.从事各种工作的通用性能。

C.具有感知、学习、计算、决策等不同程度的智能。

D.相对独立性。

1.3 机器人发展历史

1.3.1 古代机器人

早在机器人一词出现之前，人们就一直幻想和追求制造一种像人一样的机器，以替代人类完成各种工作，现今有记载的如：春秋时期鲁班设计制作的木鸟能在空中飞行；三国时期诸葛亮制作的木牛流马，是一种通过仿造牛的行走来为军队运送粮草的古代木制运输工具；1662年，日本人竹田近江利用钟表技术发明了能进行表演的自动机器玩偶；18世纪法国天才技师发明制作了机器鸭。这些都是人类在机器人从梦想到现实的漫长探索。

1.3.2 近代机器人

20世纪40年代后期，机器人的研究与发明得到了更多人的关心或关注。

20 世纪 50 年代以后，美国橡树岭国家实验室开始研究能搬运核原料的遥控操纵机械手，如图 1-1 所示。这是一种主从型控制系统，系统中加入力反馈，可使操作者获知施加力的大小，主、从机械手之间由防护墙隔开，操作者可通过观察窗或闭路电视对从机械手操作机进行有效的监视。主、从机械手系统的出现为机器人的产生及近代机器人的设计与制造做了铺垫。

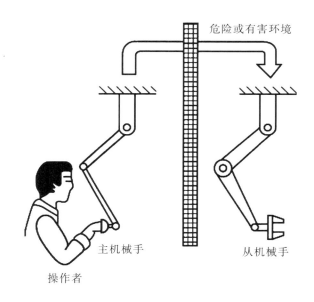

图 1-1　遥控操纵机械手

"主机械手"由操作者引导做动作，而"从机械手"则尽可能准确地模仿主机械手的动作。后来，人们将力的反馈加入"机械耦合主从机械手"的动作中，使操作者能够感觉到环境物体给机械手的作用力。20 世纪 50 年代中期，机械手中的机械耦合装置被液压装置取代，如通用电气公司的"巧手人"机器人和通用制造厂的"怪物"I 型机器人。

1.3.3 现代机器人

机器人是以控制论和信息论为指导，综合了机械学、微电子技术、计算机、传感技术等学科的成果而诞生的。因此，随着这些学科，特别是计算机技术的发展，现代机器人的出现已经是顺理成章的事了。

1954 年，美国人乔治·德沃尔(G.C.Devol)制造出世界上第一台可编程的机器人。当年,他提出了"通用重复操作机器人"的方案,并在 1961 年获得了专利。1958 年,被誉为"工业机器人之父"的约瑟夫·英格伯格(Joseph F.Engel Berger)创建了世界上第一个机器人公司——Unimation(意为"Universal Automation")公司。1959 年,德沃尔与美国发明家约瑟夫·英格伯格联手制造出全球第一台工业机器人 Unimate,如图 1-2 所示。这是一台用于压铸的五轴液压驱动机器人,手臂的控制由一台计算机完成。它采用了分离式固体数控元件,并装有存储信息的磁鼓,具有能够完成 180 个工作步骤的记忆。

 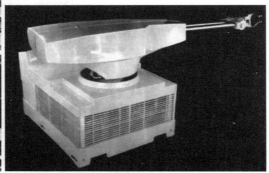

图 1-2 Unimate 机器人

与此同时,另一家美国公司——AMF 公司也开始研制工业机器人,即 Versatran(意为"Versatile Transfer")机器人,如图 1-3 所示。它采用液压驱动,主要用于机器之间的物料运输。该机器人的手臂可以绕底座回转,沿垂直方向升降,也可以沿半径方向伸缩。一般认为,Unimate 和 Versatran 机器人是世界上最早的工业机器人。

1978 年,德国徕斯(Reis)机器人公司开发了首款拥有独立控制系统的六轴机器人 RE15,如图 1-4 所示。1979 年,Unimation 公司推出了 PUMA 系列工业机器人,它们是由全电动驱动、关节式结构、多 CPU 二级微机控制,采用 VAL 专用语言,可配置视觉和触觉的力觉感受器且技术较为先进的机器人。同年,日本山梨大学的牧野洋研制成具有平面关节的 SCARA 型机器人。

20 世纪 70 年代出现了更多的机器人商品,并在工业生产中逐步获得推广应用。随着计算机科学技术、控制技术和人工智能的发展,机器人的研究开发,

图 1-3　Unimate 机器人

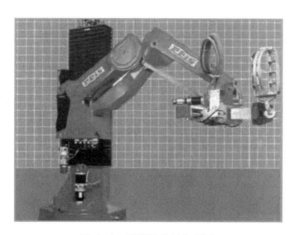

图 1-4　RE15 六轴机器人

无论在水平和规模上,都得到了迅速发展。据国外统计,到 1980 年全世界约有 2 万余台机器人在工业中应用。

　　近十几年来,世界上的机器人无论是从技术水平上,还是从已装备的数量上,其优势集中在以日、欧、美为代表的少数几个发达的工业化国家和地区。2019 年和 2020 年全球工业机器人装机量分别为 275 万台和 315 万台,并以每年超过 10% 的速度快速增长,工业机器人的四大家族 ABB、KUKA、FANUC、YASKAWA 更是占有全球机器人市场 75% 以上。目前在工业领域中以六轴机器人应用最为广泛,带有六个关节的工业机器人与人类的手臂极为相似,六个自由度适合于几乎任何轨迹或角度的工作。其次应用较多的是三轴、四轴和五轴双臂的工业机器人，具体轴数的选择通常取决于实际应用的灵活性、经济成本和速度等。

1.4 机器人发展方向

1.4.1 工业机器人正在向智能化、模块化、仿生化方向发展

1.4.1.1 智能化

工业机器人智能化，即让机器人有感觉、有知觉，能够迅速、准确地检测及判断各种复杂的信息。随着执行与控制、自主学习与智能发育等技术进步，机器人将从预编程、示教再现控制、直接控制、遥操作等被操纵作业模式，逐渐向自主学习、自主作业方向发展。

人工智能是关于人造物的智能行为，它包括知觉、推理、学习、交流和在复杂环境中的行为，人工智能的长期目标是发明出可以像人类一样能更好地完成以上行为的机器。

1.4.1.2 模块化

通过标准化模块组装制造工业机器人将成为趋势。当前，各个国家都在研究、开发和发展组合式机器人，这种机器人将由标准化的伺服电机、传感器、手臂、手腕与机身等工业机器人组件标准化组合件拼装制成。

研究新型机器人结构是未来发展趋势。新型微动机器人结构可以提升工业机器人的作业精度、改善工业机器人的作业环境。研制新型工业机器人结构将成为适应工作强度高、环境复杂作业的需求。

1.4.1.3 仿生化

直至今天，大多数机器人才被认为属于生物纲目之一。工具型机器人保持了机器人应有的基本元素，如装备了爪形机械、抓具和轮子，但不管怎么看，它都像是台机器。相比之下，类人形机器人则最大限度地与创造它们的人类相似，它们的运动臂上有自己的双手，下肢有真正的脚，有人类一样的脸，最重要的是还能像人一样活动。有的类人机器人，不仅在外观上像人，甚至自己还会

去"想",会思考,有智慧,可以代替人类从事更加复杂的工作。

1.4.2 工业机器人总体发展趋势

1.4.2.1 技术发展趋势

在技术发展方面,工业机器人正向结构轻量化、智能化、模块化和系统化的方向发展。未来主要的发展趋势如下。

A.机器人结构的模块化和可重构化。

B.控制技术的高性能化、网络化。

C.控制软件架构的开放化、高级语言化。

D.伺服驱动技术的高集成度和一体化。

E.多传感器融合技术的集成化和智能化。

F.人机交互界简单化、协同化。

1.4.2.2 应用发展趋势

自工业机器人诞生以来,汽车行业一直是其应用的主要场合。2014 年北美机器人工业协会在年度报告中指出,截至 2013 年底,汽车行业仍然是北美机器人最大的应用市场,但其在电子/电气行业、金属加工行业、化工行业、食品等行业的出货量却增速迅猛。由此可见,未来工业机器人的应用依托汽车产业,并迅速向各行业延伸。对于机器人行业来讲,这是一个非常积极的信号。

1.4.2.3 产业发展趋势

国际机器人联合会(IFR)最新发布的《2020 年世界机器人报告》显示,全球范围内,2019 年工业机器人年度安装量排名前五的市场分别是中国、日本、美国、韩国和德国。中国是目前世界上最大,增长最快的机器人市场。相信在不久的将来,工业机器人的时代将很快来临,并将在智能制造领域掀起一场变革。

第 2 章 工业机器人基本组成及技术参数

2.1 工业机器人基本组成

工业机器人通常由执行机构、驱动系统、控制系统和传感系统四部分组成，如 2-1 所示。

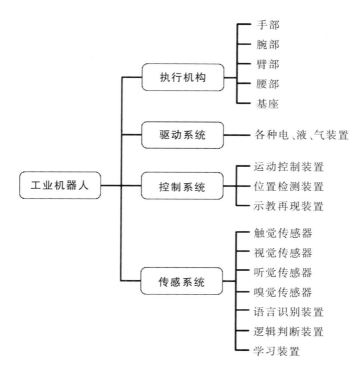

图 2-1 工业机器人组成示意图

2.1.1 执行机构

执行机构是机器人赖以完成工作任务的实体，通常由一系列连杆、关节或其他形式的运动副组成。从功能的角度可分为手部、腕部、臂部、腰部和机座，每个部位相连处由伺服电机驱动以实现转动，俗称关节，如图2-2所示。

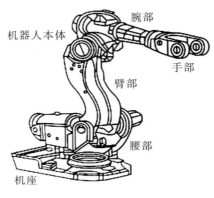

图 2-2 工业机器人

2.1.2 驱动系统

工业机器人的驱动系统是向执行系统各部件提供动力的装置，包括驱动器和传动机构两部分，它们通常与执行机构连成一体。驱动器通常有电动、液压、气动装置以及把它们结合起来应用的综合系统。常用的传动机构有谐波传动、螺旋传动、链传动、带传动以及各种齿轮传动等。

2.1.2.1 气力驱动

气力驱动系统通常由气缸、气阀、气罐和空压机等组成，以压缩空气来驱动执行机构进行工作。其优点是空气来源方便、动作迅速、结构简单、造价低、维修方便、防火防爆、漏气对环境无影响，缺点是操作力小、体积大，又由于空气的压缩性大、速度不易控制、响应慢、动作不平稳、有冲击。因起源压力一般只有60MPa左右，故此类机器人适宜对抓举力要求较小的场合。

2.1.2.2 液压驱动

液压驱动系统通常由液动机(各种油缸、油马达)、伺服阀、油泵、油箱等组成，以压缩机油来驱动执行机构进行工作，其特点是操作力大、体积小、传动平稳且动作灵敏、耐冲击、耐振动、防爆性好。相对于气力驱动，液压驱动的机器人具有大得多的抓举能力，可高达上百千克或者更高。但液压驱动系统对密封的要求较高，且不宜在高温或低温的场合工作。

2.1.2.3 电力驱动

电力驱动是利用电动机产生的力或力矩直接或经过减速机构驱动机器人，以获得所需的位置、速度和加速度。电力驱动具有电源易取得，无环境污染，响应快，驱动力较大，信号检测、传输、处理方便，可采用多种灵活的控制方案，运动精度高，成本低，驱动效率高等优点，是目前机器人使用最多的一种驱动方法。驱动电动机一般采用步进电动机、直流伺服电动机以及交流伺服电动机。

2.1.3 控制系统

工业机器人的位置控制方式有点位控制和连续路径控制两种。其中，点位控制方式只关心机器人末端执行器的起点和终点位置，而不关心这两点之间的运动轨迹，这种控制方式可完成无障碍条件下的点焊、上下料、搬运等操作。连续路径控制方式不仅要求机器人以一定的精度达到目标点，而且对移动轨迹也有一定的精度要求，如机器人喷漆、弧焊等操作。实质上这种控制方式是以点位控制方式为基础，在每两点之间用满足精度要求的位置轨迹插补算法实现轨迹连续化的。

2.1.4 传感系统

传感系统是机器人的重要组成部分，按其采集信息的位置，一般可分为内部和外部两类传感器。内部传感器是完成机器人运动控制所必需的传感器，如位置、速度传感器等，用于采集机器人内部信息，是构成机器人不可缺少的基本元件。外部传感器可检测机器人所处环境、外部物体状态或机器人与外部物体的关系。常用的外部传感器有力觉传感器、触觉传感器、接近觉传感器、视觉传感器等。机器人传感器的分类如表 2-1 所示。

表 2-1　机器人传感器分类

传感器分类	用途	机器人的精确控制
内部传感器	监测的信息	位置、角度、速度、加速度、姿态、方向等
	所用传感器	微动开关、光电开关、差动变压器、编码器、电位计、旋转变压器、测速发电机、加速度计、陀螺、倾角传感器、力(或力矩)传感器等
外部传感器	用途	了解工件、环境或机器人在环境中的状态,对工件的灵活、有效的操作
	检测的信息	工件和环境:形状、位置、范围、质量、姿态、运动、速度等机器人与环境:位置、速度、加速度、姿态等
		对工件的操作:非接触(间隔、位置、姿态等)、接触(障碍检测、碰撞检测等)、触觉(接触觉、压觉、滑觉)、夹持力等
	所用传感器	视觉传感器、光学测距传感器、超声测距传威器、触觉传感器、电容传感器、电磁感应传感器、限位传感器、压敏导电橡胶、弹性体加应变片等

　　传统的工业机器人仅采用内部传感器,用于对机器人运动、位置及姿态进行精确控制。使用外部传感器,使得机器人对外部环境具有一定程度的适应能力,从而表现出一定程度的智能。

　　一般驱动器和控制器等其他控制部件放到同一个柜子里,称之为控制柜。进行人机交互的操作单元指的就是示教器,由硬件和软件组成,其本身就是一个完整的计算机。因此工业机器人的系统组成我们可以简单地理解为三大件,如图 2-3 所示:工业机器人本体 A、控制柜 B(包含主计算机控制模块、轴计算

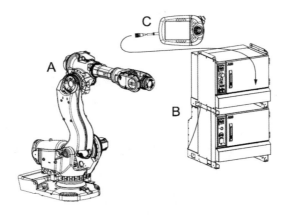

图 2-3　工业机器人基本组成

机板、轴伺服驱动、连接伺服轴编码器的SMB测量板、I/O板等)、机器人示教器C(手持式操作员装置)。

2.2 基本组成部分

2.2.1 机器人本体

工业机器人的机械主体,是完成各种作业的执行机构。一般包含：
- 机械臂；
- 驱动及传动装置；
- 各种内外部传感器。

工作时通过末端夹具也称末端执行器用于实现机器人对工作目标的夹取、搬用等动作。

2.2.1.1 机械臂

关节型机器人的机械臂是由若干个机械关节连接在一起的集合体，主要有以下组成部分。

A.基座：是构成机器人的支撑部分,内部安装有机器人的执行机构和驱动装置。

B.腰部：是连接机器人机座和大臂的中间支撑部分。工作时,腰部可以通过关节1在机座上转动。

C.臂部：6关节机器人的臂部一般由大臂和小臂构成,大臂通过关节2与腰部相连,小臂通过肘关节3与大臂相连。工作时,大、小臂各自通过关节电机转动,实现移动或转动。

D.手腕：连接小臂和末端执行器的部分,主要用于改变末端执行器的空间位姿,联合机器人的所有关节实现机器人预期的动作和状态。

2.2.1.2 本体驱动及传动装置

机器人运动时,每个关节必须有驱动装置和传动机构完成,如图2-4所示为交流伺服电机和精密减速机组成的典型驱动。

驱动装置是向机器人各机械臂提供动力和运动的装置。不同类型的机器人,驱动采用的动力源不同,驱动系统的传动方式也不同。驱动系统的传动方

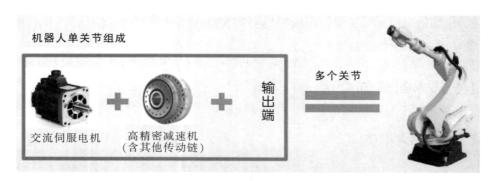

图 2-4　典型的驱动或运动关节组成

式主要有四种:液压式、气压式、电气式和机械式。

2.2.1.3 本体所用传感器

本体所用传感器为内部传感器,包括微动开关、光电开关、差动变压器、编码器、电位计、旋转变压器、加速度计、陀螺、倾角传感器、力矩传感器等。

2.2.2 控制系统

控制系统是构成工业机器人的神经中枢,由主计算机单元、驱动单元、轴计算机单元、输入输出接口和一些专用电路等构成。工作时,根据编写的指令以及传感信息控制机器人本体完成一定的动作或路径,主要用于处理机器人工作的全部信息,其正面内部结构如图 2-5 所示。

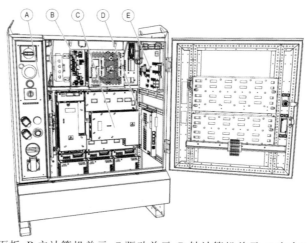

A.控制面板;B.主计算机单元;C.驱动单元;D.轴计算机单元;E.安全面板

图 2-5　机器人控制柜

2.2.3 示教器

示教器是人机交互的一个接口,也称示教盒或示教编程器,主要由液晶屏和可供触摸的操作按键组成,如图 2-6 所示。操作时由控制者手持设备,通过按键将需要控制的全部信息通过与控制器连接的电缆送入控制柜中的存储器中,实现对机器人的控制。示教器是机器人控制系统的重要组成部分,操作者可以通过示教器进行手动示教,控制机器人到达不同位姿,并记录各个位姿点坐标,也可以利用机器人语言进行在线编程,实现程序回放,让机器人按编写好的程序完成轨迹运动。

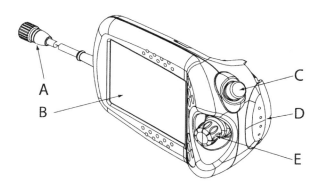

A.连接器;B.触摸屏;C.紧急停止按钮;D.使动装置;E.控制杆

图 2-6 示教器

2.3 工业机器人主要技术参数

工业机器人技术参数如下。
- 自由度
- 分辨率
- 定位精度
- 重复定位精度
- 作业范围
- 运动速度
- 承载能力

各技术参数详解如下。

(1) 自由度

自由度是指机器人所具有的独立坐标轴运动的数目，不包括末端执行器的开合自由度。一般情况下机器人的一个自由度对应一个关节，所以自由度与关节的概念是相等的。自由度是表示机器人动作灵活程度的参数，自由度越多就越灵活，但结构也越复杂，控制难度越大，所以机器人的自由度要根据其用途设计，一般在 3~6 个之间，如图 2-7 所示为六自由度机器人，每个轴的旋转方向如箭头所示。

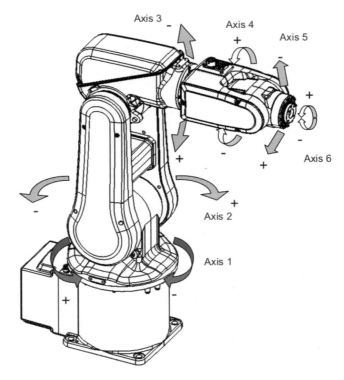

图 2-7　六自由度机器人

(2) 分辨率

分辨率是指机器人每个关节所能实现的最小移动距离或最小转动角度。工业机器人的分辨率分编程分辨率和控制分辨率两种。

- 编程分辨率：指控制程序中可以设定的最小距离，又称基准分辨率。当机器人某关节电机转动 0.1°,机器人关节端点移动直线距离为 0.01mm,其基准分辨率即为 0.01mm。

- **控制分辨率**:是系统位置反馈回路所能检测到的最小位移,即与机器人关节电机同轴安装的编码盘发出单个脉冲电机转过的角度。

(3)定位精度

定位精度是指机器人末端执行器的实际位置与目标位置之间的偏差,由机械误差、控制算法与系统分辨率等部分组成。典型的工业机器人定位精度一般在±0.02~±5mm 范围。

(4)重复定位精度

重复定位精度是指在同一环境、同一条件、同一目标动作、同一命令之下,机器人连续重复运动若干次时,其位置的分散情况,是关于精度的统计数据。因重复定位精度不受工作载荷变化的影响,故通常用重复定位精度这一指标作为衡量示教—再现工业机器人水平的重要指标。

(5)作业范围

作业范围是指机器人运动时手臂末端或手腕中心所能到达的位置点的集合,如图 2-8 所示为 ABB IRB120 机器人工作范围。作业范围的大小不仅与机器人各连杆的尺寸有关,而且与机器人的总体结构形式有关。作业范围的形状和大小是十分重要的,机器人在执行某作业时可能会因存在手部不能到达的盲区而不能完成任务。因此在选择机器人执行任务时,一定要合理选择符合当前作业范围的机器人。

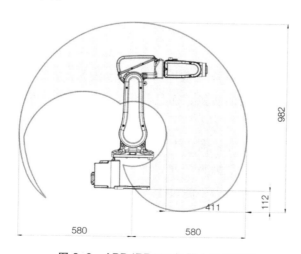

图 2-8 ABB IRB120 机器人工作范围

图 2-9 所示为直角坐标机器人的工作范围示意图,该结构具有容易通过计算机控制实现、容易达到、高精度等优点;缺点是占地面积大、妨碍工作、运动速度低、密封性不好。

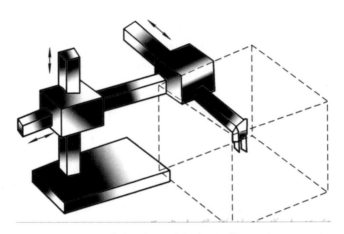

图 2-9　直角坐标机器人的工作范围示意图

(6)运动速度

运动速度影响机器人的工作效率和运动周期,它与机器人所提取的重力和位置精度均有密切的关系。运动速度高,机器人所承受的动载荷增大,必将承受着加减速时较大的惯性力,影响机器人的工作平稳性和位置精度。就目前的技术水平而言,通用机器人的最大直线运动速度大多在1000mm/s以下,最大回转速度一般不超过120°/s。

(7)承载能力

承载能力是指机器人在作业范围内的任何位姿上所能承受的最大质量。承载能力不仅取决于负载的质量,而且与机器人运行的速度和加速度的大小和方向有关。

根据承载能力不同工业机器人大致分为:

微型机器人——承载能力为 1 N 以下;

小型机器人——承载能力不超过 105N;

中型机器人——承载能力为 105~106N;

大型机器人——承载能力为 106~107N;

重型机器人——承载能力为 107N 以上。

2.4 工业机器人的分类及典型应用

工业机器人的种类很多,其功能、特征、驱动方式、应用场合等参数不尽相同。目前,国际上还没有形成机器人的统一划分标准。一般可从机器人的结构特征、控制方式、驱动方式、应用领域等几个方面进行分类。

(1)按结构特征可划
- 直角坐标机器人
- 柱面坐标机器人
- 极坐标机器人
- 多关节型机器人
- 并联机器人

①直角坐标机器人:直角坐标机器人是指在工业应用中,能够实现自动控制的、可重复编程的、在空间上具有相互垂直关系的三个独立自由度的多用途机器人,如图 2-10 所示。机器人在空间坐标系中有三个相互垂直的移动关节 X/Y/Z,每个关节都可以在独立的方向移动。特点是直线运动、其控制简单。缺点是灵活性较差,自身占据空间较大。

直角坐标机器人可以非常方便的用于各种自动化生产线中,可以完成诸

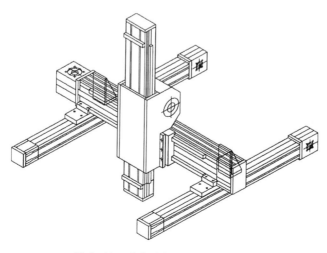

图 2-10 直角坐标机器人示意图

如焊接、搬运、上下料、包装、码垛、检测、探伤、分类、装配、贴标、喷码、打码、喷涂、目标跟随、排爆等一系列工作。

②柱面坐标机器人：

柱面坐标机器人是指轴能够形成圆柱坐标系的机器人。其结构主要由一个旋转机座形成的转动关节和垂直、水平移动的两个移动关节构成，如图2-11。柱面坐标机器人末端执行器的姿态由(x,R,θ)决定。

柱面坐标机器人具有空间结构小，工作范围大，末端执行器速度高、控制简单、运动灵活等优点。缺点在于工作时，必须有沿R轴线前后方向的移动空间，空间利用率低。目前，圆柱坐标机器人主要用于重物的装卸、搬用等工作作业。著名的Versatran机器人就是一种典型的柱面坐标机器人。

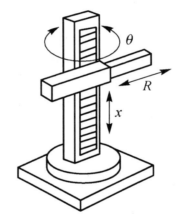

图2-11 柱面坐标机器人示意图

③极坐标机器人：

极坐标型机器人一般由两个回转关节和一个移动关节构成，其轴线按极坐标配置，R为移动坐标，β是手臂在铅垂面内的摆动角，θ是绕手臂支撑底座垂直的转动角。这种机器人运动所形成的轨迹表面是半球面，所以又称为球坐标型机器人，如图2-12所示。

极坐标机器人同样具有占用空间小、操作灵活且范围大的优点，但运动学模型较复杂，难以控制。

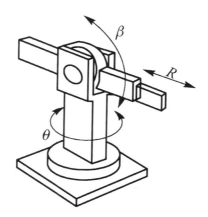

图 2-12　极坐标机器人运动示意图

(4) 多关节型机器人

多关节型机器人,也称关节手臂机器人或关节机械手臂,是当今工业领域中应用最为广泛的一种机器人。多关节型机器人按照关节的构型又可分为垂直多关节型(如图 2-13 所示)和水平多关节型机器人。

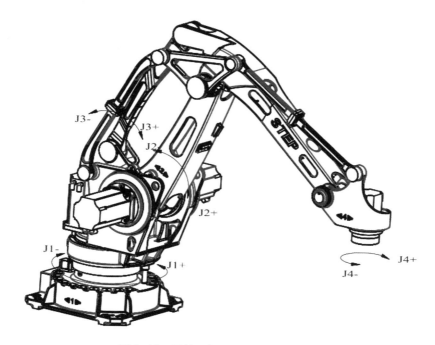

图 2-13　四轴码垛机器人运动示意图

多关节机器人同样占用空间小,操作灵活且范围大,但运动学模型较复杂,难以控制。机器人由多个旋转和摆动关节组成,其结构紧凑、工作空间大、动作接近人类,工作时能绕过机座周围的一些障碍物,对装配、喷涂、焊接等多种作业都有良好的适应性,且适合电机驱动,关节密封,防尘比较容易。目前,瑞士 ABB、德国 KUKA、日本安川、国内的一些公司都在推出这类产品。

水平多关节机器人,也称 SCARA (Selective Copmliance Assembly Robot Arm 选择顺应性装配机器手臂),是一种水平多关节机器人,具有四个轴和四个运动自由度:绕 A、B、C 轴的旋转和 Z 向的平动,可实现 X、Y、Z 方向的平动自由度和绕 Z 轴的转动自由度,如图 2-14 所示。

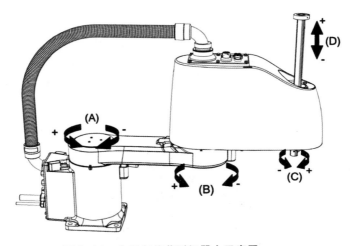

图 2-14 水平多关节型机器人示意图

⑤并联机器人

并联机器人是近年来发展起来的一种由固定机座和具有若干自由度的末端执行器以不少于两条独立运动链连接形成的新型机器人,如图 2-15 所示为 Delta 并联机器人。

并联机器人具有以下特点:

- 无累积误差,精度较高;
- 驱动装置可置于定平台上或接近定平台的位置,运动部分重量轻,速度高,动态响应好;
- 结构紧凑,刚度高,承载能力大;

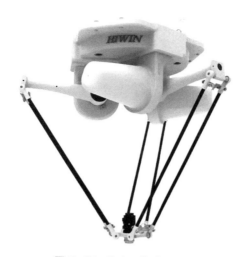

图 2-15 Delta 并联机器人

- 具有较好的各向同性；
- 工作空间较小。

(2) 从机器人的控制方式、驱动方式、应用领域等几个方面进行分类

① 按控制方式可划分

- 伺服控制机器人
- 非伺服控制机器人

② 按驱动方式可划分

- 液压驱动机器人
- 气压驱动机器人
- 电力驱动机器人
- 新型驱动机器人

③ 按结构特征或应用领域可划分

- 焊接机器人(如图 2-16 所示)
- 搬运机器人(如图 2-17 所示)
- 装配机器人(如图 2-18 所示)
- 喷涂机器人(如图 2-19 所示)

图 2-16　焊接机器人

图 2-17　搬运机器人

图 2-18　装配机器人

图 2-19　喷涂机器人

第 3 章 工业机器人安装与调试

工业机器人是精密机电设备,其运输和安装有着特别的要求,每一个品牌的工业机器人都有自己的安装与连接指导手册,但大同小异。工业机器人一般的安装流程如图 3-1 所示,每个步骤操作时要认真参阅指导手册相关部分。

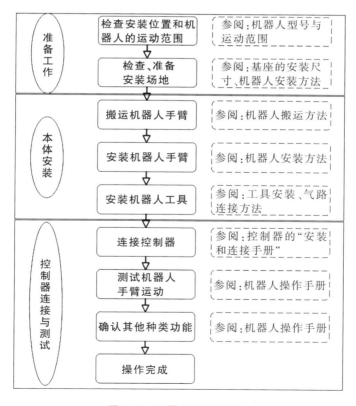

图 3-1 机器人安装流程

3.1 机器人拆包装与固定

工业机器人都是按照标准流程打包好才发送到客户现场的,拆包与安装的工作具体如下:

A.机器人到达现场后,箭头朝上摆放,如图 3-2 所示。第一时间检查外观是否有破损,是否有进水等等异常情况。如果有问题需要马上联系厂家或供应商进行处理。

图 3-2 机器人到货状态

B.首先使用工具将木箱顶盖拆除,然后依次取出配件包装纸箱,如图 3-3 所示,以免拆除木箱侧板时损坏配件。

C.清点工业机器人装箱物品:以 ABB 机器人 IRB1200 为例,主要包括机器人本体、示教器、线缆配件及控制柜等 4 大部分。具体应参照厂家的装箱单进行物品清点,随机文档还包括:SMB 电池安全说明、出厂清单

图 3-3 机器人拆箱示意图

和机器人操作说明书。

D.使用扳手拆掉将固定机器人底座的四颗螺丝,如图 3-4 所示。

图 3-4　机器人底盘

E.如图 3-5 所示,将机器人本体转运至安装位置时,应安装并使用厂家提供的专用提升配件,否则可能导致机器人工作部件受损;严禁在转运过程中或机器人底盘未紧固的状态下改变机器人本体出厂时的标准姿态,因为移动手臂会使机器人重心移位进而可能导致机器人翻倒。

图 3-5 机器人本体转运标准姿态

F.准备好动力电缆、SMB 电缆和示教器电缆,仔细阅读 ABB IRC5 Compact 控制柜的接线图,如图 3-6 所示。

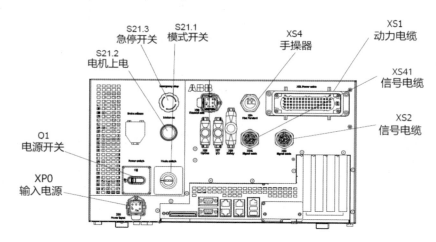

图 3-6 IRC5 Compact 控制柜的接线图

G.将动力电缆标注为 XP1 的插头接入控制柜 XS1 接口,将动力电缆标为 R1.MP 的插头接入机器人本体底座的插头上。

H.将 SMB 电缆(直头)接头插入到控制柜 XS2 接口,将 SMB 电缆(弯头)接头插入到机器人本体底座 SMB 端口。

I.将示教器电缆(红色)的接头插入到控制柜 XS4 接口。

J.根据 ABB IRB1200 的供电参数,准备电源线并且制作控制柜端的接头。控制柜端电源接头定义说明如图 3-7 所示。

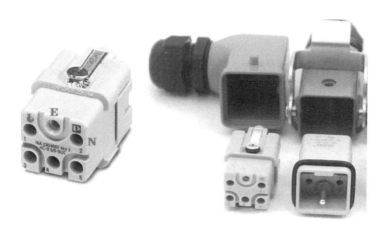

图 3-7 控制柜输入电源接头

K.再次确认以上所有接线正确无误后,将电源接头按插入控制柜 XP0 端口并锁紧,至此机器人本体与控制器的基本连接已经完成,效果如图 3-8 所示。如图 3-8 所示。

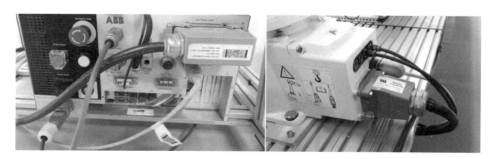

图 3-8 机器人完成接线效果图

L.将示教器支架安装到合适的位置(要求便于示教器的取放,且不易被操作人员碰到),然后将示教器放好。

M.确认所有连接正确后,即可打开电源开关进行试运行,只要将控制柜上的总电源旋钮从【OFF】扭转到【ON】即可开机;将机器人控制柜上的总将电源旋钮逆时针从【ON】扭转到【OFF】即可切断系统电源。

3.2 机器人调试流程

3.2.1 设置语言

第一次通电开机时,默认的语言是英语,需要更改为汉语,方便操作。

A.依次点击示教器屏幕左上角"ABB"菜单图标——"Control Panel"选项,如图3-9所示。

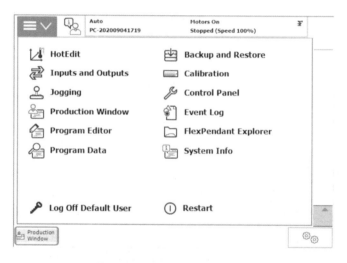

图3-9 "Control Panel"选项

B.点击"language"选项,选择"Chinese",点击"OK"确认,如图3-10所示。

第 3 章　工业机器人安装与调试

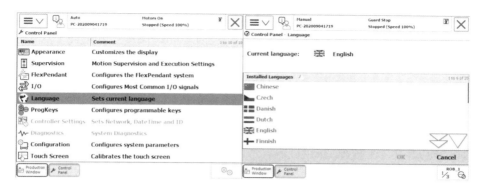

图 3-10　语言设置

C.点击弹出的提示对话框"Yes"按钮,机器人重启,语言设置就完成了,如图 3-11 所示。

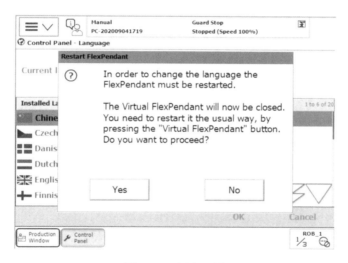

图 3-11　语言设置

3.2.2 备份与恢复

定期对机器人进行备份,是保证机器人正常工作的良好习惯。备份文件可以放在机器人内部的存储器上,也可以备份到 U 盘上。

备份文件包含运行程序和系统配置参数等内容。当机器人系统出错,可以通过备份文件快速的恢复备份前的状态。平时在程序更改之前,一定要做好备份。需要注意的是,备份恢复数据是具有唯一性的,不能将一台机器人的备份数据恢复到另一个机器人上。

3.2.3 零点检查

ABB 机器人每个关节轴都有一个机械原点的位置,出厂时已对各个关节轴的机械原点进行了校准,并在机器人本体六个关节轴上有标记位。该零点作为各关节轴运动的基准,因此机器人安装之后需对机器人进行回零操作,检查零点是否丢失,新出厂机器人只要零点没有丢失则无需校准可以直接使用。

3.2.4 系统 I/O 配置及接线

ABB 机器人提供了丰富的 I/O 通信接口,可以轻松地实现与周围设备的通信。下面以 ABB 最常用的标准 I/O 版 DSQC651 为例做详细讲解。ABB 的标准 I/O 板提供的常用信号处理有:数字输入 DI、数字输出 DO、模拟输入 AO、以及输送链跟踪等。

3.2.5 检查信号

点击"ABB"图标进入系统菜单,点击"输入输出",对 I/O 信号进行监控。0 表示没信号,1 表示有信号。检查配置的信号与实际信号是否对应正确。例如 doGripperA 和 doGripperB 分别代表机器人两个夹具气缸,点击其中的一个再点击 0 或 1 即可更改夹具状态,强制进行夹具松开、闭合的操作,查看电磁阀接线是否错误。

3.2.6 导入程序

点击"ABB"图标,选择"程序编辑器",点击"模块",在模块界面,选择"加载模块",从存放程序模块的路径加载你所需要加载的程序模块,模块通常存放在 PROGMOD 文件夹下,可以用记事本打开。

3.2.7 工件坐标系设定

3.2.7.1 机器人的坐标系统

设定工件坐标是进行示教的前提,所有的示教点都在必须在对应的工件坐标中建立。如果在wobj0上建立示教点,如果机器人在搬动以后就必须重教所以的点。如果是在对应的工件左边上示校的话就可以只修改一下工件坐标,而无需重教所有的点。

3.2.7.2 正确设定工件坐标的必要性

不准确的工件坐标,使机器人在工件对象上的X/Y方向移动变得困难。

3.2.7.3 设定坐标

在示教器创建一个wobj1项目,定义工件坐标,验证工件坐标准确度。

3.2.8 较基准点

点击"ABB"图标进入主系统界面,点击"程序数据",点击"robtarget",选择需要修改的工位,pPick1和pPkck2分别对应1工位和2工位的基础位置,点击工位后出现下拉菜单,选择编辑,在编辑选项中选择"修改位置"。修改完成后,机器人就会自动记录下新的位置。

3.2.9 调整参数

对机器人运行程序参数进行精准调整,例如码垛机器人需微调所搬运纸箱的长宽高、纸箱数量、抓取位置和码垛摆放位置等,确保机器人所有动作稳定可靠。

3.2.10 手动调试

调试好的程序在自动运行前必须经过手动运行检验,在手动运行调试过程中,一旦发现问题,松开控制器机器人就会立即停止。

3.2.11 自动运行

手动运行调试完成无误后才能自动运行,且前期一定要设置为较低的速度。自动运行过程中按暂停键可以停止机器人运行,此时电机还是开启的,按下启动键,机器人会继续运行。

调试过程注意事项:

- 送电前一定要确保电源接线正确、牢靠,并且有效接地;
- 示教器要断电插拔;

- 断电后再重启,一定要等待完全关机后约 1 分钟再启动,防止数据丢失;
- 对程序内容等进行修改后,一定要复查一遍;
- 对程序参数等修改后,一定要先手动低速运行程序,再自动运行;
- 改动前,要及时做好备份。

第 4 章 工业机器人的基本操作

4.1 认识示教器

机器人示教器是工业机器人重要的一个外设,它是我们操作者与工业机器人进行"对话"交流的一个手持输出设备。通过机器人示教器可以手动操作机器人,也可以对工业机器人进行手动编程、调试程序、修改机器人系统参数等,因此示教器在工业机器人设备中占有举足轻重的地位。ABB IRB120 示教器如图 4-1 所示,图中各部分分别为:

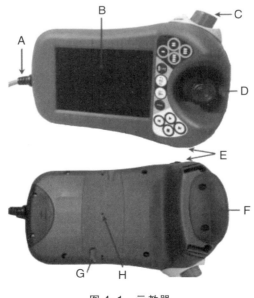

图 4-1 示教器

- A 链接电缆
- B 触摸屏
- C 急停开关
- D 手动操纵摇杆
- E USB 端口
- F 使能器按钮
- G 触摸屏用笔
- H 示教器复位按钮

正确使用示教器时,右手持示教笔进行屏幕和按钮的操作,左手放在使能键上,这样就能舒适地将示教器放在左手上,然后用右手进行屏幕和按钮的操作了。ABB 的这款示教器是按照人体工程学进行设计的,同时适合左手操作(如图 4-2 所示),只要在屏幕中进行切换就能适应左手的操作习惯。

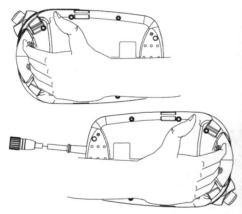

图 4-2 示教器握法示意图

系统开机之后示教器界面如图 4-3 所示,图纸标记各区域说明如下。
- 1——主菜单:显示机器人各个功能主菜单界面。
- 2——操作员窗口:机器人与操作员交互界面显示当前状态信息。
- 3——状态栏:显示机器人当前状态如工作模式、电机状态、报警信息等。
- 4——关闭按钮:关闭当前窗口按钮。
- 5——快速设置菜单:快速设置机器人功能界面如速度、运行模式、增

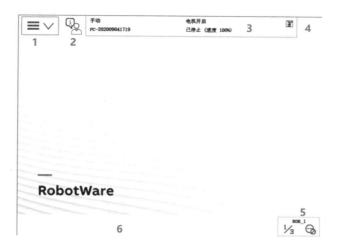

图 4-3 示教器开机界面

量等。

• 6——任务栏:当前打开界面的任务列表,最多支持打开 6 个界面。

示教器屏幕左上角红色"ABB"按钮为主菜单,单击后展开菜单如图 4-4 所示,各选项功能说明详见表 4-1。

图 4-4 示教器主菜单

表 4-1　主菜单选项功能表

选项名称	功能说明
HotEdit	用于程序模块下轨迹点位置的补偿设置
输入输出	用于设置及查看 I/O 信号
手动操纵	用于查看并配置手动操作属性
自动生产窗口	在自动模式下,可直接调试程序并运行
程序编辑器	用于对机器人进行编程调试
程序数据	用于查看并配置变量数据
备份与恢复	用于对系统数据进行备份和恢复
校准	用于对机器人机械零点进行校准
控制面板	用于对系统参数进行配置
事件日志	查看系统出现的各种提示信息
资源管理器	用于对系统资源、备份文件等进行管理
系统信息	用于查看系统控制器属性以及硬件和软件信息
注销	退出当前用户权限
重新启动	重新启动示教器

4.2 配置必要的操作环境

4.2.1 设定机器人系统的时间

为了方便进行文件的管理和故障的查阅与管理,在进行各种操作之前要将机器人系统的时间设定为本地时区的时间,具体操作如下。

点击示教器屏幕左上角的 ABB 菜单按钮,依次选择"控制面板"—"日期和时间"选项,进入日期和时间设定界面,如图 4-5 所示。日期和时间修改完成后,单击"确定"按键完成设置。

4.2.2 正确使用使能器按钮

使能器按钮是工业机器人为保证操作人员人身安全而设置的。只有在按下使能器按钮,并保持在"点击开启"的状态,才可对机器人进行手动的操作与程序的调试。当发生危险时,人会本能地将使能器按钮松开或按紧,机器人则会马上停下来,保证安全。

A.使能器按钮位于示教器手动操作摇杆的右侧,操作者应用左手的四个

第 4 章 工业机器人的基本操作

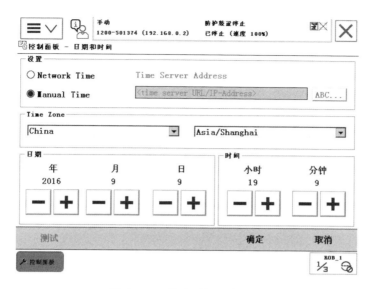

图 4-5 日期和时间设定界面

手指进行操作,如图 4-6 所示。

B.ABB 机器人使能器是一种特殊类型的装置,又称为三位使动装置,必须将按钮按下一半才能激活,在完全按下和完全弹出位置是无法操作机器人的。在手动状态下轻按使能器至中间挡位,机器人处于电机开启状态,完全按下去以后,机器人就会处于防护装置停止状态,如图 4-7 所示。

4.2.3 查看 ABB 机器人常用信息与事件日志

操作者可以通过示教器画面上的状态栏进行 ABB 机器人常用信息及事

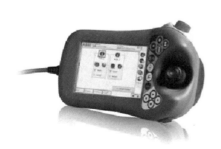 

图 4-6 使能器按钮操作示意图

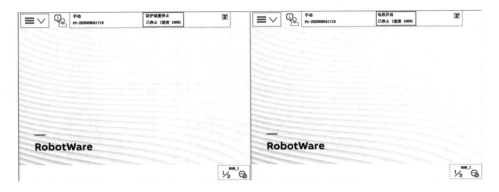

图 4-7 使能器按下机器人状态变化图

件日志的查看。如图 4-3 标记 3 所示,状态栏包含信息如下。

- 机器人的状态(手动、全速手动和自动)。
- 机器人的系统信息。
- 机器人的电机状态。
- 机器人的程序运行状态。
- 当前机器人或外轴的使用状态。

单击画面中的状态栏就可以查看机器人的事件日志,如图 4-8 所示。

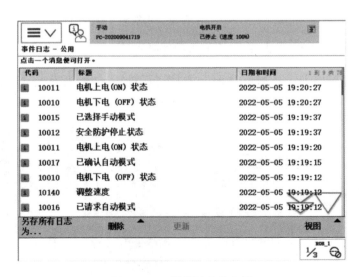

图 4-8 示教器屏幕状态栏

4.2.4 ABB 机器人数据的备份与恢复

定期对 ABB 机器人的数据进行备份，是保证 ABB 机器人正常工作的良好习惯。ABB 机器人数据备份的对象是所有正在系统内存运行的 RAPID 程序和系统参数。当机器人系统出现错乱或者重新安装新系统以后，可以通过备份快速地把机器人恢复到备份时的状态。

4.2.4.1 ABB 机器人数据的备份

A.点击示教器屏幕左上角的 ABB 菜单按钮，选择"备份与恢复"选项，如图 4-9 所示。

图 4-9 备份与恢复选项

B.单击"备份当前系统…"选项，如图 4-10 所示。

C.进入备份路径设置界面如图 4-11 所示，单击"ABC…"按钮，进行存放备份数据目录名称的设定；单击"…"按钮，选择备份存放的位置（机器人硬盘或 USB 存储设备）；最后单击"备份"按钮进行备份的操作。

4.2.4.2 ABB 机器人数据的恢复

A.单击图 4-10 界面中的"恢复系统…"按钮进入图 4-12 所示的恢复路径选择界面，单击"…"选择备份存放的目录，最后单击"恢复"。

图 4-10 备份当前系统选项

图 4-11 备份路径设置

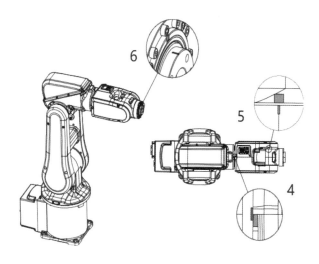

图 4-12 恢复路径选择

B.系统弹出恢复确认对话框如图 4-13 所示,单击"是"之前需再次确认恢复文件夹选择是否正确。在进行恢复时,要注意:备份的数据具有唯一性,不能将一台机器人的备份恢复到另一台机器人中去,否则会造成系统故障。但是,也常会将程序和 I/O 的定义做成通用的,方便批量生产时使用。这时,可以通过分别导入程序和 EIO 文件来解决实际的需要。

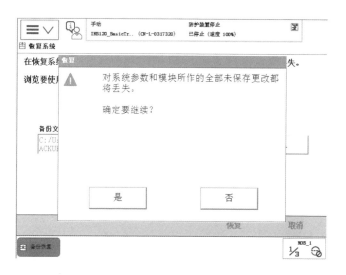

图 4-13 恢复确认对话框

4.2.4.3 什么情况下需要为机器人进行备份？

A.新机器第一次上电后.

B.在做任何修改之前.

C.在完成修改之后.

D.如果机器人重要,定期备份(如每周 1 次)。

E.最好在 U 盘也做备份.

F.太旧的备份定期删除,腾出硬盘空间。

4.2.5 导入程序

导入程序可以将离线编程生成的程序或其他机器人的程序快速导入当前机器人,省去手动编程的时间,提高工业机器人的利用率。

A.点击示教器屏幕左上角的 ABB 菜单按钮,选择"程序编辑器"选项,如图 4-14 所示。

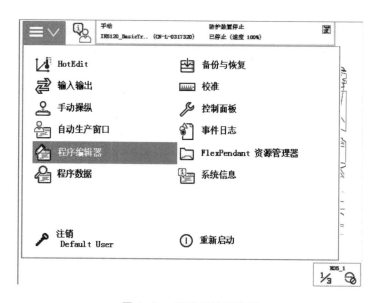

图 4-14 程序编辑器选项

B.单击"模块"标签,如图 4-15 所示。

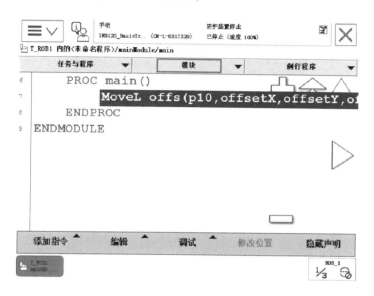

图 4-15　程序编辑器模块选项

C.打开"文件"菜单,点击"加载模块…",从"备份目录/RAPID"路径下加载所需要的程序模块,如图 4-16 所示。

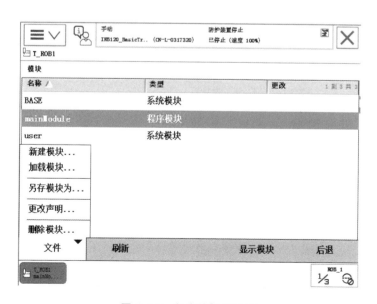

图 4-16　程序编辑器选项

4.2.6 导入 EIO 文件

A.点击示教器屏幕左上角的 ABB 菜单按钮,选择"控制面板"选项,选择"配置"选项,如图 4-17 所示。

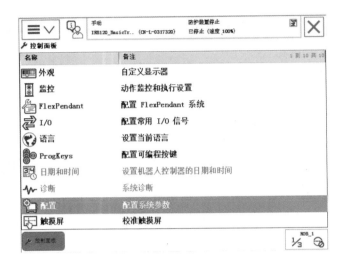

图 4-17 配置系统参数选项

B.打开"文件"菜单,单击"加载参数",如图 4-18 所示。

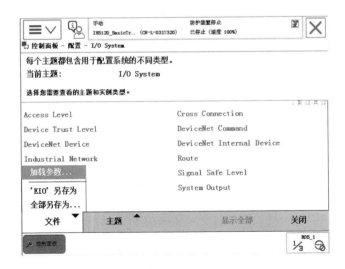

图 4-18 加载参数选项

C.选择"删除现有参数后加载",单击"加载…",如图 4-19 所示。

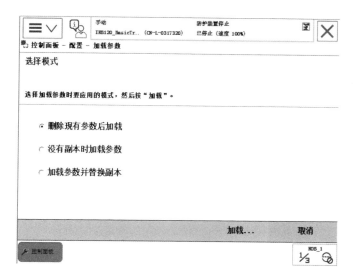

图 4-19　加载参数选项

D.在"备份目录/SYSPAR"路径下找到 EIO.cfg 文件,单击"确定",如图 4-20 所示。

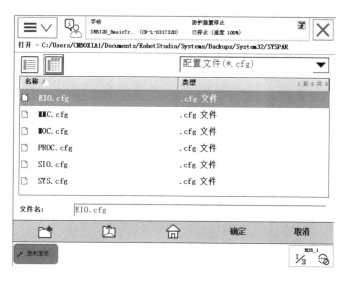

图 4-20　选择 EIO 文件

E.系统弹出如图 4-21 所示重新启动画面,单击"是"重启机器人。重启完成后新导入 EIO 生效。

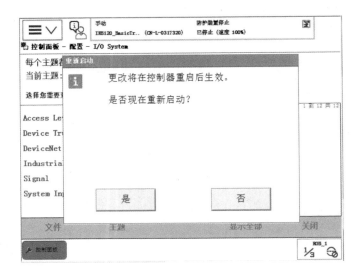

图 4-21　EIO 导入完成画面

4.3　单轴运动的手动操纵

手动操纵机器人运动一共有三种模式:单轴运动、线性运动和重定位运动。下面介绍如何手动操纵机器人进行三种运动。

4.3.1　单轴运动

一般地,ABB 机器人是由六个伺服电机分别驱动机器人的六个关节轴(图3-9),那么每次手动操纵一个关节轴的运动,就称为单轴运动。以下就是手动操纵单轴运动的方法。

为了方便进行文件的管理和故障的查阅与管理,在进行各种操作之前要将机器人系统的时间设定为本地时区的时间,具体操作如下。

A.选择"手动操纵",如图 4-22 所示。

第 4 章 工业机器人的基本操作

图 4-22 手动操纵选项

B.单击"动作模式",如图 4-23 所示。

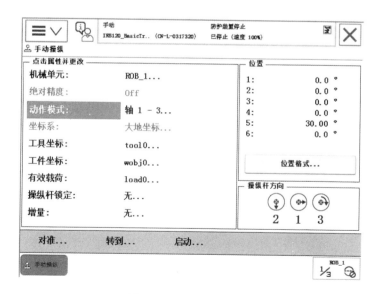

图 4-23 运动模式选项

C.进入动作模式选择界面,单击"动作模式",如图 4-24 所示。其中"轴 1—3"选项可以操纵轴 1—3,而操作轴 4—6 则需选中"轴 4—6"选项。

图 4-24 运动模式选项

D.返回手动操纵界面,按下使能按钮,进入"电机开启"状态,如图 4-25 所示。当前状态下使用操纵杆即可控制机器人轴 1—3 做旋转运动,操纵杆的上下、左右、旋转分别对应轴 2、轴 1、轴 3,右下角的箭头代表各轴运动正方向。

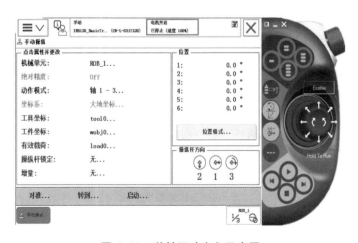

图 4-25 单轴运动方向示意图

E.操纵杆的使用技巧:可以将机器人的操纵杆比作汽车的节气门,操纵杆的操纵幅度是与机器人的运动速度相关的。操纵幅度较小,则机器人运动速度较慢;操纵幅度较大,则机器人运动速度较快。所以在操作时,尽量以小幅度操纵使机器人慢慢运动来开始手动操纵学习。

4.3.2 线性运动

机器人的线性运动是指安装在机器人第六轴法兰盘上工具的 TCP 在空间中做线性运动。以下就是手动操纵步骤。

A.点击示教器屏幕左上角的 ABB 菜单按钮,依次单击"手动操纵"——"动作模式",选择"线性"选项,如图 4-26 所示。

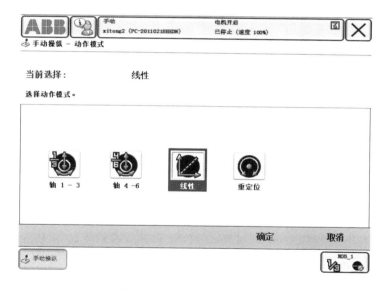

图 4-26 线性运动模式选项

B.单击"工具坐标"选项,如图 4-27 所示。

C.选中对应的工具"tool1"单击"工具坐标"选项,如图 4-28 所示。系统默认的工具坐标为机器人法兰坐标,用户建立的工具坐标以法兰坐标为基准。

D.返回手动操纵界面,按下使能按钮,进入"电机开启"状态,如图 4-29 所示。操纵杆的上下、左右、旋转分别对应 X、Y、Z,右下角的箭头代表各轴运动正方向。

图 4-27 工具坐标选项

图 4-28 工具选项

第 4 章 工业机器人的基本操作

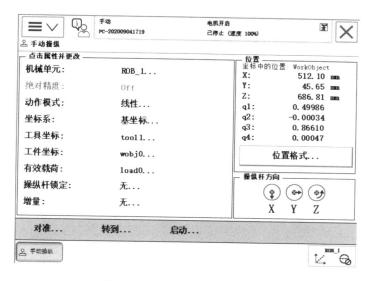

图 4-29 线性运动方向示意图

E.当前状态下操纵示教器上的操纵杆,机器人工具的 TCP 点在机器人参考坐标系中做线性运动,如图 4-30 所示。

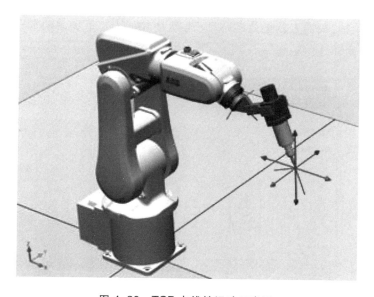

图 4-30 TCP 点线性运动示意图

F.增量模式的使用：如果对使用操纵杆通过位移幅度来控制机器人运动的速度不熟练，那么可以使用"增量"模式来控制机器人的运动，在"手动操作"界面选择"增量选项"，如图4-31所示。

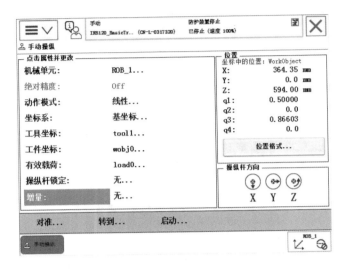

图4-31　增量模式选项

在增量模式下，操纵杆每位移一次，机器人就移动一步。如果操纵杆持续一秒或数秒钟，机器人就会持续移动(速率为10步/s)，每步运动量在增量模式中选择，如图4-32所示。

图4-32　增量大小选项

4.3.3 重定位运动

机器人的重定位运动是指机器人第六轴法兰盘上的工具 TCP 点在空间中绕着坐标轴旋转的运动，也可以理解为机器人绕着工具 TCP 点做姿态调整的运动。以下就是手动操纵重定位运动的方法。

A.点击示教器屏幕左上角的 ABB 菜单按钮，依次单击"手动操纵"—"动作模式"，选择"重定位"选项，如图 4-33 所示。

图 4-33　重定位选项

B.返回手动操纵界面，选择"坐标系"选项，如图 4-34 所示。

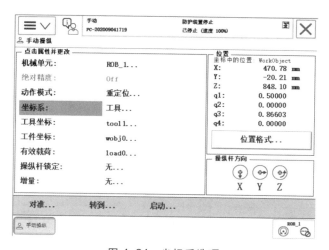

图 4-34　坐标系选项

C.选择"工具"选项并确认,如图4-35所示。(其中"基坐标"为机器人本体坐标,大地坐标为工作车间内某一固定参考坐标)

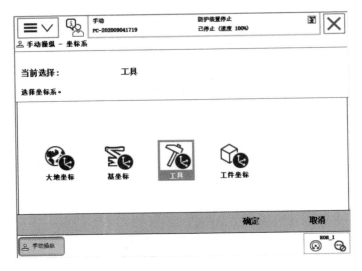

图4-35 坐标系选择界面

D.确认当前工具坐标为目标TCP点所在工具坐标后,按下使能按钮,进入"电机开启"状态,如图4-36所示。

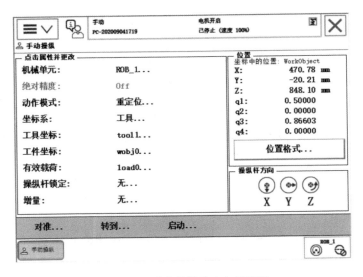

图4-36 重定位运动方向示意图

E.当前状态下操纵示教器上的操纵杆,机器人绕着工具TCP点做姿态调整的运动,如图4-37所示。

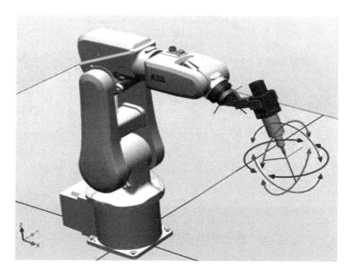

图4-37 TCP点线性运动示意图

4.3.4 手动操纵快捷方式

为便于手动操纵的快捷设置,ABB示教器上提供了4个快捷键按钮和快捷菜单。

A.示教器上快捷键按钮共4个,如图4-38所示,分别对应如下功能。

- A 机器人/外轴的切换
- B 线性运动/重定位运动的切换
- C 关节运动轴1-3/轴4-6的切换
- D 增量开/关

B.示教器屏幕右下角为快捷菜单按钮,单击后弹出快捷菜单,单件菜单第一项"手动操纵"展开菜单,如图4-39所示。

C.单击"显示详情"显示所有菜单选项如图4-40所示,功能分别如下。

图4-38 TCP点线性运动示意图

图 4-39 示教器快捷菜单

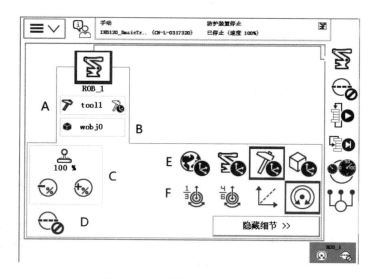

图 4-40 手动操纵快捷菜单选项

- A 选择当前使用的工具数据
- B 选择当前使用的工件坐标
- C 操纵杆速率
- D 增量开/关

- E 坐标系选择
- F 动作模式选择

D.单击快捷菜单第二项"增量模式"按钮,弹出增量选项如图 4-41 所示。自定义增量值的方法:选择"用户模块",然后单击"显示值"就可以进行增量值的自定义了。

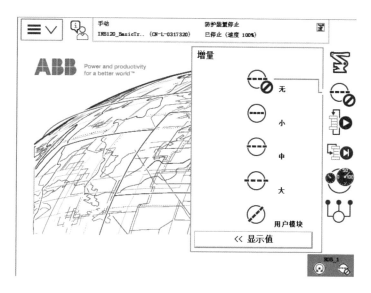

图 4-41 增量模式快捷选项

4.4 机器人的零点校准

机器人零点信息是指机器人各轴处于机械零点时各轴电机编码器对应的读数,零点信息存储在本体串行测量板上,数据需要供电才能保存,掉电后数据会丢失。遇到下列情况时,需要对机械原点的位置进行转数计数器的更新操作。

A.更新伺服电动机转数计数器电池后。

B.当转数计数器发生故障,修复后。

C.转数计数器与测量板之间断开过以后。

D.断电后,机器人关节轴发生了移动。

E.当系统警报提示"10036 转数计数器未更新"。

4.4.1 机器人零点位置

不同品牌、型号机器人的零点标记位置会有所不同,ABB IRB1200 工业机器人正确回零后的姿态如图 4-42 所示,放大图 4、5、6 分别是轴 4、轴 5、轴 6 的零点标志位置。

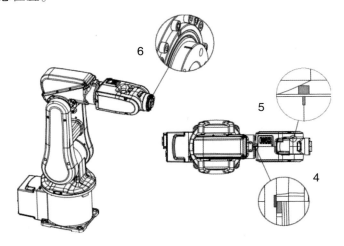

图 4-42　IRB1200 机器人零位示意图

图 4-43 所示为轴 1—3 的零点标记位置。

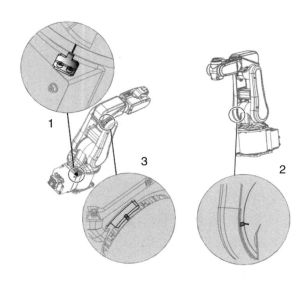

图 4-43　零点标记位置示意图

4.4.2 转数计数器更新

转数计数器更新的操作需使用手动操纵,使机器人各关节轴运动到机械原点刻度的位置,一般顺序是:4—5—6—1—2—3。各个型号的机器人机械原点刻度位置会有所不同,可参考机器人使用说明书和机器人本体进行对照。机器人各轴零点标记位置调整好之后即可操纵手操器进行转速计数器更新,详细步骤如下。

A.点击示教器屏幕左上角的 ABB 菜单按钮,单击"校准"选项,进入校准机械单元选择界面,如图 4-44 所示。

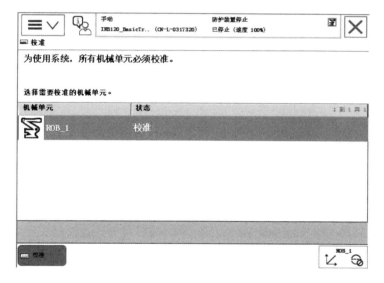

图 4-44 校准单元选择界面

B.单击"ROB_1",进入校准设计界面并选择"校准参数"选项,如图 4-45 所示。

C.单击"编辑电极校准偏移..."选项,进入电机偏移值设置界面,如图 4-46 所示。机器人本体侧面有记录电机校准偏移值的铭牌,仔细检查示教器中显示的数值,修改至与机器人本体上的标签数值一致,单击"确认"按钮。

D.系统提示重新启动机器人控制器,单击"是"重启,如图 4-47 所示。

E.重启后重新进入"校准"设置界面"转数计数器"选项,单击"更新转速计数器..."选项,如图 4-48 所示。

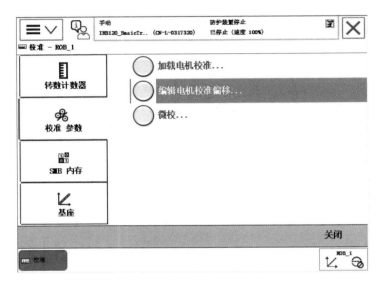

图 4-45 校准单元选择界面

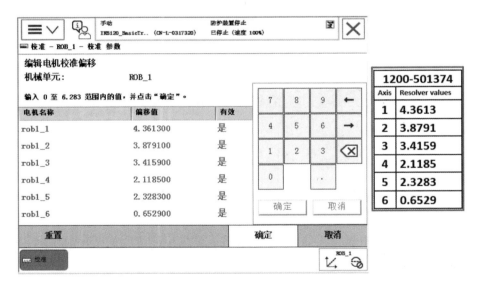

图 4-46 电机偏移值设置界面

图 4-47 重启确认对话框

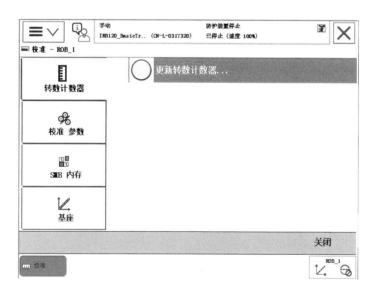

图 4-48 更新转数计数器选项

F.在弹出的确认对话框中单击"是"按钮,进入"机械单元"确认界面,按系统提示单击"确认",进入如图4-49所示"轴"选择界面。全选六个轴并单击"更新"按钮,等待更新完成后各轴会显示"转数计数器已更新",说明转数计数器更新已完成。(注意:如果机器人由于安装位置的关系,无法六个轴同时到达机械原点刻度位置,则可以逐一选取对应关节轴进行转数计数器更新)

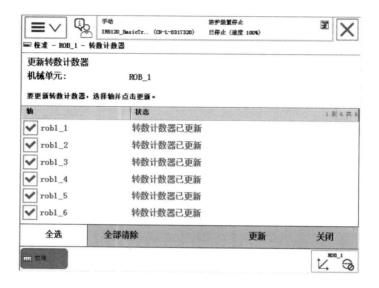

图4-49 转数计数器更新界面

4.5 机器人的重新启动

ABB机器人系统可以长时间进行工作,无须定期重新启动运行。但出现以下情况时需要重新启动机器人系统。

A.安装了新的硬件。

B.更改了机器人系统配置参数。

C.出现系统故障(SYSFAIL)。

D.RAPID程序出现程序故障。

单击示教器屏幕左上角的ABB菜单按钮,单击"重新启动"选项,重新启动选项界面,如图4-50所示。

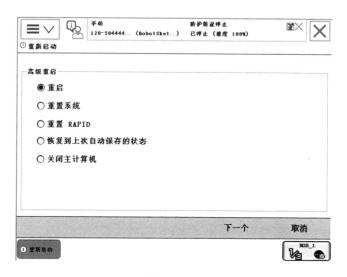

图 4-50 转数计数器更新界面

重新启动的类型包括重启、重置系统、重置 RAPID、恢复到上次自动保存的状态和关闭主计算机。各选项详细说明如下。

• 重启:使用当前的设置重新启动当前系统。

• 重置系统:重启并将丢弃当前的系统参数设置和 RAPID 程序,将会使用原始的系统安装设置。

• 重置 RAPID:重启并将丢弃当前的 RAPID 程序和数据,但会保留系统参数设置。

• 恢复到上次自动保存的状态:重启并尝试回到上一次自动保存的系统状态。一般在从系统崩溃中恢复时使用。

• 关闭主计算机:关闭机器人控制系统,应在控制器 UPS 故障时使用。

4.6 机器人的紧急停止与恢复

在机器人的手动操纵过程中,操作者因为操作不熟练容易引起碰撞或者发生其他突发状况时,会选择按下紧急停止按钮(控制柜和示教器上各有一个),启动机器人安全保护机制,停止机器人。紧急停止优先于任何其他 ABB 机器人控制操作,它会断开 ABB 机器人电机的驱动电源,停止所有运转部件,并切断由 ABB 机器人系统控制、存在潜在危险的功能部件的电源。

紧急停止状态意味着断开了 ABB 机器人中除手动制动闸释放电路外的所有电源，机器人无法通过示教器进行动作，在操作其运动前，必须执行恢复步骤。即重置紧急停止按钮，并按下"电机开启"按钮，才能返回至正常操作。可在 ABB 机器人系统中进行相应的配置，使紧急停止处于以下状态。

A.非受控停止：断开 ABB 机器人电机的电源，立刻停止 ABB 机器人运行。

B.受控停止：停止 ABB 机器人运行，但为了保留 ABB 机器人路径，不断开 ABB 机器人电机电源。操作完成后，电源断开。

系统默认设置是非受控停止。但是，受控停止可最小化 ABB 机器人额外的、不必要的磨损，以及使 ABB 机器人返回生产状态的必要操作。紧急停止不应用作正常程序停止，因为这会造成 ABB 机器人额外的、不必要的磨损。

第 5 章 RobotStudio 软件应用

5.1 RobotStudio 软件介绍

RobotStudio 是一款由 ABB 集团研发生产的计算机仿真软件,它的主要功能就是帮助用户在电脑上模拟构建一个机器人系统,并对它的程序和功能进行调试,从而找到最合适的设计和工作方案,确保用户能够顺利完成现实的机器人系统部署。RobotStudio 适用于机器人系统的各个周期阶段,是大家设计部署机器人的必备软件,其特色功能如下。

5.1.1 CAD 导入

RobotStudio 可方便地导入各种主流 CAD 格式的数据,包括 IGES、STEP、VRML、VDAFS、ACIS 及 CATIA 等。机器人程序员可依据这些精确的数据编制精度更高的机器人程序,从而提高产品质量。

5.1.2 AutoPath

RobotStudio 中最能节省时间的功能之一。该功能通过使用待加工零件的 CAD 模型,仅在数分钟之内便可自动生成跟踪加工曲线所需要的机器人位置(路径),而这项任务以往通常需要数小时甚至数天。

5.1.3 程序编辑器

程序编辑器(Program Maker)可生成机器人程序,使用户能够在 Windows 环境中离线开发或维护机器人程序,可显著缩短编程时间、改进程序结构。

5.1.4 路径优化

如果程序包含接近奇异点的机器人动作，RobotStudio 可自动检测出来并发出报警，从而防止机器人在实际运行中发生这种现象。仿真监视器是一种用于机器人运动优化的可视工具，红色线条显示可改进之处，以使机器人按照最有效方式运行。可以对 TCP 速度、加速度、奇异点或轴线等进行优化，缩短周期时间。

5.1.5 Autoreach

Autoreach 可自动进行可到达性分析，使用十分方便，用户可通过该功能任意移动机器人或工件，直到所有位置均可到达，在数分钟之内便可完成工作单元平面布置验证和优化。

5.1.6 虚拟示教台

虚拟示教台是实际示教台的图形显示，其核心技术是 VirtualRobot。从本质上讲，所有可以在实际示教台上进行的工作都可以在虚拟示教台(Quick-Teach™)上完成，因而是一种非常出色的教学和培训工具。

5.1.7 事件表

一种用于验证程序的结构与逻辑的理想工具。程序执行期间，可通过该工具直接观察工作单元的 I/O 状态。可将 I/O 连接到仿真事件，实现工位内机器人及所有设备的仿真。该功能是一种十分理想的调试工具。

5.1.8 碰撞检测

碰撞检测功能可避免设备碰撞造成的严重损失。选定检测对象后，RobotStudio 可自动监测并显示程序执行时这些对象是否会发生碰撞。

5.1.9 Visual Basic for Applications

可采用 VBA 改进和扩充 RobotStudio 功能，根据用户具体需要开发功能强大的外接插件、宏，或定制用户界面。

5.1.10 PowerPac's

ABB 协同合作伙伴采用 VBA 进行了一系列基于 RobotStudio 的应用开

发,使 RobotStudio 能够更好地适用于弧焊、弯板机管理、点焊、CalibWare(绝对精度)、叶片研磨以及 BendWizard(弯板机管理)等应用。

5.1.11 接上传和下载

整个机器人程序无需任何转换便可直接下载到实际机器人系统。

5.2 RobotStudio 安装

A.获得 RobotStudio 软件的渠道有以下两种。

a)登录 www.robotstudio.com 网站,在页面内找到图 5-1 所示点击进入个人信息填写页面,认真填写个人邮箱完成提交,收到邮件链接打开后可以开始下载。安装试用许可证可获得 30 天的 RobotStudio 试用权限。

图 5-1　RobotStudio 下载按钮

b)登录 https://new.abb.com/products/robotics/robotstudio/downloads 网站链接,页面内有最新 RobotStudio 的版本信息,如图 5-2 所示,点开后操作与 a)相同。

图 5-2　RobotStudio 下载按钮

B.解压缩安装包后双击setup.exe文件开始安装,如图5-3所示。

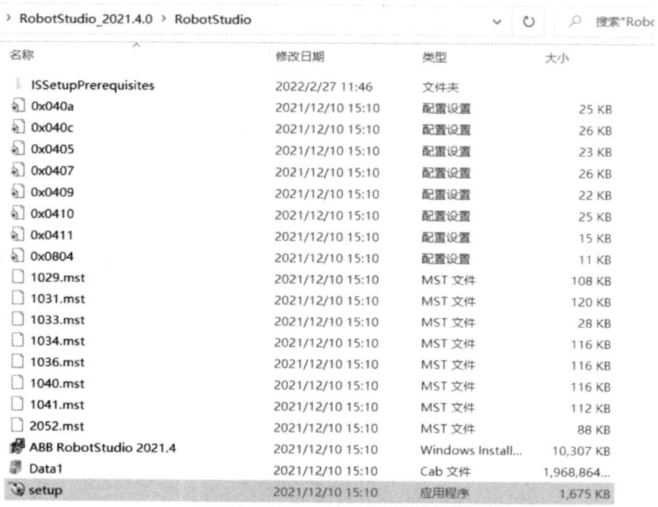

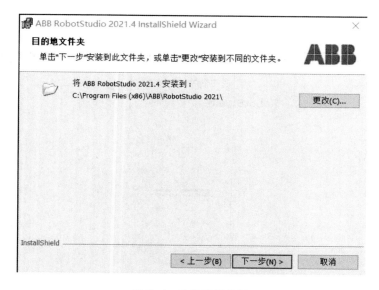

图5-3　安装文件

C.按提示勾选接受许可协议并点击"下一步",确认软件安装路径,如图5-4所示。

图5-4　安装路径设置

D.选择安装类型如图 5-5 所示。完整安装:安装运行完整所需的所有功能,可以使用基本版和高级版的所有功能;自定义安装:安装用户自定义的功能可以选择不安装不需要的机器人库文件和 CAD 转换器等。

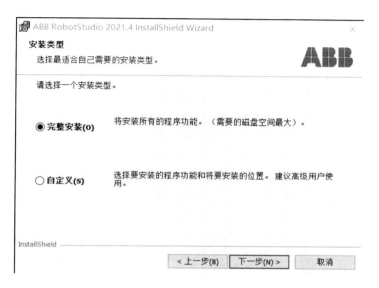

图 5-5 安装类型选择

E.点击下一步开始安装,安装完成界面如图 5-6 所示,按提示重启计算机。

图 5-6 安装完成界面

F.第一次打开软件会自动弹出激活对话框,如图 5-7 所示,否则应依次单击"文件—帮助—管理授权—激活向导"手动打开激活界面。选项"我希望激活单机许可证密匙"需输入正版软件 License,选项"我希望申请试用许可"可申请 30 天试用期(注意:每个试用许可证只能在一台电脑上安装一次)。

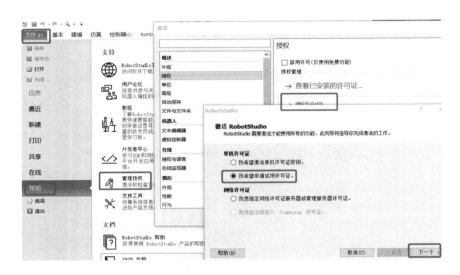

图 5-7 试用许可申请

G.确保安装的电脑能正常连接 Internet,激活完成后界面如图 5-8 所示。

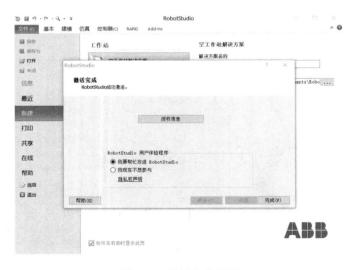

图 5-8 激活完成界面

H. 试用版安装完成后输出栏会提示许可将在 30 天后过期,如图 5-9 所示,过期后软件只可以使用基本功能。

图 5-9　试用许可期限

5.3 RobotStudio 软件界面

初次打开 RobotStudio 软件界面如图 5-10 所示,七个菜单分别为:文件、基本、建模、仿真、控制器、RAPID、Add-Ins。

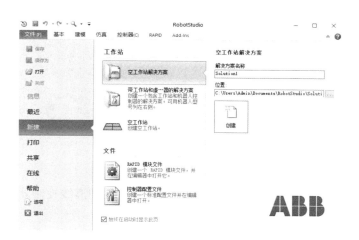

图 5-10　RobotStudio 软件界面

各菜单主要功能如下。

A."文件"菜单会打开 RobotStudio 后台视图,可以对工作站进行保存(另存为),打开,新建,查看,关闭,退出。对工作站进行打包后共享,通过帮助查看消息和手册文档等。注意:"在线" 就是通过 RobotStudio 连接真实的控制器,"离线"使用虚拟示教器、控制面板等。

B."基本"菜单如图 5-11 所示,可进行工作站创建、系统搭建、路径编程、基本设置及机器人的基本控制等。

图 5-11 "基本"菜单

C."建模"菜单如图 5-12 所示,可进行创建、分组组件、模型导入、创建一些简单的部件,测量以及进行 2D、3D 模型的一些控制等。

图 5-12 "建模"菜单

D."仿真"菜单如图 5-13 所示,主要包括创建、配置、仿真控制、监控、信号分析和录制仿真等功能。

图 5-13 "仿真"菜单

E."控制器"菜单如图 5-14 所示,主要用于在线和离线控制器的功能。用于对控制器的同步、配置和分配任务及控制措施。

图 5-14 "控制器"菜单

F."RAPID"菜单如图 5-15 所示,主要是对应 RAPID 程序相关的一些功能。(比如:创建程序,编辑、管理)

图 5-15 "RAPID"菜单

G."Add-ins"菜单如图 5-16 所示,主要用二次开发的相关功能。在连网的情况下,在线添加其他版本的 RobotWare。RobotWare 是机器人系统软件,缺少 RobotWare 则无法创建虚拟控制器。RobotWare 既可以在线添加也可以下载安装包,如"ABB.RobotWare-6.13.0164.rspak"点击"安装文件包"按钮进行离线添加。

图 5-16 "Add-ins"菜单

RobotStudio 软件界面更详细的介绍也可以查阅系统自带的 HTML 帮助文件,打开方法为依次单击菜单"文件—帮助—RobotStudio 帮助",打开帮助文件如图 5-17 所示。

图 5-17 HTML 帮助文档

5.4 创建虚拟工作站

RobotStudio 是针对 ABB 机器人开发的专用软件,它不仅可以用来开发新的机器人程序,更强大的是在电脑内创建一个虚拟的机器人,帮助用户进行离线编程和仿真测试。RobotStudio 包含了 ABB 所有的机器人,有模拟示教器,以及跟真实的示教器一样的操作、功能,非常便于 ABB 机器人的操作和编程学习。创建机器人虚拟工作站的步骤如下。

A.依次单击菜单"文件—新建—空工作站—创建"以创建一个新的工作站工程,如图 5-18 所示。

B.依次单击菜单"基本—ABB 模型库—IRB 1200",选择要添加的机器人型号,如图 5-19 所示。

C.进入机器人版本确认界面,如图 5-20 所示,确认参数后点击确认完成机器人本体添加。

D.依次单击菜单"基本—虚拟控制器—从布局..."如图 5-21 所示。

E.进入 RobotWare 机器人控制系统选择界面如图 5-22 所示,若系统未能正常安装 RobotWare 软件,则下拉选项为空,无法进行虚拟控制器的添加。

第 5 章　RobotStudio 软件应用

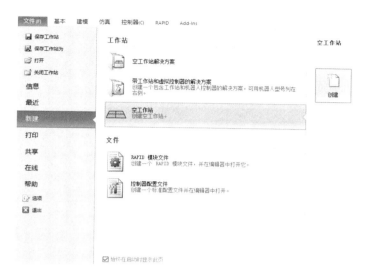

图 5-18　新建工作站

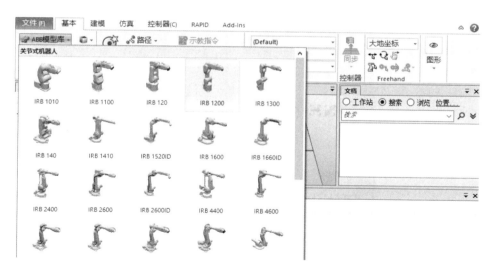

图 5-19　机器人选择

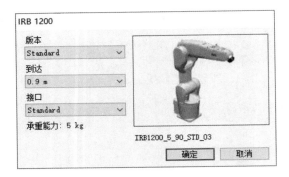

图 5-20 机器人版本确认

图 5-21 添加虚拟控制器

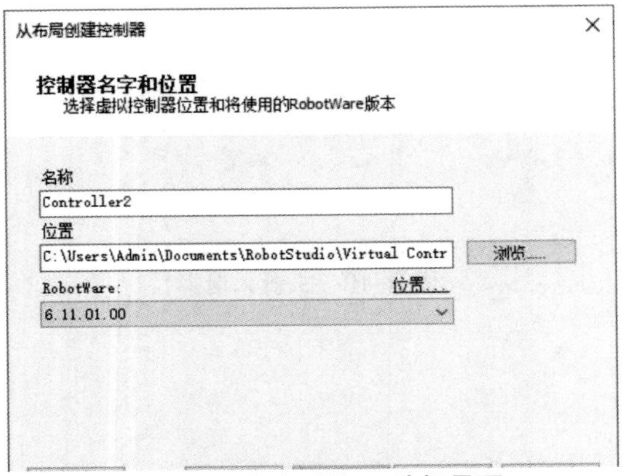

图 5-22 RobotWare 选择界面

F.单击"下一步菜单"进入控制器选项设置界面,依次单击"选项—Default Language—Chinese",可将虚拟控制器的语言改为中文,如图5-23所示。

图5-23 控制器选项设置界面

G.依次单击菜单"基本—导入模型库—设备—myTool"进行工具添加,如图5-24所示。

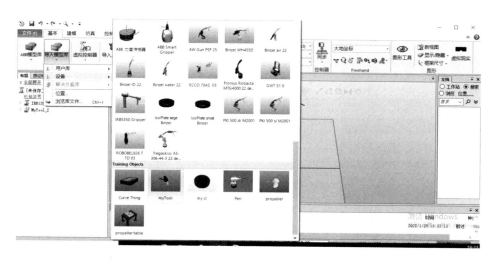

图5-24 添加工具

H. 添加完成后左侧目录树会新增"MyTool"项目，单击后该工具高亮显示，其默认位置为机器人本体坐标系原点位置。将工具正确安装到机器人法兰末端，需右键单击"MyTool"项目条，在弹出的编辑菜单中依次选择"安装到—IRB1200_5_90_STD_03(T_ROB1)"，如图 5-25 所示。

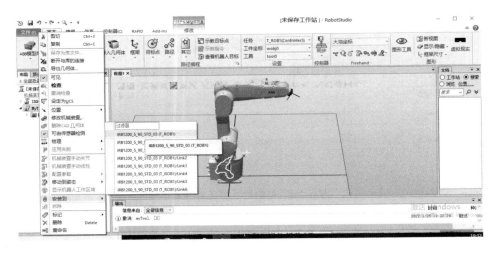

图 5-25　工具重定位

I. 确认后"MyTool"工具被正确安装到机器人法兰坐标上，重定位前后工具位置变化，如图 5-26 所示。

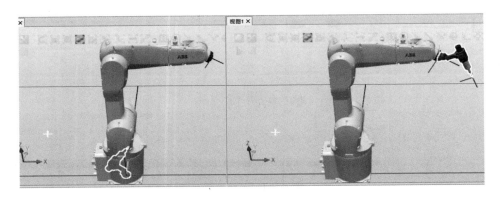

图 5-26　工具位置变化对比图

J.依次单击菜单"控制器—示教器"打开虚拟示教器,如图 5-27 所示。

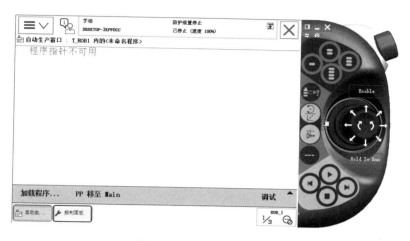

图 5-27　虚拟示教器按钮

K.至此虚拟机器人工作站创建完毕,用户可按照第四章所述工业机器人基本操作的办法,使用虚拟示教器控制虚拟机器人进行基本操作的练习,如图 5-28 所示。

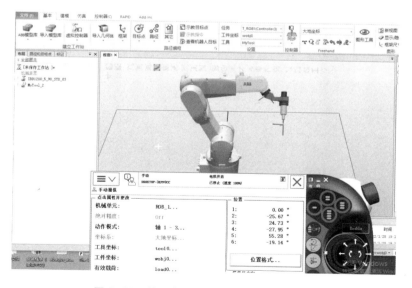

图 5-28　使用虚拟示教器控制机器人运动

第 6 章

ABB 机器人的 I/O 通讯

标准的 ABB IRC5 控制器由一个机柜组成,如图 6-1 中 C 所示,也可以选择分为两个模块(图中 A 和 B):Control Module 和 Drive Module,这种控制器称为 Dual Cabinet Controller。Drive Module 包含所有为机器人电机供电的电源电子设备,Control Module 包含所有的电子控制装置,例如主机、I/O 电路板和闪存。I/O 是 Input/Output 的缩写,即输入输出端口,机器人可通过 I/O 与外部设备进行交互,例如:

A.数字量输入:各种开关信号反馈,如按钮开关、转换开关、接近开关等;传感器信号反馈,如光电传感器、光纤传感器;还有接触器,继电器触点信号反馈;另外还有触摸屏里的开关信号反馈。

B.数字量输出:控制各种继电器线圈,如接触器、继电器、电磁阀;控制各种指示类信号,如指示灯、蜂鸣器。

ABB 机器人的标准 I/O 板的输入输出都是 PNP 类型。

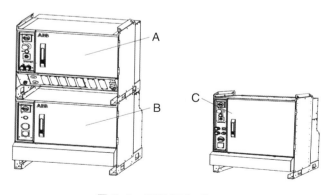

图 6-1　ABB IRC5 控制器

6.1 ABB 机器人 I/O 通信的种类

ABB 机器人提供了丰富 I/O 通信接口,如表 6-1 所示,有 ABB 的标准通信,与 PLC 的现场总线通信,还有与 PC 机的数据通信,可以轻松地实现与周边设备的通信。

表 6-1 ABB 机器人 I/O 通信种类

ABB 标准	现场总线	PC
标准 I/O 板	Device Net	RS232
PLC	Profibus	OPC Server
	Profibus-DP	Socket Message
	Profinet	
	EtherNet IP	

ABB 的标准 I/O 板提供的常用信号处理有数字输入 DI、数字输出 DO、组输入 GI、组输出 GO、模拟输入 AI、模拟输出 AO。

ABB 机器人可以选配标准 ABB 的 PLC,省去了原来与外部 PLC 进行通信设置的麻烦,并且在机器人的示教器上就能实现与 PLC 的相关操作。

IRC5 控制柜的 I/O 模块如图 6-2 所示,这些模块在内部与 I/O(DSQC651 或 DSQC652)连接,具体对应关系详见机器人产品说明书。在本章中,以最常用

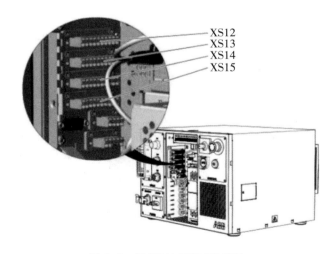

图 6-2 IRC5 控制柜 I/O 模块

的 ABB 标准 I/O 板 DSQC651 和 Profibus-DP 为例，对如何进行相关参数设定进行详细的讲解。

6.2 DSQC651 介绍

常用的 ABB 标准 I/O 板如表 6-2 所示。

表 6-2　ABB 标准 I/O 板

型号	说明
DSQC 651	分布式 I/O 模块 DI8\DO8\AO2
DSQC 652	分布式 I/O 模块 DI16\DO16
DSQC 653	分布式 I/O 模块 DI8\DO8 带继电器
DSQC 355A	分布式 I/O 模块 AI4\AO4
DSQC 377A	输送链跟踪单元

本章着重介绍的 ABB 标准 I/O 板 DSQC651 如图 6-3 所示，主要提供 8 个数字输入信号、8 个数字输出信号和 2 个模拟输出信号的处理。其主要组成部分有：

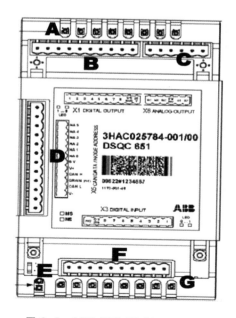

图 6-3　ABB 标准 I/O 板 DSQC651

- 数字输出信号指示灯。
- X1 数字输出接口。
- X6 模拟输出接口。
- X5 是 DeviceNet 接口。
- 模块状态指示灯。
- X3 数字输入接口。
- 数字输入信号指示灯。

(1)X1 模块接口连接说明如表 6-3 所示

表 6-3 X1 端子连接说明

X1 端子编号	使用定义	地址分配
1	OUTPUT CH1	32
2	OUTPUT CH2	33
3	OUTPUT CH3	34
4	OUTPUT CH4	35
5	OUTPUT CH5	36
6	OUTPUT CH6	37
7	OUTPUT CH7	38
8	OUTPUT CH8	39
9	0V	
10	24V	

(2)X3 模块接口连接说明如表 6-4 所示

表 6-4 X3 端子连接说明

X3 端子编号	使用定义	地址分配
1	INPUT CH1	0
2	INPUT CH2	1
3	INPUT CH3	2
4	INPUT CH4	3
5	INPUT CH5	4
6	INPUT CH6	5
7	INPUT CH7	6
8	INPUT CH8	7
9	0V	
10	未使用	

(3) X5 模块接口连接说明如表 6-5 所示。

表 6-5 X5 端子连接说明

X5 端子编号	使用定义
1	0V BLACK(黑色)
2	CAN 信号线 low BLUE(蓝色)
3	屏蔽线
4	CAN 信号线 high WHITE(白色)
5	24V RED(红色)
6	GND 地址选择公共端
7	模块 ID bit 0(LSB)
8	模块 ID bit 1(LSB)
9	模块 ID bit 2(LSB)
10	模块 ID bit 3(LSB)
11	模块 ID bit 4(LSB)
12	模块 ID bit 5(LSB)

(4) X6 模块接口连接说明如表 6-6 所示,模拟输出的范围:0~10V

表 6-6 X6 端子连接说明

X6 端子编号	使用定义	地址分配
1	未使用	
2	未使用	
3	未使用	
4	0V	
5	模拟输出 AO1	0~15
6	模拟输出 AO2	16~31

ABB 标准 I/O 板是挂在 DeviceNet 网络上的,所以要设定模块在网络中的地址。端子 X5 的 6-12 的跳线用来决定模块的地址,地址可用范围在 10~63。如果想要获得 10 的地址,可将第 8 脚和第 10 脚的跳线剪去,如图 6-4 所示,"2+8=10"就可以获得 10 的地址。

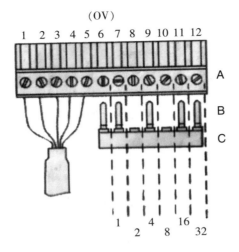

图 6-4 端子 X5 的跳线示意图

6.3 DSQC651 配置

ABB 标准 I/O 板 DSQC651 是最为常用的模块,下面以创建数字输入信号 DI、数字输出信号 DO、组输入信号 GI、组输出信号 GO 和模拟输出信号 AO 为例,详细讲解其配置过程。

ABB 标准 I/O 板都是下挂在 DeviceNet 现场总线下的设备,通过 X5 端口与 DeviceNet 现场总线进行通信,定义 DSQC651 板的总线连接的相关参数详见表 6-7。

表 6-7 参数定义说明

参数名称	设定值	说明
Name	d651	设定 I/O 板在系统中的名字,系统默认名称 d651
Type	d651	设定 I/O 板的类型
Network	DeviceNet	I/O 板连接的总线
Address	10	设定 I/O 板在总线中的地址

6.3.1 RobotStudio 机器人工作站配置

使用 RobotStudio 软件建立虚拟机器人工作站时,ABB 标准 I/O 板

第 6 章　ABB 机器人的 I/O 通讯

DSQC651 默认是不加载的，在加载 RobotWare 软件时，需点击如 6-5 所示"选项"按钮进行配置。

图 6-5　控制器选型按钮

A.点击类别"Industrial Networks"，选中"709-1 DeviceNet Master/Slave"和"969-1 PROFIBUS Controller"选项如图 6-6 所示。

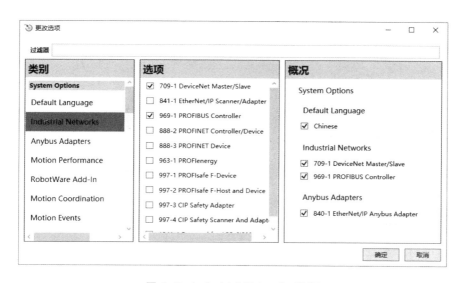

图 6-6　Industrial Networks 选项

89

B.点击类别"Anybus Adapters",选中"840-2 PROFIBUS Anybus Device"选项如图6-7所示。

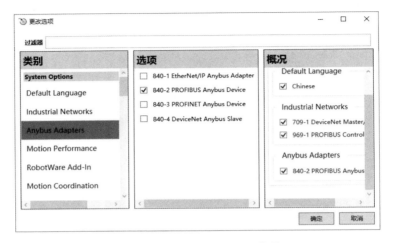

图6-7　Anybus Adapters选项

6.3.2　DeviceNet Device设置

标准I/O板DSQC651需要在DeviceNet Device模块进行设置,具体步骤如下。

A.点击示教器屏幕左上角的ABB菜单按钮,选择"控制面板"选项,选择"配置"选项,如图6-8所示。(注意:配置模块只有手动模式才能进入)

图6-8　DeviceNet Device选项

B.双击"DeviceNet Device"选项,进入设置界面如图6-9所示。

图6-9 DeviceNet Device 界面

C.点击"添加"按钮,选择"DSQC 651 Combi I/O Device"设备模板,如图6-10所示。

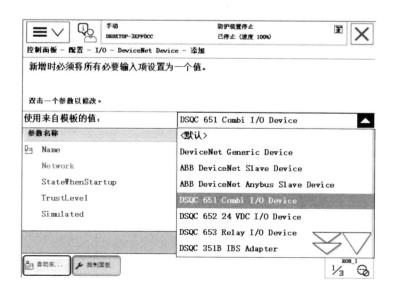

图6-10 I/O 设备模板

D.选择模板后系统默认 I/O 设备名称为"d651",双击选项可修改各项参数的数值,如图 6-11 所示,设置地址"Address"为 10。

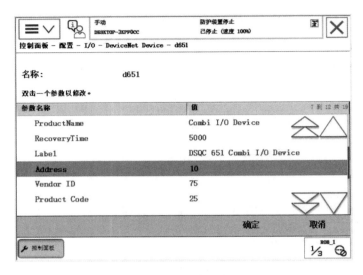

图 6-11　I/O 设备参数配置

E.参数设置完成后点击"确认"键,弹出如图 6-12 所示重启对话框,点击"是"重启控制器使 I/O 板设置生效。

图 6-12　重启确认对话框

6.3.3 I/O 信号配置

A.点击示教器屏幕左上角的 ABB 菜单按钮,选择"控制面板"选项,选择"配置"选项,双击"Signal"选项,如图 6-13 所示。

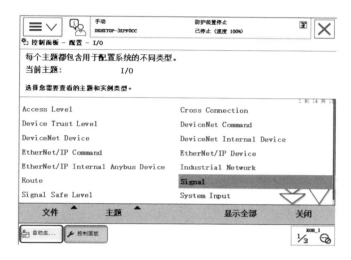

图 6-13　Signal 配置选项

B.进入 Signal 设置界面如图 6-14 所示,点击"添加"按钮定义新的 Signal 变量。

图 6-14　Signal 设置界面

C.进入 Signal 参数设置界面,所需定义参数名称和说明如表 6-8 所示,点击参数名可对其进行修改,数字输入信号 Di1 设置完成效果如图 6-15 所示。

表 6-8 数字输入信号参数表

参数名称	设定值	说明
Name	Di1	设定数字输入信号的名字
Type of Signal	Digital Input	设定信号的类型
Assigned to Device	d651	设定信号所在的 I/O 模块
Device Mapping	0	设定信号所占用的地址

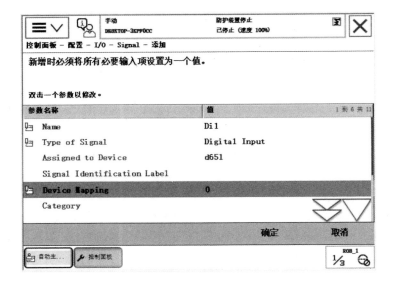

图 6-15 Signal 参数设置界面

D.设置完成后点击确定弹出如图 6-16 所示重启确认界面,点击"是"重启控制器使新设置的参数生效,点击"否"可以继续设置新的 Signal 变量,待所有设置完成后再重新启动控制器。

E.数字输出信号的定义与数字输入信号操作方法相同,其参数名称和说明如表 6-9 所示,Do1 信号设置完成效果如图 6-17 所示。

第 6 章 ABB 机器人的 I/O 通讯

图 6-16 Signal 设置界面

表 6-9 数字输出信号参数表

参数名称	设定值	说明
Name	Do1	设定数字输出信号的名字
Type of Signal	Digital Output	设定信号的类型
Assigned to Device	d651	设定信号所在的 IO 模块
Device Mapping	32	设定信号所占用的地址

图 6-17 数字输出信号定义

95

F.定义组输入信号 Gi1:组输入信号就是将几个数字输入信号组合起来使用,用于接受外围设备输入的 BCD 编码的十进制数。其参数名称和说明如表 6-10 所示,"Device Mapping"地址"1-4"代表组输入信号 Gi1 共占用 4 位数字输入端口,所组成的 4 位二进制数可以代表十进制数 0-15。如此类推,如果占用地址 5 位的话,可以代表十进制数 0-31。组输入信号 Gi1 定义完成效果如图 6-18 所示。

表 6-10 组输入信号参数表

参数名称	设定值	说明
Name	Gi1	设定组输入信号的名字
Type of Signal	Group Input	设定信号的类型
Assigned to Device	d651	设定信号所在的 IO 模块
Device Mapping	1-4	设定信号所占用的地址

图 6-18 组输入信号定义

G.定义组输出信号 G01:组输出信号的定义与组输入信号操作方法类似,区别为数据类型为"Group Output",且映射地址必须为数字输出端口地址,组输出信号 Go1 定义完成效果如图 6-19 所示。

第 6 章 ABB 机器人的 I/O 通讯

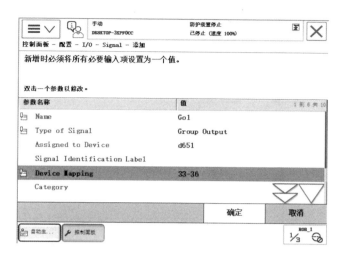

图 6-19 组输出信号定义

H.定义模拟输出信号 Ao1：模拟输出信号常见应用于控制设备工作电压，进而控制电机转速或设备功率，模拟输出参数名称和说明如表 6-11 所示。

表 6-11 模拟输出信号

参数名称	设定值	说明
Name	Ao1	设定模拟输出信号的名字
Type of Signal	Analog Output	设定信号的类型
Assigned to Device	d651	设定信号所在的 IO 模块
Device Mapping	0-15	设定信号所占用的地址
Default Value	12	默认值，不得小于最小逻辑值
Analog Encoding Type	Unsigned	默认值，不得小于最小逻辑值
Maximum Logical Value	24	最大逻辑值，如最大输出电压 24V
Maximum Physical Value	10	最大物理值，如最大输出电压时所对应 IO 板卡最大输出电压值
Maximum Physical Value Limit	10	最大物理限值，IO 板卡端口最大输出电压值
Maximum Bit Value	65535	最大逻辑位值，16 位
Minimum Logical Value	12	最小逻辑值，如最小输出电压 12V
Minimum Physical Value	0	最小物理值，焊机最小输出电压时所对应 IO 板卡最小输出电压值
Minimum Physical Value Limit	0	最小物理限值，IO 板卡端口最小输出电压
Minimum Bit Value	0	最小逻辑位值

模拟量输出信号 Ao1 定义完成效果如图 6-20 所示,所有信号定义完成后点击"确认"重启控制器使所有设置生效。

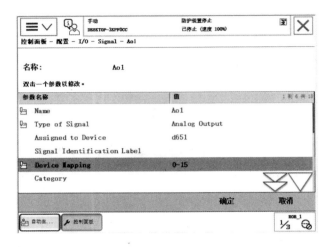

图 6-20　模拟输出信号定义

6.3.4 I/O 信号监控

所有定义的 I/O 信号可以在"输入输出"模块进行信号监控和仿真操作,具体操作方法如下。

A.点击示教器屏幕左上角的 ABB 菜单按钮,选择"输入输出"选项,如图 6-21 所示。

图 6-21　输入输出选项

B.进入"输入输出"设置界面如图 6-22 所示,单击屏幕右下角"视图"按钮弹出视图分类选项菜单,选择"IO 设备"。

图 6-22　输入输出设置界面

C.视图显示切换为所有"IO 设备"如图 6-23 所示,选中"d651"设备并单击底部的"信号"按钮。

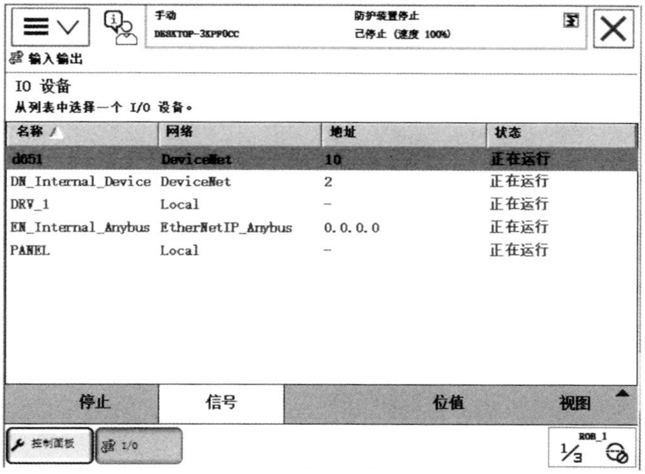

图 6-23　I/O 设备视图

D.进入 I/O 信号列表界面如图 6-24 所示,可以看到 6.3.3 节中定义的所有信号,这时可以对信号进行监控,改变外部输入信号可以看到相应信号的变化。

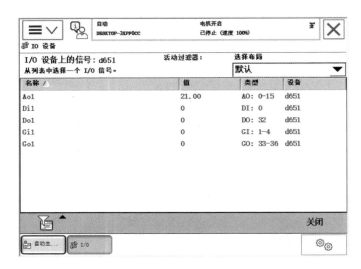

图 6-24　I/O 信号列表

E.也可以强制修改输出信号并观察外接设备的动态,还可以进行仿真操作。如图 6-25 所示,选中要修改的输出信号点击仿真,点击 0 或 1 的按钮即可修改对应信号的值。(注意:仿真功能只有手动模式才能使用)

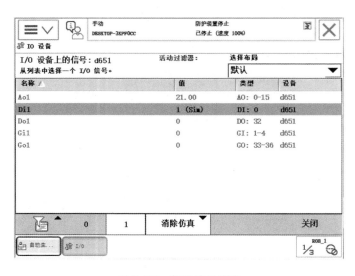

图 6-25　数字信号仿真

F.输入信号的数值也可以强制修改以便调试程序,组输入信号的数值设置如图 6-26 所示,点击"123…"按钮打开输入键盘,系统会根据组输入信号的位数提示允许设置的最大值和最小值。

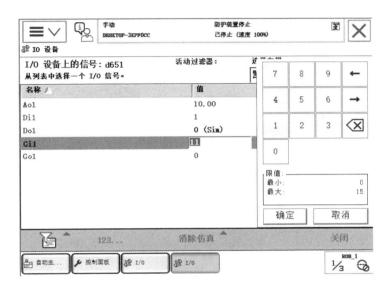

图 6-26　组输入信号仿真

G.模拟量信号数值的设置与组输入、输出信号类似,不同的是其最大值和最小值的数值取决于模拟量定义时的设置。

6.4 Profibus 通信

除了通过 ABB 机器人提供的标准 I/O 板进行与外围设备进行通信,ABB 机器人还可以使用 DSQC667 模块通过 Profibus 与 PLC 进行快捷和大数据量的通信。Profibus 通讯示意如图 6-27 所示,图中红框部分为安装在电柜中的主机上的 DSQC667 模块,最多支持 512 个数字输入和 512 个数字输。其他部分分别代表:

- A:PLC 的主站
- B:总线上的从站
- C:机器人的 Profibus 适配器 DSQC667
- D:机器人控制柜

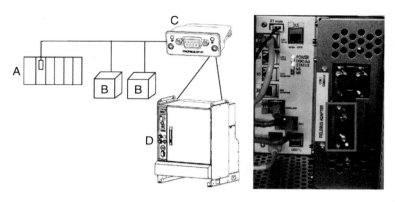

图 6-27 Profibus 通讯示意图

设置的机器人端 Profibus 地址，需要与 PLC 端添加机器人站点时设置的 Profibus 地址保持一致，从站机器人端 Profibus 地址参数设置详见表 6-12 所示。

表 6-12 Profibus 地址参数表

参数名称	设定值	说明
Name	PROFIBUS_Anybus	总线网络（不可编辑）
Identification Label	PROFIBUS Anybus Network	识别标签
Address	8	总线地址
Simulated	No	模拟状态

从站机器人端 Profibus 输入输出字节参数设置如表 6-13 所示，字节大小设置为"4"代表 32 位，表示机器人与 PLC 通信支持 32 个数字输入和 32 个数字输出。该参数允许设置的最大值为 64，意味着最多支持 512 个数字输入和 512 个数字输出。

表 6-13 Profibus 输入输出参数表

参数名称	设定值	说明
Name	PB_Internal_Anybus	板卡名称
Network	PROFIBUS_Anybus	总线网络
VendorName	ABB Robotics	供应商名称
ProductName	PROFIBUS Internal Anybus Device	产品名称
Label		标签
Input Size(bytes)	4	输入大小（字节）
Output Size(bytes)	4	输出大小（字节）

Profibus 通信连接设置操作如下。

A.点击示教器屏幕左上角的 ABB 菜单按钮,依次选择"控制面板""配置""Industrial Network"选项,如图 6-28 所示。

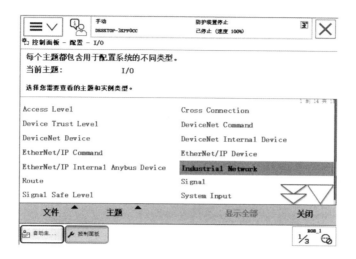

图 6-28　Industrial Network 选项

B.进入"Industrial Network"设置界面,如图 6-29 所示。

图 6-29　Industrial Network 设置界面

C.双击"PROFIBUS_Anybus"选项,进入 Profibus 参数设置界面如图 6-30 所示,双击"Address"选项,将从站机器人端 Profibus 地址设置为 8。

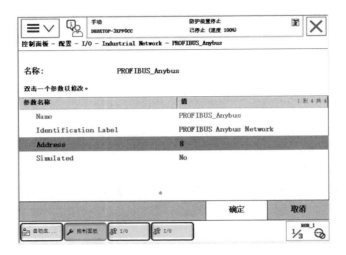

图 6-30　Profibus 地址设置界面

D.点击"确认"完成设置并弹出重新启动问询对话框,选择"否"继续进行 I/O 信号配置。点击"回退"按钮返回至"配置"选项界面,双击"PROFIBUS Internal Anybus Device"选项,如图 6-31 所示。

图 6-31　Profibus 配置选项

E.进入 Profibus 参数设置界面,双击"PB_Internal_Anybus"选项,如图 6-32 所示。

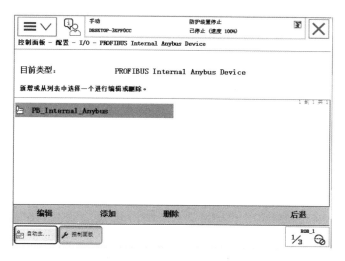

图 6-32　Profibus 参数设置界面

F.将"Input Size(bytes)"和"Output Size(bytes)"设定为"4",对应 32 个数字输入信号和 32 个数字输出信号,如图 6-33 所示。设置完成后单击"确定",在弹出的对话框点击"是"重启控制器以使所有设置生效。

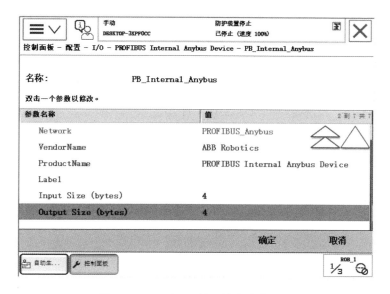

图 6-33　Profibus 输入输出字节设置

G.基于 Profibus 设定信号的方法和 ABB 标准 I/O 板上设定信号的方法基本一样,要注意的区别就是在 I/O 设备选择"PB_Internal_Anybus",如图 6-34 所示。

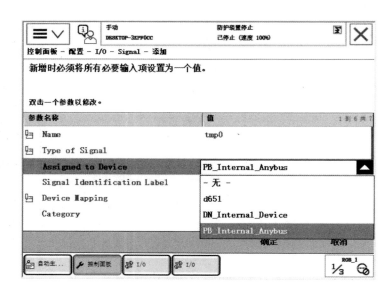

图 6-34　Profibus 信号设定

H.在完成了 ABB 机器人上的 Profibus 从站的设定后,需要在 PLC 端完成以下操作。

• 将 ABB 机器人的 DSQC667 配置文件安装到 PLC 组态软件中(进入"FlexPendant"资源管理器,按照路径 PRODUCTS/RobotWare_6** /utility/service/GSD/HMS_***.gsd 即可获取配置文件)。

• 在组态软件中将新添加的"Anybus-CC PROFIBUS DP-V1"加入到工作站中并设定 Profibus 地址(例如:8)。

• 添加输入输出模块(例如:添加总数各 4 字节的输入输出模块)。

• ABB 机器人中设置的信号与 PLC 端设置的信号是一一对应的。

6.5 示教器可编程按键设定

ABB 机器人示教器上有四个可编程快捷键,如图 6-35 所示。在调试的过程中通过配置这四个快捷键,可以模拟外围的信号输入或者对信号进行强制输出,灵活地使用可编程按键可以大大提高调试的效率。为可编程按键 1 配置数字输出信号 Do1 的详细操作步骤如下。

A.点击示教器屏幕左上角的 ABB 菜单按钮,依次选择"控制面板""ProgKeys"选项,如图 6-36 所示。

B.进入可编程按键配置界面,选择"按键 1"选项卡,类型选择"输出",如图 6-37 所示。

C. 所有数字输出信号会显示在右侧列表框,选中 Do1,选择按下按键后的选项,如图 6-38 所示。各选项功能区别如下。

图 6-35 Profibus 信号设定

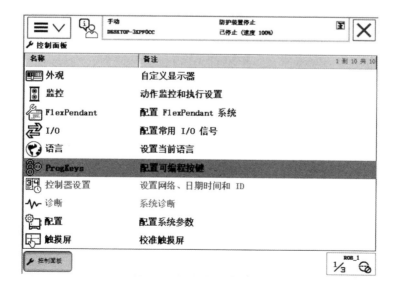

图 6-36 ProgKeys 选项

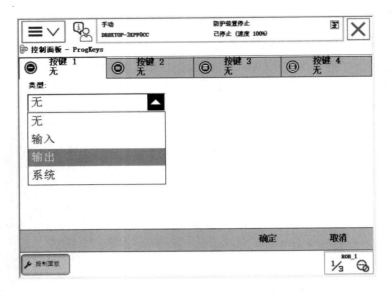

图 6-37 可编程按键配置界面

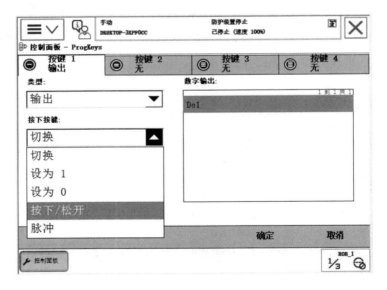

图 6-38 可编程按键配置界面

- 切换(Toggle):按下按键后 Do1 的值在 0 和 1 之间切换。
- 设为 1(Set to 1):按下按键后 Do1 的值被置 1,相当于置位。
- 设为 0(Set to 0):按下按后 Do1 的值被置 0,相当于复位。

- 按下/松开(Press/Release):按下按键后 Do1 的值被置 1,松开后 Do1 的值被置 0。
- 脉冲(Pulse):按下按键的上升沿 Do1 的值被置 1。

D.设定完成后可以进入 I/O 信号监控界面,操作按键 1 查看对应信号是否变化,或直接运行程序进行测试。

第 7 章 ABB 机器人编程基础

ABB 机器人编程基础主要包括机器人系统相关术语、ABB 机器人程序结构和程序数据三个方面。

7.1 机器人系统术语

机器人工作站主要由机器人及其控制系统、辅助设备以及其他周边设备所构成,它包括操作机器人所必需的所有硬件和软件。机器人系统的主要概念和术语如下。

7.1.1 工具(TOOL)

工具是能够直接或间接安装在机器人转动盘上,或能够装配在机器人工作范围内固定位置上的物件。如图 7-1 所示,以机器人法兰面为界,A 侧为工

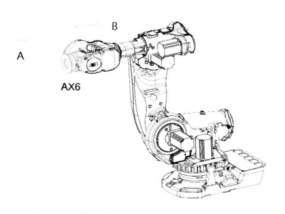

图 7-1 使用虚拟示教器控制机器人运动

具侧,B侧为机器人侧。为完成各种作业任务,工业机器人末端安装的工具形状、大小各不相同,如喷枪、抓手、焊枪等。在更换或者调整工具之后,机器人的实际工作点相对于机器人末端的位置会发生变化,因此所有工具必须用TCP(工具中心点)定义。为了获取精确的工具中心点位置,必须测量机器人使用的所有工具并保存测量数据。

7.1.2 工具中心点(Tool Center Point, TCP)

为了定义机器人末端的工具,通常的做法是在机器人工具上建立一个工具坐标系,其原点被称为工具中心点(Tool Center Point, TCP)。机器人在此坐标系内进行编程,当工具调整后,只需重新标定工作坐标系的位姿,即可使机器人重新投入使用。

同一个机器人可以因为挂载不同的工具,而定义不同的工具中心点,但是同一时刻,机器人只能只能存在一个有效TCP。TCP有两种基本类型:移动式工具中心点(Moving TCP)和静态工具中心点(Stationary TCP)。

• 移动式工具中心点比较常见,它的特点是会随着机器人手臂的运动而运动。比如焊接机器人的焊枪、搬运机器人的夹具等。

• 静态工具中心点是以机器人本体以外的某个点作为中心点,机器人携带工件围绕该点做轨迹运动。例如使用固定的点焊枪时,TCP要参照静止设备而不是移动的操纵器来定义。

机器人系统默认TCP:无论是何种品牌的工业机器人,事先都定义了一个默认工具坐标系,无一例外地将这个坐标系XY平面绑定在机器人第六轴的法兰盘平面上,坐标原点与法兰盘中心重合。显然,这时TCP就在法兰盘中心。不同品牌的机器人有不同的称呼,ABB机器人把这个工具坐标系称为tool0。

7.1.3 坐标系

坐标系是为确定机器人的位置和姿态而在机器人或其他空间上设定的位姿指标系统,通常坐标系从一个被称为原点的固定点通过轴定义平面或空间,机器人目标和位置通过沿坐标系轴的测量来定位。工业机器人定义了多个坐标系,每一坐标系都适用于特定类型的微动控制或编程。

A.大地坐标系(World Coordinate System):也称为世界坐标系,是固定在空间上的标准直角坐标系,它被固定在事先确定的位置,用户坐标系是基于该坐标系而设定的。大地坐标系在工作单元或工作站中的固定位置有其相应的零点,这有助于处理若干个机器人或由外轴移动的机器人。如图 7-2 所示,坐标系 A、C 为机器人基坐标系,坐标系则为大地坐标系。在默认情况下,大地坐标系与基坐标系是一致的。

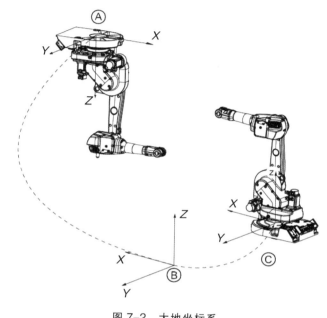

图 7-2 大地坐标系

B.基坐标系(Base Coordinate System):也称为本体坐标系,如图 7-3 所示,是位于机器人底座基点上的直角坐标系,该坐标系是机器人其他坐标系的基础。基坐标系使固定安装的机器人的移动具有可预测性,因此它对于将机器人从一个位置移动到另一个位置很有帮助。

C.关节坐标系(Joint Coordinate System):关节坐标系是设定在机器人关节中的坐标系,它是每个轴相对其原点位置的绝对角度。若将机器人末端移动到期望位置,在关节坐标系下操作,可以依次驱动各关节运动,从而引导机器人末端到达指定的位置。

D.工具坐标系(Tool Coordinate System):工具坐标系用来确定工具的位姿,它由工具中心点(TCP)与坐标方位组成,如图 7-4 所示。工具坐标系必须事先进行设定,在没有定义的时候,将由默认工具坐标系来替代该坐标系。执

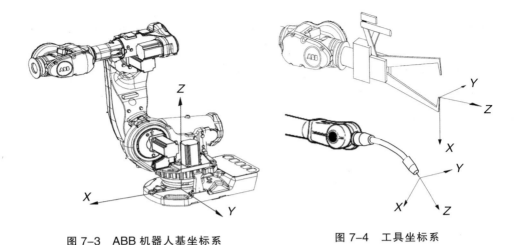

图 7-3　ABB 机器人基坐标系　　　　图 7-4　工具坐标系

行程序时,机器人就是将 TCP 移至编程位置。这意味着,如果要更改工具(以及工具坐标系),机器人的移动将随之更改,以便新的 TCP 到达目标。ABB 机器人的默认工具坐标系为 tool0,而新的工具坐标系则定义为 tool0 的偏移值。微动控制机器人时,如果不想在移动时改变工具方向(例如移动焊枪时保持平动),工具坐标系就显得非常有用。

E.工件坐标系(Work Object Coordinate System):工件坐标系用来确定工件的位姿,定义工件相对于大地坐标系(或其它坐标系)的位置,它由工件原点与坐标方位组成。工件坐标系可采用三点法确定:点 X_1 与点 X_2 连线组成 X 轴,通过点 Y_1 向 X 轴作的垂直线为 Y 轴,Z 轴方向以右手定则确定。如图 7-5 所示,B、C 为工件坐标系。机器人可以拥有若干工件坐标系,或者表示不同工件,或者表示同一工件在不同位置的若干副本。对机器人进行编程时就是在工件坐标系中创建目标和路径,这带来很多优点:重新定位工作站中的工件时,只需更改工件坐标系的位置,所有路径将即刻随之更新。

F.用户坐标系(User Coordinate Sys-

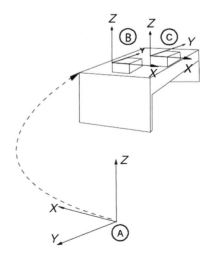

图 7-5　工件坐标系

tem):用户坐标系是用户对每个作业空间进行自定义的直角坐标系,它用于位置寄存器的示教和执行、位置补偿指令的执行等。如图 7-6 所示,其中 A 为用户坐标系,B 为大地坐标系,C 为基坐标系,D 为移动后的用户坐标系,E 为工件坐标系随用户坐标系一起移动。在没有定义的时候,将由大地坐标系来替代该坐标系。

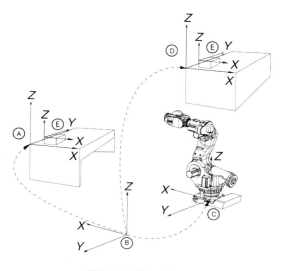

图 7-6 用户坐标系

7.2 ABB 机器人的程序结构

在 ABB 机器人中,机器人所运行的程序被称为 RAPID,RAPID 程序中包含了一连串控制机器人的指令,执行这些指令可以实现对机器人的控制操作。RAPID 下面又划分了 Task(任务),任务下面又划分了 module(程序模块),模块是机器人的程序与数据的载体,模块又分为 System modules(系统模块)与 Task modules(任务模块)。图 7-7 所示为 ABB 机器人的程序结构示意图。

A.Task(任务):通常每个任务包含了一个 RAPID 程序和系统模块,并实现一种特定的功能(例如点焊或操纵器的运动)。一个 RAPID 应用程序包含一个任务。如果安装了 Multitasking 选项,则可以包含多个任务。

B.Program(程序):每个程序通常都包含具有不同作用的 RAPID 代码的程

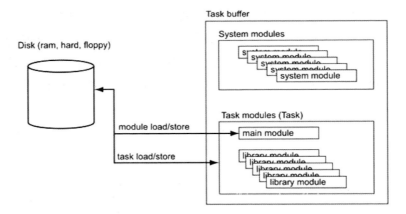

图 7-7　ABB 机器人的程序结构示意图

序模块。所有程序必须定义可执行的录入例行程序。

C.Module(程序模块):每个程序模块都包含特定作用的数据(Program Data)和例行程序(Routine)。将程序分为不同的模块后,可改进程序的结构,且使其便于处理。每个模块表示一种特定的机器人动作或类似动作。从控制器程序内存中删除程序时,也会删除所有程序模块,程序模块通常由用户编写。

D.Routine(例行程序):例行程序包含一些指令集,它定义了机器人系统实际执行的任务,也包含指令需要的数据。

E.main(主程序/录入例行程序):在英文中有时称为"main"的特殊例行程序,被定义为程序执行的起点。每个程序必须含有名为"main"的录入例行程序,否则程序将无法执行。

F.Instruction(指令):指令是对特定事件的执行请求,例如"运行操纵器 TCP 到特定位置 MoveJ"或"设置特定的数字化输出 Set"。

在程序编辑和测试过程中,需了解以下概念。

A.程序指针(PP):程序指针指的是无论按 FlexPendant 上的"启动""步进"或"步退"按钮都可启动程序的指令。程序将从"程序指针"指令处继续执行。但是,如果程序停止时光标移至另一指令处,则程序指针可移至光标位置(或者光标可移动至程序指针),程序执行也可从该处重新启动。"程序指针"在"程序编辑器"和"运行时窗口"中的程序代码左侧显示为黄色箭头。

B.动作指针(MP):动作指针是机器人当前正在执行的指令,通常比"程序指针"落后一个或几个指令,因为系统执行和计算机器人路径比执行和计算机器人移动更快。"动作指针"在"程序编辑器"和"运行时窗口"中的程序代码左侧显示为小机器人。

C.光标:可表示一个完整的指令或一个变元,它在"程序编辑器"中的程序代码处以蓝色突出显示。

D.程序编辑器:如果在"程序编辑器"和其他视图之间切换并再次返回,只要程序指针未移动,"程序编辑器"将显示同一代码部分。如果程序指针已移动,"程序编辑器"将在程序指针位置显示代码。

7.3 ABB 机器人程序数据

在开始编程前,需确保已在机器人系统安装过程中设置了基坐标系和大地坐标系,根据需要定义工具坐标系、工件坐标系和有效载荷三个程序数据。所谓程序数据就是程序内声明的数据,由同一个模块或其他模块中的指令进行引用。如图 7-8 所示,MoveJ 调用了 4 个程序数据,分别是:

p01:机器人运动目标位置数据(robtarget);

v1000:机器人运动速度数据(speeddata);

z50:机器人运动转弯数据(zonedata);

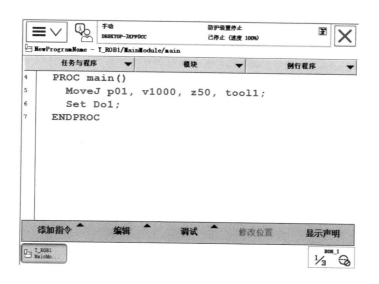

图 7-8 程序指令示例

tool1：机器人工具数据 TCP(tooldata)。

ABB 机器人控制器数据类型多达 100 余种，其中常见的数据类型包括基本数据、I/O 数据、运动相关数据，如表 7-1 所示。

表 7-1 ABB 常用数据类型表

类型	名称	数据类型	使用说明
基本数据	bool	逻辑值	逻辑状态下赋予的真或者假。逻辑值有两种情况：成立和不成立。若逻辑值为真，使用 true 或 1，表示不成立则逻辑值为假，使用 false 或 0。
	byte	字节值	用于计量存储容量的一种计量单位，取值范围为 (0–255)
	num	数值	变量、可存储整数或小数整数取值范围 (−8388607~8388608)
	dnum	双数值	可存储整数和小数，整数取值范围 (−4503599627370495~+4503599627370496)
	string	字符串	字符串是由数字、字母、下划线组成的一串字符。它在编程语言中表示文本的数据类型。
	stringdig	只含数字的字符串	可处理不大于 4294967295 的正整数
I/O 数据	dionum	数字值	取值为 0 或 1 用于处理数字 i/o 信号，数字 i/o 信号中 0 作为低电平 0~0.7v，1 作为高电平 3.4~5.0v
	signaldi/do	数字输入/输出	二进制值输入、输出 如开关接通是 1，断开是 0
	signalgi/go	数字量输入/输出信号组	多个数字量输入或输出组合配合使用
	signalai	模拟量输入	例如通过温度采样器采集到一个温度值，就要经过变送器转换，转换成 PLC 能够识别的二进制数数据−变送器−执行机构
	signalao	模拟量输出	
运动数据	robtarget	位置数据	定义机械臂和附加轴的位置
	jointtarget	关节数据	定义机械臂和外轴的各单独轴位置
	robjoint	关节数据	定义机械臂各关节位置
	speeddate	速度数据	定义机械臂和轴移动速率，共包含四个参数
	zonedata	区域数据	一般也称为转弯半径，用于定义机器人轴在朝向下一个移动位置前如何接近编程位置
	tooldata	工具数据	用于定义工具的特征，包含工具中心点(TCP)的位置和方向，以及工具的负载
	wobjdata	工件数据	用于定义工件的位置及状态
	loaddata	负载数据	用于定义机械臂安装界面的负载

7.3.1 程序数据建立方法

程序数据的建立一般可以分为两种形式，一种是直接在示教器中的程序数据画面中建立程序数据，另一种是在建立程序指令时，同时自动生成对应的程序数据。示教编程通常使用 FlexPendant 示教器来编程，最合适用于修改程序中的各种数据。以 bool 类型为例，在示教器中的程序数据画面中建立程序数据的方法如下。

A.点击示教器屏幕左上角的 ABB 菜单按钮，选择"程序数据"选项，如图 7-9 所示。

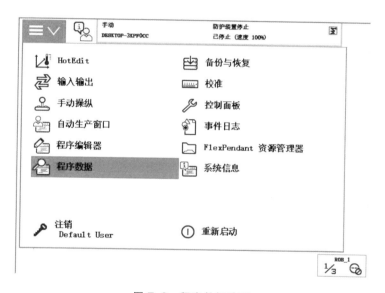

图 7-9　程序数据选项

B.进入数据类型显示界面如图 7-10 所示，选择数据类型"bool"，双击或单击底部的"显示数据"按钮。（注意：如果当前界面找不到数据类型 bool，原因是当前任务范围没有定义 bool 类型的数据，单击右下角的"视图"按钮，切换选项为"全部数据类型"即可）

C.进入 bool 类型数据列表界面如图 7-11 所示，单击"新建"按钮定义新数据。

D.依次设置变量名称、应用范围、所属任务和模块，也可以单击"确认"完

图 7-10　数据类型显示界面

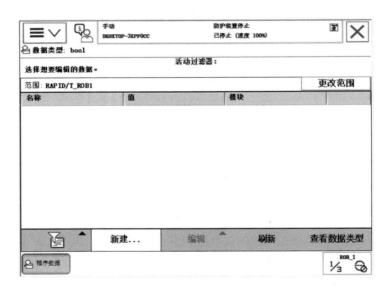

图 7-11　新建 bool 数据

成新数据的定义,如图 7-12 所示。其中范围有三个选项:全局可用于任何一个模块,本地作用于特定的模块,任务作用于机器人具体几个任务,单台机器人默认的一个任务。任务和模块选项则可以设定该数据的使用范围,便于数据管理。

图 7-12　bool 数据定义

E.存储类型有三个选项：变量（VAR）、可变量（PERS）、常量（CONST），主要区别如下。

• 变量型数据：在程序执行的过程中和停止时，会保持当前的值。但如果程序指针复位或者机器人控制器重启，数值会恢复为声明变量时赋予的初始值。（示例：VAR num Distance:=0;）

• 可变量 PERS：无论程序的指针如何变化，无论机器人控制器是否重启，可变量型的数据都会保持最后赋予的值。（示例：PERS num Height:=0;）

• 常量 CONST：常量的特点是在定义时已赋予了数值，并不能在程序中进行修改，只能手动修改。（示例：CONST num Pi:=3.14159;）

F.系统默认所有新建 bool 变量的初始值为 false，可单击左下角的"初始值"按钮进行修改，如图 7-13 所示。

7.3.2　工具数据 TOOLDATA 的设定

机器人工具坐标系、TCP 点的概念在 7.1 节已有描述，工具数据 Tooldata 则是用于描述安装在机器人第六轴上工具的 TCP、姿态、质量、重心等参数。默认 ABB 机器人工具（tool0）的工具中心点（Tool Center Ponit）位于机器人安装法兰的中心，其 XYZ 方向如图 7-14 所示，Z 轴垂直于法兰安装面，X 轴与基坐

图 7-13 bool 数据初始值设置

图 7-14 tool0 工具坐标系

标系 Z 轴相反,坐标原点即 TCP 点位于法兰盘中心。

　　Tooldata 数据会影响机器人控制算法(例如计算加速度、速度、加速度监控、力矩监控、碰撞监控、能量监控等),因此机器人的工具数据需要正确设置。新工具具有质量、框架、方向等初始默认值,这些值在工具使用前必须进行定义,所有新定义工具坐标系的位置可以理解为在 tool0 的基础上偏移和旋转,详细设置步骤如下。

A.点击示教器屏幕左上角的 ABB 菜单按钮,选择"程序数据"选项,选中"tooldata"数据类型选项单击"显示数据"或者直接双击,如图 7-15 所示。

图 7-15 tooldata 数据类型选项

B.或者进入手动操纵界面,单击"工具坐标"选项,如图 7-16 所示。

图 7-16 工具坐标选项

C.进入tooldata显示界面,如图7-17所示,单击"新建"按钮。则可在"编辑"菜单进行工具坐标的删除、更改声明、更改值、复制和定义。

图7-17 tooldata显示界面

D.进入tooldata定义界面,如图7-18所示,系统会自动将工具按顺序命名为"tool+数字"。为使用方便应更改为具体的名称,例如焊枪、夹具等;工具的范围应该始终保持全局状态,以便用于程序中的所有模块;工具数据必须始终是持久变量。最后点击"确认"完成新建。

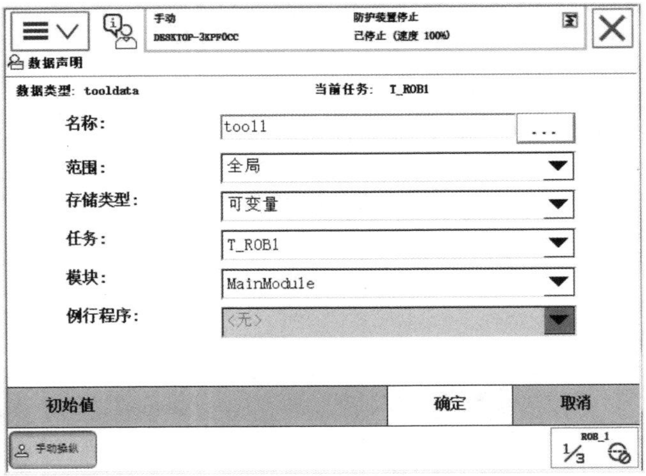

图7-18 tooldata定义界面

E.新建坐标系的各项参数默认与 tool0 一致,对于工具坐标系数据已知的也可以点击"初始值"按钮,进入 tooldata 参数设置界面进行各项参数的设定,如图 7-19 所示。

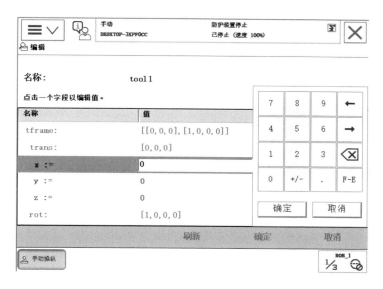

图 7-19　tooldata 参数设定界面

表 7-2 所示为 tooldata 数据的详细参数表,详细说明了各个参数的含义和取值。(注:在图形学中,最常用的空间位姿表示方法便是四元数和欧拉角两种。欧拉角由三个角度组成,直观、容易理解,但存在死锁问题;四元数不存在万向节死锁问题,存储空间小、计算效率高,但其四元数表示不够直观。欧拉角和四元数的相互转换可以借助专用软件或网站(https://quaternions.online/)。

F.返回 tooldata 显示界面,选中 tool1,单击"编辑"菜单的"定义"选项,可以采用示教标定法对工具坐标系 TCP 进行设定,如图 7-20 所示。

G.进入如图 7-21 所示的工具坐标定义界面,可使用三种不同的方法。所有这三种方法都需要定义基于 TCP 的笛卡尔坐标,不同的方法对应不同的工具姿态定义方式。工具 TCP 通过 N 种不同姿态同某定点相碰,得出多组解,通过计算得出当前 TCP 与默认工具坐标系 tool0 的相应位置。三种方法说明如下。

• TCP(默认方向):工具坐标系方向与 tool0 一致时使用。

表 7-2 tooldata 数据参数表

参数	说明
robhold	全称：robot hold 数据类型：bool 定义机械臂是否夹持工具： • TRUE：机械臂正夹持着工具。 • FALSE：机械臂未夹持工具，即为固定工具。
tframe	全称：tool frame 数据类型：pose 定义工具坐标系位置和姿态： • trans[x y z]：TCP 相对于 tool0 的位置，单位：mm。 • rot[q1 q2 q3 q4]：代表工具坐标系相对于 tool0 姿态的四元数，作为工具坐标系姿态的表示方式。
tload	全称：tool load 数据类型：loaddata 定义机械臂夹持工具的负载，即： • mass：工具的质量，单位：kg。 • cog[x y z]：工具负载相对于 tool0 的重心（x、y、z），单位：mm。 • aom[q1 q2 q3 q4]：工具力矩主惯性轴相对于 tool0 的方位。 • [ix iy iz]：围绕力矩惯性轴的惯性矩，单位：kg·m^2。如果将所有惯性部件定义为 0，则将工具作为一个点质量来处理。

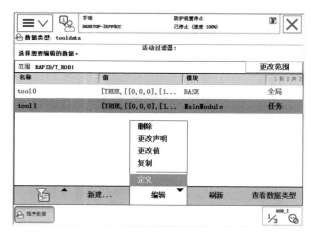

图 7-20 工具坐标系定义选项

第 7 章 ABB 机器人编程基础

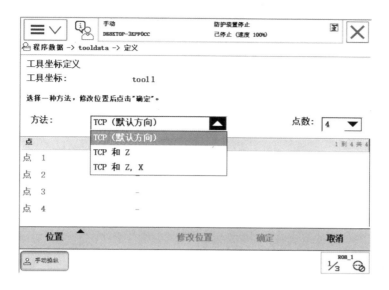

图 7-21 工具坐标定义界面

- TCP & Z：工具坐标 Z 轴方向与 ool0 Z 轴方向不一致时使用。
- TCP & Z, X：工具坐标方向需要更改 Z 轴和 X 轴方向时使用。

H.定义 TCP 时通常 4 个接近点就足够了,如果为了获得更精确的结果而选取了更多的点数,则应在定义每个接近点时均同样小心。选择四点法定义 TCP 时,四个姿态差别应尽可能大以便提高拟合精度,如图 7-22 所示。

I.具体操作方法是：

- 将机器人移至合适的位置 A,取得第一个接近点。使用示教器手动操纵,小幅增量,尽量将工具顶点的位置接近参照点。点击修改位置如图 7-23 所示，完成第一点的设定。

- 重复上述步骤,定义其他的

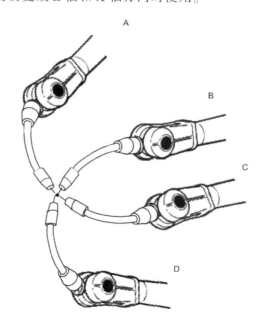

图 7-22 TCP 4 点法定义

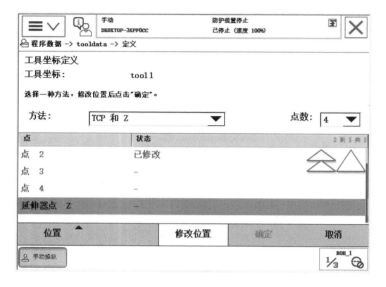

图 7-23 修改接近点位置

接近点,得到位置 B、C、D。注意移动机器人时,使其远离固定大地坐标点,以便获得最佳效果。仅修改工具方向不会获得良好的效果。

• 如果使用的方法是 TCP & Z 或 TCP & Z, X,则还必须对方向进行定义。延伸器点的定义是在不改变工具方向的情况下移动机器人,使参考大地坐标点成为所需旋转工具坐标系正轴上的某个点。

• 如果出于某种原因需要重新设定接近点,点击"位置"按钮"全部重置"选项,重新进行标定。

• 将所有点都定义好后,点击"确认"按钮完成工具坐标系定义。

J.tool1 标定完成之后系统会弹出如图 7-24 所示的误差信息,操作人员需对误差进行确认,虽说误差越小越好,但人工操作很难达到高精度,应根据实际应用场景和经验进行评定。

K.最后,还要根据实际情况设定工具的重量 mass 和重心位置数据(相对于 tool0),完成 tooldata 数据的设定。

7.3.3 工件坐标数据 WOBJDATA 的设定

工件坐标系的概念在 7.1 节已有描述,对编程使用工件坐标系,可以在重新定位工件时,只需更改工件坐标系的定义,所有路径将即刻随之更新。如图

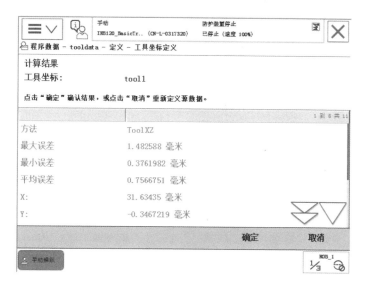

图 7-24　工具坐标系设定误差

7-25 所示,如果在工作台上还有一个相同工件 C 需要与工件 A 相同的轨迹,只需建立工件坐标 D,将工件坐标 B 中的程序复制一份,然后将新程序的工件坐标从 B 改为 D,无需重复轨迹编程即可得到新的轨迹。

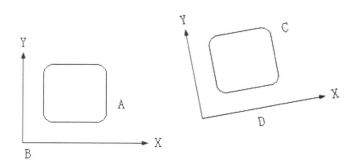

图 7-25　程序轨迹与工件坐标系关系示意图

工件坐标数据 WOBJDATA 定义方法如下。

A.点击示教器屏幕左上角的 ABB 菜单按钮,选择"程序数据"选项,选中"wobjdata"数据类型选项单击"显示数据"或者直接双击,如图 7-26 所示。

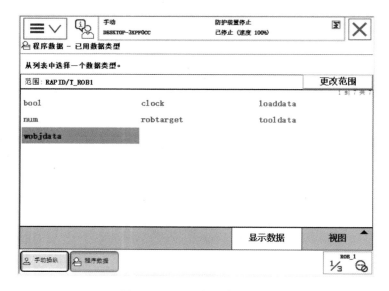

图 7-26 wobjdata 类型选项

B.或者在"手动操纵"界面点击"工件坐标"选项,进入工件坐标系显示界面如图 7-27 所示,默认工件坐标系 wobj0 与大地坐标相同,用户定义工件坐标系需点击"新建"按钮。

图 7-27 wobjdata 数据显示界面

C.进入 wobjdata 定义界面如图 7-28 所示,系统会自动按顺序命名为"wobj+数字",可以改变命名使之更明确(工件坐标系一经创建不可轻易改名,否则所有用到该工件坐标系的程序都得进行修改)。所有模块中,工件都应该是程序中的全局变量;工件数据必须始终是持久变量。最后点击"确认"完成新建。

图 7-28 wobjdata 定义界面

D.新建坐标系的各项参数默认与 wobj0 一致,对于工件坐标系数据已知的也可以点击"初始值"按钮,进入 wobjdata 参数设置界面进行各项参数的设定,如图 7-29 所示。

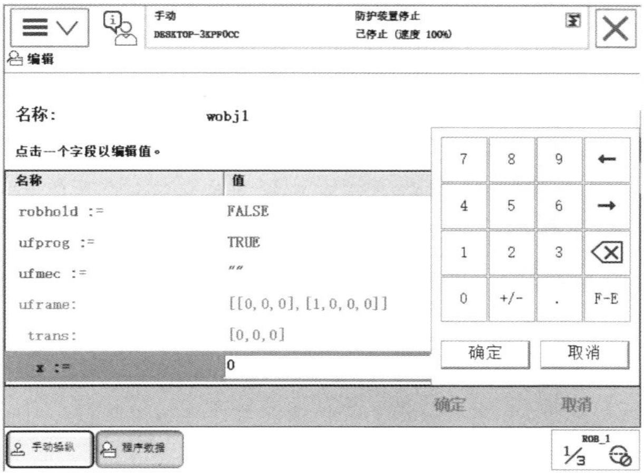

图 7-29 wobjdata 参数设置界面

表7-3所示为wobjdata数据的详细参数表,详细说明了各参数的含义和取值。

表7-3 wobjdata数据参数表

参数	说明
robhold	全称:robot hold 数据类型:bool 定义机械臂是否夹持工件: • TRUE:机械臂正夹持着工件,即使用了固定工具。 • FALSE:机械臂未夹持工件,即机械臂夹持工具。
ufprog	全称:user frame programmed 数据类型:bool 定义是否使用固定的用户坐标系: • TRUE:固定的用户坐标系。 • FALSE:可移动的用户坐标系,即使用协调外轴。也用于MultiMove系统的半协调或同步协调模式。
ufmec	全称:user frame mechanical unit 数据类型:string 定义与机械臂协调移动的机械单元。仅在可移动的用户坐标系中进行指定(ufprog为FALSE)时,指定系统参数中所定义的机械单元名称,如orbit_a。
uframe	全称:user frame 数据类型:pose 定义用户坐标系,即当前工作面或固定装置的位置: • trans[x y z]:用户坐标系原点的位置,单位:mm。 • rot[q1 q2 q3 q4]:代表用户坐标系的旋转的四元数。 说明:如果机械臂正夹持着工具,则在大地坐标系中定义用户坐标系(如果使用固定工具,则在腕坐标系中定义)。对于可移动的用户坐标系(ufprog为FALSE),由系统对用户坐标系进行持续定义。
oframe	全称:object frame 数据类型:pose 在用户坐标系中定义目标坐标系,即当前工件的位置: • trans[x y z]:目标坐标系原点的位置,单位:mm。 • rot[q1 q2 q3 q4]:代表目标坐标系的旋转的四元数。

E.也可以返回 wobjdata 显示界面,选中 wobj1,单击"编辑"菜单的"定义"选项,采用示教法对工件坐标系进行设定,如图 7-30 所示。

图 7-30 wobjdata 数据定义选项

F.进入如图 7-31 所示的工件坐标定义界面,在"用户方法"下拉菜单列表中选择"3 点",系统显示用户点 X1、X2、Y1 选项。

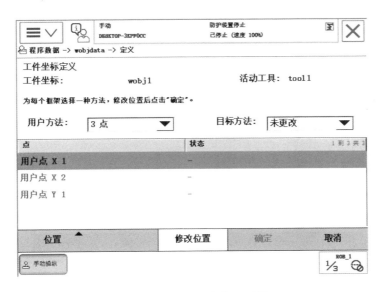

图 7-31 wobjdata 参数设置界面

G. 在工件对象的平面上，只需要定义三个点，确定原点和 X、Y 轴正方向，就可以依据右手定则建立一个工件坐标，如图 7-32 所示。其中 X1 点确定工件坐标的原点，X1、X2 点确定工件坐标 X 正方向，Y1 确定工件坐标 Y 正方向。

H. 手动操作机器人的工具参考点依次靠近工件上的 X1、X2 和 Y1 位置，分别点击"修改位置"记录坐标，完成后点击"确认"完成 wobjdata 数据的创建。

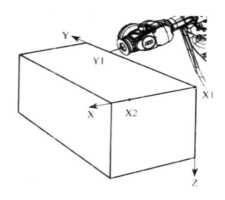

图 7-32　三点法设定工件坐标示意图

7.3.4　有效载荷 loaddata 设定

对于搬运工业机器人，必须正确设定夹具 tooldata 的质量、重心以及搬运对象的质量和重心数据 loaddata，其重心 tooldata 数据是基于 tool0 设定的。载荷数据是定义机器人的有效负载或抓取物的负载（通过指令 GripLoad 或 MechUnitLoad 来设置），即机器人夹具所夹持的负载。同时将 loaddata 作为 tooldata 的组成部分，以描述工具负载。详细操作方法如下。

A. 点击示教器屏幕左上角的 ABB 菜单按钮，选择"程序数据"选项，选中"loaddata"数据类型选项单击"显示数据"或者直接双击，如图 7-33 所示。

图 7-33　loaddata 类型选项

B.或者在"手动操纵"界面点击"有效载荷"选项,进入有效载荷显示界面如图 7-34 所示,点击"新建"按钮。

图 7-34 loaddata 数据显示界面

C.进入 loaddata 定义界面如图 7-35 所示,系统会自动按顺序命名为"load+数字",可以改变命名使之更明确(loaddata 一经创建不可轻易改名,否则所有用到该有效载荷的程序都得进行修改)。所有模块中,有效载荷都应该是程序中的全局变量,有效载荷变量必须是持续变量。

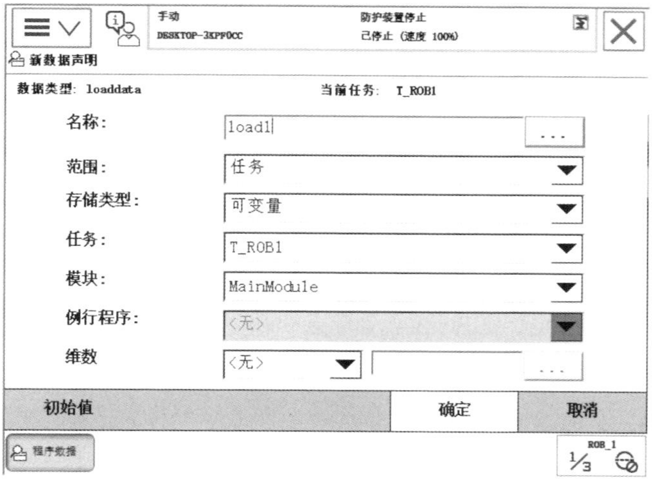

图 7-35 loaddata 参数设置界面

D.点击"初始值"按钮,进入 loaddata 参数设置界面如图 7-36 所示,图中定义的 load1 为:重量 5kg、重心坐标[50,0,40]、有效负载为一个点质量。

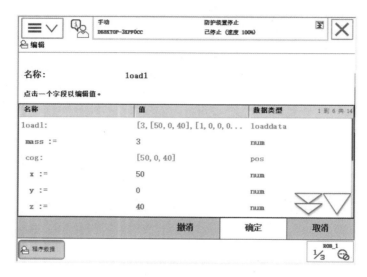

图 7-36　loaddata 参数设置界面

表 7-4 所示为 loaddata 数据的详细参数表,详细说明了各参数的含义和取值。

E.在编程过程中,需要根据负载变化对有效载荷的情况进行实时的调整,如图 7-37 所示。当夹具夹紧时,加载有效载荷 load1;当夹具松开时,切换为默认有效载荷 load0。

7.4　ABB 机器人常用指令

在 ABB 机器人的程序中,所有的机器人的行为都是通过 RAPID 语言或者 RAPID 指令来进行描述与控制的,不同的操作对应不同的指令,不同的指令语法和用法也不同。ABB 机器人的指令数量很多,本书主要介绍一些常用指令。

7.4.1　基本运动指令

ABB 工业机器人基本运动指令格式分为 5 个部分:运动方式,目标位置,

表 7-4 loaddata 数据参数表

组件	描述
mass	数据类型:num 负载的重量,单位:kg。
cog	全称:center of gravity 数据类型:pos 定义负载的重心位置,单位:mm:如果机械臂正夹持着工具,则有效负载的重心是相对于工具坐标系;如果使用固定工具,则有效负载的重心是相对于机械臂上的可移动的工件坐标系。
aom	全称:axes of moment 数据类型:orient 定义矩轴的方向姿态:是指处于 cog 位置的的有效负载惯性矩的主轴。如果机械臂正夹持着工具,则方向姿态是相对于工具坐标系;如果使用固定工具,则方向姿态是相对于可移动的工件坐标系。
ix	全称:inertia x 数据类型:num 定义负载绕着 X 轴的转动惯量,单位:$kg \cdot m^2$。 转动惯量的正确定义,有利于合理利用路径规划器和轴控制器。当处理大块金属板等时,该参数尤为重要。所有等于 0 的转动惯量 ix、iy 和 iz 均指一个点质量。
iy	全称:inertia y 数据类型:num 定义负载绕着 Y 轴的转动惯量
iz	全称:inertia z 数据类型:num 定义负载绕着 Z 轴的转动惯量

运行速度,转弯半径,工具中心点。机器人转弯半径示意如下图 7-38 所示,图中轨迹对应的指令如下。

- MoveL p10,v1000,z50,tool0;！机器人 TCP 到达距离目标点 p10 的 50mm 处绕过目标点。
- MoveL p20,v1000,fine,tool0;！机器人 TCP 到达目标点 p20 时速度降为零。

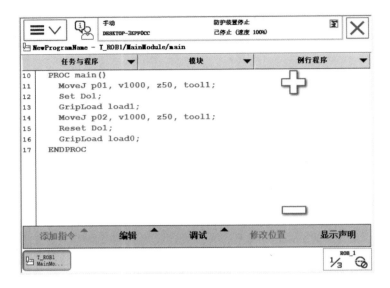

图 7-37 有效载荷使用示例

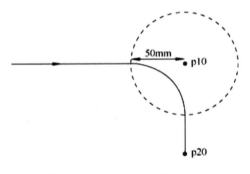

图 7-38 机器人转弯半径示意

程序指令中 fine 是指机器人 TCP 到达目标点时速度降为零，机器人动作有所停顿然后继续运动。zone 数据 z50 是指机器人 TCP 不达到目标点，而是在距离目标点所设置的数值长度（50mm）处圆滑绕过目标点。一般情况下，一段路径的最后一个点，设置转弯区尺寸为参数 fine。

在机器人基本运动中，大部分是由直线或圆弧运动轨迹组成的，比较复杂的运动轨迹也是由这些基本的运动轨迹组合而成，常用指令如下。

A.MoveJ（关节运动指令）：用于将机械臂迅速地从一点移动至另一点。机械臂和外轴沿非线性路径运动至目的位置，如图 7-39 所示，所有轴均同时达到目的位置，因此运动轨迹精度不高。

其指令语法如下，表 7-5 详细说明了 MoveJ 指令各参数的含义和取值。

MoveJ [\Conc] ToPoint [\ID] Speed [\V] | [\T] Zone [\Z] [\Inpos] Tool [\WObj] [\TLoad]

第 7 章 ABB 机器人编程基础

图 7-39 MoveJ 关节运动轨迹示意图

表 7-5 MoveJ 指令参数表

参数	参数
[\Conc]	全称：Concurrent 数据类型：switch 当机械臂正在运动时，执行后续指令。通常不使用参数，但是，当使用飞越点时，可使用参数以避免由过载 CPU 所引起的多余停止。当编程点在高速度下极为接近时，该参数适用。当不需要同外设备与机械臂运动之间的外设和同步进行通信时，参数亦适用。 运用参数 \Conc，将连续运动指令的数量限制为 5。在包含 StorePath-RestoPath 运动指令以及参数的程序段中，不允许 \Conc。 如果省略该参数，且 ToPoint 并非停止点，则在机械臂达到编程区之前，执行后续指令一段时间，不能将该参数用于 MultiMove 系统中的协调同步移动。
ToPoint	数据类型：robtarget 机器人和外部轴的目标点。定义为已命名的位置或直接存储在指令中（在指令中加 * 标记）。
[\ID]	全称：Synchronization id 数据类型：identno 如果协调同步运动，且在任何其他情形下均不被允许时，则必须在 MultiMove 系统中使用该参数。在所有合作程序任务中，指定 id 编号必须相同。id 编号能确保各运动不会在运行时混淆。
Speed	数据类型：speeddata 适用于运动的速度数据。速度数据规定了关于工具中心点、工具方位调整和外轴的速率。

（待续）

表 7-5(续)

参数	参数
[\V]	全称:Velocity 数据类型:num 该参数用于规定指令中 TCP 的速率,以 mm/s 计。用于取代速度数据中指定的相关速率。
[\T]	全称:Time 数据类型:num 该参数用于规定机械臂运动的总时间,以秒计。用于取代相关的速度数据。
Zone	数据类型:zonedata 定义相关移动的区域数据,区域数据描述了所生成拐角路径的大小。
[\Z]	全称:Zone 数据类型:num 该参数用于规定指令中机械臂 TCP 的位置精度。角路径的长度以 mm 计,用于替代区域数据中指定的相关区域。
[\Inpos]	全称:In position 数据类型:stoppointdata 该参数用于规定停止点中机械臂 TCP 位置的收敛准则,停止点数据用于取代 Zone 参数中的指定区域。
Tool	数据类型:tooldata 移动机械臂时正在使用的工具,工具中心点是指移动至指定目的点的点。
[\WObj]	全称:Work Object 数据类型:wobjdata 指令中机器人位置关联的工件坐标系。若省略该参数,则位置与世界坐标系相关。另一方面,如果使用固定式 TCP 或协调的外轴,则必须指定该参数。
[\TLoad]	全称:Total load 数据类型:loaddata \TLoad 主动轴描述了移动中使用的总负载。如果使用了 \TLoad 自变数,那么就不考虑当前 tooldata 中的 loaddata。 如果 \TLoad 自变数被设置成 load0,那么就不考虑 \TLoad 自变数,而是以当前 tooldata 中的 loaddata 作为代替。

• MoveJ 指令使用示例 1：

MoveJ *，v2000\V:=2200，z40 \Z:=45，grip3；

工具的 TCP grip3 沿非线性路径运动至指令中储存的位置。将数据设置为 v2000 和 z40 时，开始运动。TCP 的速率和区域半径分别为 2200 mm/s 和 45 mm。

• MoveJ 指令使用示例 2：

MoveJ p5，v2000，fine \Inpos :=inpos50，grip3；

工具的 TCP grip3 沿非线性路径运动至停止点 p5。当满足关于停止点 fine 的 50%的位置条件和 50%的速度条件时，机械臂认为该工具位于点内。其最多等待 2 秒，以满足各条件。

B.MoveL(TCP 直线运动指令)：用于将工具中心点沿直线移动至给定目的，运动轨迹保持直线如图 7-40 所示。当 TCP 保持固定时，则该指令亦可用于调整工具方位。

图 7-40　MoveL 直线运动轨迹示意图

其指令语法如下，各参数的含义和取值与 MoveJ 指令相似。

MoveL [\Conc] ToPoint [\ID] Speed [\V] | [\T] Zone [\Z] [\Inpos] Tool [\WObj] [\Corr] [\TLoad]

• MoveL 指令使用示例 1：

MoveL *，v2000 \V:=2200，z40 \Z:=45，grip3；

工具的 TCP grip3 沿直线运动至指令中储存的位置。将数据设置为 v2000 和 z40 时，开始运动。TCP 的速率和区域半径分别为 2200 mm/s 和 45 mm。

• MoveL 指令使用示例 2：

MoveL p5，v2000，fine \Inpos :=inpos50，grip3；

工具的 TCP grip3 沿直线运动至停止点 p5。当满足关于停止点 fine 的 50%的位置条件和 50%的速度条件时，机械臂认为该工具位于点内。其最多等待 2 秒，以满足各条件。

• MoveL 指令使用示例 3：

MoveL start，v2000，z40，grip3 \WObj:=fixture；

工具的 TCP grip3 沿直线运动至在 fixture 的工件坐标系中指定的位置 start。

C.MoveC(TCP 圆弧运动指令)：用于将工具中心点(TCP)沿圆周移动至给定目的地。移动期间,该周期的方位通常相对保持不变。众所周知,三点确定一段圆弧,MoveC 指令运动的实际轨迹与起点密切相关,图 7-41 所示代码如下。

MoveL p40,v200,z50,tool1；

MoveC p50,p60,v200,fine,tool1；

上述代码中机器人 TCP 从前一位置以直线运动到达圆弧起始点 p40，接着从起始点 p40 到终点 p60 做圆弧运动,p50 为圆弧中间点,运行速度为 200mm/s,转弯区半径是 fine,使用的

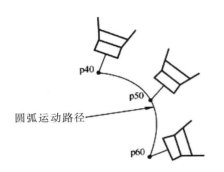

图 7-41 MoveC 圆弧运动轨迹示意图

工具是 tool1。由此可以看出,若想圆弧轨迹确定,必须有严格规定圆弧起点的运动指令和 MoveC 组合出现。

其指令语法如下,相比 MoveJ 和 MoveL 指令多了圆弧点 CirPoint,圆弧点是指相关起点与终点间的圆弧上的某个位置。若要获得最好的准确度,则宜把该点放在相关起点与终点的正中间处。如果该点太靠近起点或终点,那么相关机器人就可能发出一条警告。其它参数的含义和取值与 MoveJ 指令相似。

MoveC [\Conc] CirPoint ToPoint [\ID] Speed [\V] | [\T] Zone [\Z] [\Inpos] Tool [\WObj] [\Corr] [\TLoad]

• MoveC 指令使用示例：

MoveL p1，v500，fine，tool1；

MoveC p2，p3，v500，z20，tool1；

MoveC p4，p1，v500，fine，tool1；

上述显示了如何通过两个 MoveC 指令(注意 MoveL 给出了圆弧起点),实施一个完整的周期,轨迹运行效果如图 7-42 所示。

D.MoveAbsJ(轴绝对角度位置运动指令)：用于将机械臂和外轴移动至轴

位置中指定的绝对位置。使用 MoveAbsJ 运动期间，机械臂的位置不会受到给定工具和工件以及有效程序位移的影响。机械臂运用该数据，以计算负载、TCP 速度和拐角路径。

其指令语法如下，其中 ToJointPos 为关节目标点数据，如果参数 \NoEOffs 得以设置，则关于 MoveAbsJ 的运动将不受外轴有效偏移量的影响。

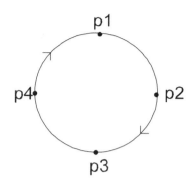

图 7-42　整圆运动轨迹示意图

MoveAbsJ [\Conc] ToJointPos [\ID] [\NoEOffs] Speed [\V] | [\T] Zone [\Z] [\Inpos] Tool [\WObj] [\TLoad]

- MoveAbsJ 指令使用示例：

MoveAbsJ p50, v1000, z50, tool2;

通过速度数据 v1000 和区域数据 z50，机械臂以及工具 tool2 得以沿非线性路径运动至绝对轴位置 p50。

7.4.2 程序内的逻辑控制指令

1.Compact IF：如果条件满足，就执行一条指令。语法如下，判断 Condition 是否为 TRUE，如果是 TRUE，则进行后面指令的操作；如果为 False，则不进行后面的操作，程序指针指向下一个语句（该选择指令没有分支选项）。

IF <conditional expression> (<instruction> | <SMT>) ';'

- Compact IF 指令使用示例：

　　IF counter > 10 Set do1;

如果 counter > 10，则设置 do1 信号。

B.IF：条件选择指令，如果满足条件，那么...；否则...。语法如下：

　　IF Condition THEN ...

　　{ELSEIF Condition THEN ...}

　　[ELSE ...]

　　ENDIF

- IF 指令使用示例：

　　IF counter > 100 THEN

```
            counter :=100;
    ELSEIF counter < 0 THEN
            counter :=0;
    ELSE
            counter :=counter + 1;
    ENDIF
```

counter 在 0~100 以内,使 counter 增加 1,但是,如果 counter 的数值超出限值 0-100,则向 counter 分配相应的限值。

C.For:循环指令,当一个或多个指令重复多次时,使用 FOR。语法如下,需要设置循环计数器 loop counter 的起始值、结束值和增量(省略时默认为 1)。

FOR Loop counter FROM Start value TO End value [STEP Step value]
DO ... ENDFOR

- FOR 指令使用示例:

```
    FOR i FROM 1 TO 10 DO
        routine1;
    ENDFOR
```

重复 routine1 无返回值程序 10 次。

D.While:循环指令,只要给定条件表达式评估为 TRUE 值,当重复一些指令时,使用 WHILE。语法如下:

WHILE Condition DO ... ENDWHILE

- While 指令使用示例:

```
    WHILE reg1 < reg2 DO
        ...
        reg1 :=reg1 + 1;
    ENDWHILE
```

其中 reg1 为条件变量,只要 reg1 < reg2,则重复 WHILE 块中的指令。

E.Test:条件选择指令,根据表达式或数据的值,当有待执行不同的指令时,使用 TEST。如果并没有太多的替代选择,也可使用 IF..ELSE 指令。应用示例如下:

```
    TEST reg1
    CASE 1,2,3 :
```

routine1;

CASE 4 :

routine2;

DEFAULT :

TPWrite "Illegal choice";

Stop;

ENDTEST

根据 reg1 的值,执行不同的指令。如果该值为 1、2 或 3 时,则执行 routine1。如果该值为 4,则执行 routine2。否则,打印出错误消息,并停止执行。

F.GOTO:跳转指令,用于将程序执行转移到相同程序内的另一线程(标签)。应用示例如下:

reg1 :=1;

next:

...

reg1 :=reg1 + 1;

IF reg1<=5 GOTO next;

程序将执行转移至 next 四次(reg1=2、3、4、5)。

G.Label:用于命名程序中的程序。常用作跳转标签与 GOTO 指令配合使用,该名称随后可用于移动相同程序内的程序执行,上面程序中的"next:"。Label 指令的语法是<identifier>':',不得与同一程序内的所有其他标记或所有数据名称相同。

7.4.3 停止程序执行

A.Stop:用于停止程序执行。在 Stop 指令就绪之前,将完成当前执行的所有移动。语法如下:

Stop [\NoRegain] | [\AllMoveTasks]

[\NoRegain]:数据类型 switch,指定下一程序的起点,无论受影响的机械单元是否应当返回停止位置。如果已设置参数 \NoRegain,则机械臂和外轴将不会返回停止位置(如果他们已远离停止位置)。如果省略该参数,且如果机械臂或外轴已从停止位置慢慢远离,则机械臂会在 FlexPendant 示教器上显示问题。随后,用户可回答机械臂是否应当返回停止位置。

[\AllMoveTasks]:数据类型 switch,指定所有运行中的普通任务以及实际任务中应当停止的程序。如果省略本参数,则仅将停止执行本指令的任务中的程序。

- Stop 指令使用示例:

MoveL p1, v500, fine, tool1;

TPWrite "Jog the robot to the position for pallet corner 1";

Stop\NoRegain;

p1_read :=CRobT(\Tool:=tool1 \WObj:=wobj0);

MoveL p2, v500, z50, tool1;

通过位于 p1 的机械臂,停止程序执行。运算符点动,机械臂移动至 p1_read。关于下一次程序起动,机械臂并未恢复至 p1,因此,可将位置 p1_read 储存在程序中。

B.EXIT:用于终止程序执行。当出现致命错误或永久地停止程序执行时,应当使用 EXIT 指令。在执行指令 EXIT 后,程序指针消失,为继续程序执行,必须设置程序指针。

- EXIT 指令使用示例:

ErrWrite "Fatal error","Illegal state";

EXIT;

程序执行停止,且无法从程序中的该位置重启。

C.Break:用于立即中断程序执行。指令立即停止程序执行,且无需等待机械臂和外轴达到当时其程序规定的运动目的点。随后,可以从下一个指令重新开始程序执行。如果某些程序事件中存在 Break 指令,则将中断程序的执行,且将不会执行任何停止程序事件。

7.4.4 赋值指令

":=":赋值指令用于向数据分配新值。该值可以是一个恒定值,亦可以是一个算术表达式。语法为 Data :=Value,使用示例如下:

reg1 :=5;

counter :=counter + 1;

7.4.5 等待指令

A.WaitTime:用于等待给定的时间,程序再往下执行。该指令亦可用于等待,直至机械臂和外轴静止。程序执行等待的最短时间(以秒计)为 0 s。最长时间不受限制,分辨率为 0.001 s,语法如下:

WaitTime [\InPos] Time

- WaitTime 指令使用示例:

WaitTime \InPos,0;

程序执行进入等待,直至机械臂和外轴已静止。

B.WaitDI:用于等待,直至已设置数字信号输入。语法如下:

WaitDI Signal Value [\MaxTime] [\TimeFlag]

表 7-6 详细说明了 WaitDI 指令各参数的含义和取值,当执行本指令时,如果信号值正确,则本程序仅仅继续以下指令;如果信号值不正确,则机械臂进入等待状态,且当信号改变为正确值时,程序继续。如果等待超出最长时间值,则程序将在指定 TimeFlag 时继续,否则将会引起错误。如果指定 TimeFlag,

表 7-6 WaitDI 指令参数表

参数	说明
Signal	数据类型:signaldi 信号的名称。
Value	数据类型:dionum 信号的期望值。
[\MaxTime]	全称:Maximum Time 数据类型:num 允许的最长等待时间,以秒计。如果在满足条件之前耗尽该时间,则将调用错误处理器,错误代码 ERR_WAIT_MAXTIME。如果不存在错误处理器,则将停止执行。
[\TimeFlag]	全称:Timeout Flag 数据类型:bool 如果在满足条件之前耗尽最长允许时间,则包含该值的输出参数为 TRUE。如果该参数包含在本指令中,则不将其视为耗尽最长时间时的错误。如果 MaxTime 参数不包括在本指令中,则将忽略该参数。

则在超出时间时设置为TRUE;否则,其将设置为FALSE。如果停止程序执行,并随后重启,则本指令评估信号的当前值,否定程序停止期间的任意改变。

- WaitDI指令使用示例:

　　WaitDI di4, 1;

仅在已设置di4输入后,继续程序执行。

C.WaitDO:用于等待,直至已设置数字信号输出。语法与用法和WaitDI类似。

D.WaitGI:用于等待,直至将一组数字信号输入信号设置为指定值。语法如下。

- WaitGI Signal [\NOTEQ] | [\LT] | [\GT] Value | Dvalue [\MaxTime] [\ValueAtTimeout] | [\DvalueAtTimeout]

表7-7详细说明了WaitGI指令各参数的含义和取值。

- WaitGI指令使用示例:

　　WaitGI gi1,\LT,1;

仅在gi1小于1之后继续程序执行。

E.WaitGO:用于等待,直至将一组数字输出信号设置为指定值。语法与用法和WaitGI类似。

F.WaitAI:用于等待,直至已设置模拟信号输入信号值。语法如下,语法与用法和WaitGI类似。

　　WaitAI Signal [\LT] | [\GT] Value [\MaxTime] [\ValueAtTimeout]

- WaitAI指令使用示例:

　　WaitAI ai1, \GT, 5;

仅在ai1模拟信号输入具有大于5的值之后,方可继续程序执行。

G.WaitAO:用于等待,直至已设置模拟信号输出信号值。语法与用法和WaitAI类似。

7.4.6 输入输出信号值设定指令

A.Reset:用于将数字信号输出信号的值重置为0。如:Reset do15;

B.Set:用于将数字信号输出信号的值设置为1。如:Set do15;

C.SetDO:用于改变数字信号输出信号的值。语法如下。

- SetDO [\SDelay][\Sync] Signal Value

表 7-7　WaitGI 指令参数表

参数	说明
Signal	数据类型：signalgi
	数字组输入信号的名称。
[\NOTEQ]	全称：NOT EQual
	数据类型：switch
	如果使用该参数，则 WaitGI 指令等待，直至数字组信号值除以 Value 中的值。
[\LT]	全称：Less Than
	数据类型：switch
	如果使用该参数，则 WaitGI 指令等待，直至数字组信号值小于 Value 中的值。
[\GT]	全称：Greater Than
	数据类型：switch
	如果使用该参数，则 WaitGI 指令等待，直至数字组信号值大于 Value 中的值。
Value	数据类型：num
	信号的期望值。必须为所用数字组输入信号工作范围内的整数值，允许值取决于该组中的信号数量。
Dvalue	数据类型：dnum
	信号的期望值，与 Value 用法一样，数据类型不同。
[\MaxTime]	全称：Maximum Time
	数据类型：num
	允许的最长等待时间，以秒计。用法与 WaitDi 相同。
[\ValueAt Timeout]	数据类型：num
	如果指令超时，则将当前信号值储存在该变量中。仅将系统变量 ERRNO 设置为 ERR_WAIT_MAXTIME 时，方才设置该变量。
[\DvalueAt Timeout]	数据类型：dnum
	如果指令超时，则将当前信号值储存在该变量中。仅将系统变量 ERRNO 设置为 ERR_WAIT_MAXTIME 时，方才设置本变量。

表 7-8 详细说明了 SetDO 指令各参数的含义和取值。

• SetDO 指令使用示例 1：

SetDO \SDelay :=0.2, weld, high;

将信号 weld 设置为 high，且时间延迟为 0.2 s。通过下一指令，继续程序执行。

表 7-8 SetDO 指令参数表

参数	说明
[\SDelay]	全称：Signal Delay 数据类型：num 延迟时间改变(以秒计,最多 2000s)。通过下一指令,直接继续程序执行。在给定的时间延迟之后,改变信号,且随后程序执行不受影响。
[\Sync]	全称：Synchronization 数据类型：switch 如果使用该参数,则程序执行将进入等待,直至从物理上将信号设置为指定值。
Signal	数据类型：signaldo 待改变信号的名称。
Value	数据类型：dionum 信号的期望值,0 或 1。
Dvalue	数据类型：dnum 信号的期望值,与 Value 用法一样,数据类型不同。
[\MaxTime]	全称：Maximum Time 数据类型：num 允许的最长等待时间,以秒计。用法与 WaitDi 相同。
[\ValueAt Timeout]	数据类型：num 如果指令超时,则将当前信号值储存在该变量中。仅将系统变量 ERRNO 设置为 ERR_WAIT_MAXTIME 时,方才设置该变量。
[\DvalueAt Timeout]	数据类型：dnum 如果指令超时,则将当前信号值储存在该变量中。仅将系统变量 ERRNO 设置为 ERR_WAIT_MAXTIME 时,方才设置本变量。

- SetDO 指令使用示例 2：

  SetDO \Sync ,do1, 0;

  将信号 do1 设置为 0。程序执行进入等待,直至从物理上将信号设置为指定值。

  D.SetGO：用于改变一组数字信号输出信号的值。语法如下：

  SetGO [\SDelay] Signal Value | Dvalue

E.SetAO：用于改变模拟信号输出信号的值。语法如下：

SetAO Signal Value

在将编程值发送至物理通道之前，对于编程值按照模拟量定义的上下限制进行计算。图7-43显示了关于如何测量模拟信号值的图表。

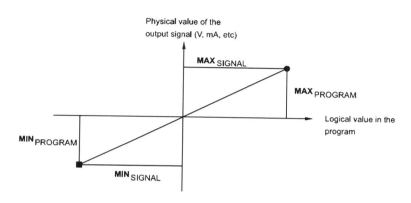

图7-43　程序值与物理值关系对应图

7.4.7 程序调用指令

ProcCall：调用新无返回值程序。过程调用用于将程序执行转移至另一个无返回值程序。当充分执行本无返回值程序时，程序将继续执行过程调用后的指令，还可以将一系列参数发送至新的无返回值程序。例行程序的定义分为两种形式。

- 不带参数的例行程序，这是使用最多的一种。

  PROC global_routine();

  ...

  ENDPROC;

- 带参数的例行程序。

  PROC routine1(num in_par, INOUT num inout_par, VAR num var_par, PERS num pers_par);

  ...

  ENDPROC;

在调用带参数的例行程序时,必须提供相应的实参。例行程序的参数有四种存取模式。

A.INPUT:通常这个参数仅仅被用来作为例行程序输入,如果参数类型省略则默认为 INPUT。在例行程序里更改变量不会影响相应的输入参数,参数可以是 CONST,VAR 和 PERS 类型。

B.INOUT:如果例行程序参数被设置为该模式,则相应的参数必须是可以被程序修改的 Var 和 PERS 数据,在调用过程中可以修改相应的输入参数。但如果是 Var 则不会更改初始值,是 Pers 则会更改初始值。

C.VAR:如果例行程序参数被设置为该模式,则相应的参数必须是可以被程序修改的 Var。实参可以被修改,但不会更改 Var 实参的初始值。

D.PERS:如果例行程序参数被设置为该模式,则相应的参数必须是可以被程序修改的 PERS 数据。实参可以被修改,同时会修改 Pers 实参的初始值。

例行程序也可以定义可选参数:一个例行程序参数可以被设置为可选参数,可选参数带有"\"标示符,在调用例行程序时,此参数可以缺省。例如:

PROC routine (num required_par \num optional_par)

Switch 则可以用来选择使用哪个可选参数,例如:

PROC routine4(\switch on|switch off)

例行程序的结束:可以通过 Return 指令终止运行,也可以通过例行程序的结束标示符(ENDPROC,BACKWARD,ERROR,UNDO)终止运行。

例行程序的调用:Procedure 的调用在示教器上是通过指令 ProCall 来完成的。在调用带有参数的例行程序时,强制参数必须指定,而且参数的顺序要正确;可选参数可以缺省。使用示例如下,当该无返回值程序就绪时,程序执行返回过程调用后的指令 Set do1。

Routine1 10;

Set do1;

...

PROC Routine1(num Count1)

  ...

ENDPROC

7.4.8 功能函数

A.offs:用于将一个机械臂位置在工件坐标系进行偏移。语法如下:

Offs (Point XOffset YOffset ZOffset)

下面的指令将机械臂移动至距位置 p2(沿 z 方向)10 mm 的位置:

MoveL Offs(p2, 0, 0, 10), v1000, z50, tool1;

B.RelTool:用于将通过有效工具坐标系表达的位移(和/或旋转)增加至机械臂位置。语法如下,其中 Dx、Dy、Dz 工具坐标系的平移量,[\Rx]、[\Ry]、[\Rz]分别为绕工具坐标系 x、y、z 轴旋转的角度。如果同时指定两次或三次旋转,则将首先围绕 x 轴旋转,随后围绕新的 y 轴旋转,最后围绕新的 z 轴旋转。

RelTool (Point Dx Dy Dz [\Rx] [\Ry] [\Rz])

下面的指令将工具围绕其 z 轴旋转 25°:

MoveL RelTool(p1, 0, 0, 0 \Rz:=25), v100, fine, tool1;

第 8 章 ABB 机器人程序编写

工业机器人编程是指为了使机器人完成某项作业而进行的程序设计。高效、合理的程序是机器人完成复杂、精确动作和自动化运行的重要保障,工业机器人编程主要有以下两种方式。

A.在线编程:在线编程也叫示教编程,是指操作员通过示教器控制机械手工具末端到达指定的姿态和位置,记录机器人姿态数据并编写机器人运动指令,以完成机器人工作的轨迹规划。是目前大多数工业机器人应用的编程方式,具有以下特点:

- 操作简单,容易使用,对于简单应用响应速度快;
- 编程效率低;
- 需要在现场操作实际机器人系统,编程作业导致机器人生产停机,降低设备利用率,且现场环境可能存在一定危险性;
- 难以实现复杂的机器人运行轨迹;
- 编程质量取决于编程者经验。

B.离线编程:离线编程是指基于CAD数据的计算机图形编程系统,通过对工作单元进行三维建模构建仿真工作场景,采用算法对机器人进行控制和轨迹规划,并对编程结果进行三维图形学动画仿真来检测编程的可靠性,最后将生成机器人实际生产的代码。相比示教编程,离线编程有以下特点:

- 离线编程不占用机器人生产时间,提高设备利用率;
- 改善了编程环境,使编程者远离危险的工作环境;
- 便于和 CAD/CAM 系统结合,编程效率高;
- 编程质量有保证,具有一致性;

- 可实现复杂的机器人运行轨迹,且便于程序修改;
- 技术复杂,对编程者要求高。

通过比较可以看出,在线编程和离线编程各有优缺点,但并不是对立存在而是互补存在的。使用者要根据实际应用领域和场景,综合评价工作效率、编程质量以及经济性,选择合理的编程方式。

8.1 示教编程

RAPID 程序是由程序模块与系统模块组成。一般来说,只通过新建程序模块来构建机器人的程序,而系统模块多用于系统方面的控制。

8.1.1 创建例行程序

创建例行程序步骤如下。

A.点击示教器屏幕左上角的 ABB 菜单按钮,选择"程序编辑器"选项,首次进入会显示图 8-1 所示 main 程序编辑界面,控制台会提示"未命名程序"。

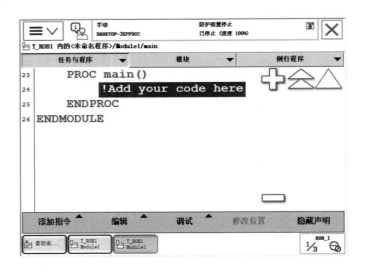

图 8-1 main 程序编辑界面

B.单击"任务与程序"选项卡进入如图 8-2 所示界面,一个任务对应一个程序名称,可通过左下角的文件菜单,对程序进行新建、加载、另存、重命名、删除等操作,如图中命名为"Transfer"。

图 8-2 任务与程序界面

C.选中任务"T_ROB1"点击右下角"显示模块"选项,进入如图 8-3 所示任务模块界面。Program 内创建多个模块有利于程序的分类管理,系统默认有三个模块:BASE、MainModule、user。左下角文件菜单可以对模块进行新建、加载、另存、更改声明、删除等操作。

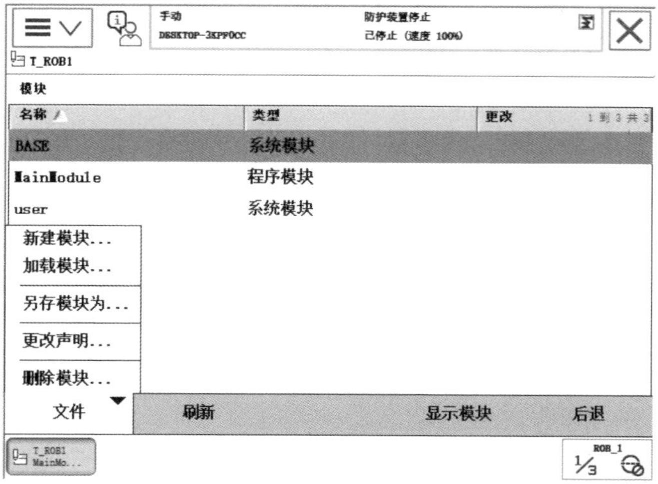

图 8-3 任务模块界面

D.选中 BASE 模块,单击"显示模块"按钮,可以看到 BASE 模块对 tool0、wobj0 和 load0 的定义(注意:系统模块、切勿修改),如图 8-4 所示。

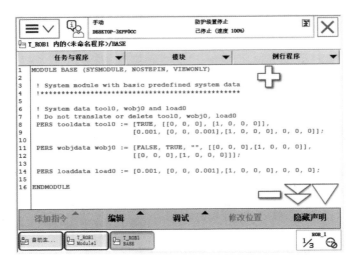

图 8-4　BASE 模块定义

E.选中 MainModule 模块,单击"显示模块"按钮,可以看到 main 程序默认在该模块下,单击"文件"菜单"新建例行程序"选项,如图 8-5 所示。

图 8-5　MainModule 模块界面

F.进入例行程序声明界面如图 8-6 所示,按需求修改名称、设置参数、选择模块,完成后单击"确定"。

图 8-6　新建例行程序界面

G.返回 MainModule 模块显示界面如图 8-7 所示,选中 Routine1()例行程序并点击"显示例行程序"按钮。

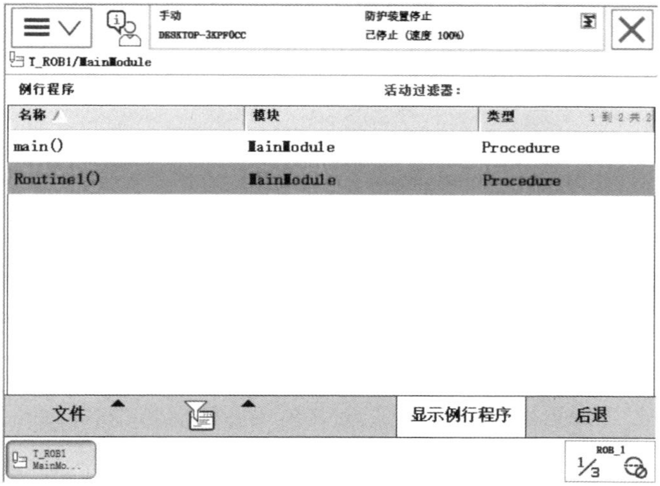

图 8-7　MainModule 模块界面

8.进入例行程序编辑界面如图 8-8 所示。

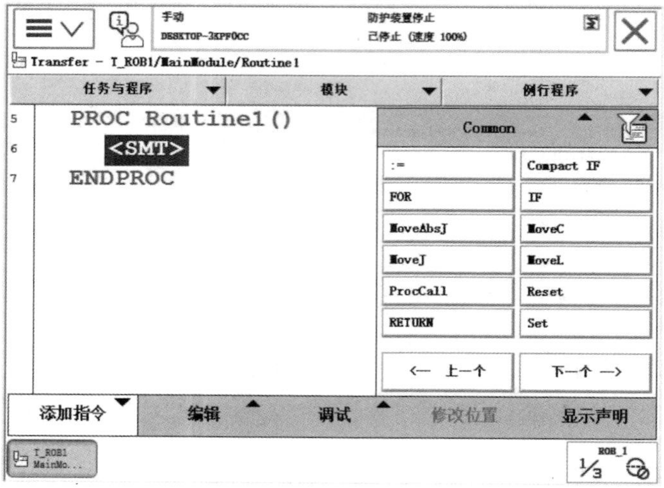

图 8-8 例行程序编辑界面

8.1.2 例行程序编写

工业机器人示教编程可采用以下三种方式：先示教点位后编程，先编程后示教点位，边示教点位边编写程序。一般来讲，先编程后示教的方式对操作人员的要求较高，需要做到思路清晰，编程熟练，但也是最省时间的方式。

假如机器人需要将一工件由 A 点搬运至 B 点，则需要定义至少三个点位：起始点、A 和 B 点，需要定义夹具夹紧的 DO 信号等。定义程序数据之后编程会更加方便，具体步骤如下。

A.首先定义 jointtarget 数据 jpos10，如图 8-9 所示。

B.一般来说，机器人程序的第一个点位最好定义为关节类型，而且采用 MoveAbsJ 运动方式，这样有利于工业机器人的姿态调整，jpos10 当前状态对应的机器人姿态如图 8-10 所示。

C.定义 robtarget 类型数据 p10、p20，对应搬运时的起点 A 和终点 B，如图 8-11 所示。新定义的数据默认为机器人当前位姿对应的[X,Y,Z,A,B,C]，先用 p10、p20 编程，之后再示教其坐标。

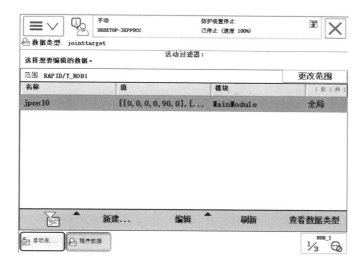

图 8-9 jointtarget 数据

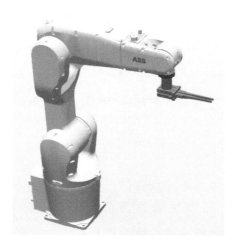

图 8-10 robjoint1 对应的机器人姿态

D.点击左下角"添加指令"按钮,添加 MoveAbsJ 指令如图 8-12 所示。

E.选中"*"字段双击进入指令参数编辑界面如图 8-13 所示,系统默认的关节数据为机器人当前姿态对应的关节角度值。

F.单击"ToJointPos"选项关节位置数据选择界面如图 8-14 所示,选中之前创建的 jpos10 完成参数更换。

图 8-11 robtarget 数据定义

图 8-12 添加 MoveAbsJ 指令

图 8-13 MoveAbsJ 指令参数界面

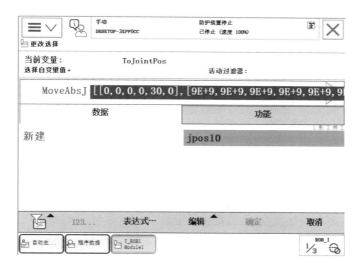

图 8-14 关节位置数据设置

G.选中 jpos10 数据之后效果如图 8-15 所示,还可以在当前界面下修改速度、工具等参数,单击确定完成参数设置。

H.继续添加 MoveL 指令,双击进入参数编辑界面,选中数据 p10,点击"功能"选项卡,点击"Offs"选项添加位置偏移,如图 8-16 所示。

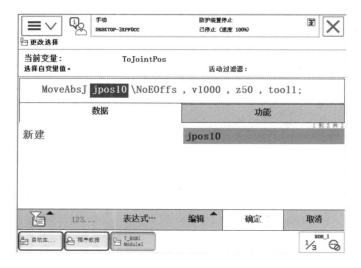

图 8-15　MoveAbsJ 指令参数设置界面

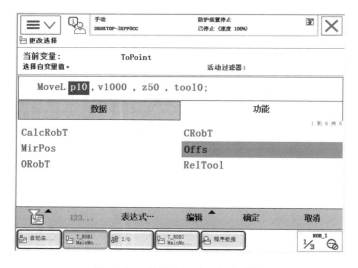

图 8-16　位置参数功能添加界面

I.进入 Offs 参数设置界面如图 8-17 所示，三个偏移量<EXP>可以用变量表示，也可以选中"编辑"菜单"仅限选定内容"直接输入数值。

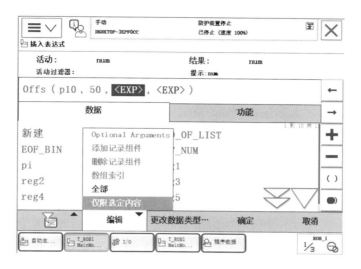

图 8-17 Offs 参数设置界面

J.按以上方法继续输入指令,完成 Routine1( )例行程序的编写如下:
　　PROC Routine1( )
　　　　MoveAbsJ jpos10\NoEOffs, v1000, z50, tool1; /程序 home 点
　　　　MoveL Offs(p10,-60,0,0), v500, fine, tool1; /接近工件位置
　　　　MoveL p10, v30, fine, tool1; /低速运动至夹紧位置
　　　　Set Do1;　　　　　　　　　/气动手爪夹紧
　　　　WaitTime 2;　　　　　　　　/等待 2s 确保夹紧动作完成
　　　　MoveL Offs(p10,0,0,100), v100, z50, tool1;/低速向上夹起工件
　　　　MoveL Offs(p20,0,0,100), v300, z50, tool1;/运动至放置点上方
　　　　MoveL p20, v20, fine, tool1; /低速运动至工件放置点
　　　　Reset Do1;　　　　　　　　/气动手爪松开
　　　　WaitTime 2;　　　　　　　　/等待 2s 确保夹紧动作完成
　　　　MoveL Offs(p20,-60,0,0), v100, z50, tool1; /机械手水平退出
　　　　MoveL Offs(p20,-60,0,100), v300, z50, tool1;/机械手向上抬起
　　　　MoveAbsJ jpos10\NoEOffs, v1000, z50, tool1;/返回 home 点
　　ENDPROC

8.1.3 程序点位示教

先编程后示教的好处是程序思路清晰,可以避免很多无用的点位示教,比如 Routine1()中利用点位偏移 Offs 实现了多个点位目标,而真正需要示教的点位只有两个。定义程序数据之后编程会更加方便,具体步骤如下。

A.手动操纵机器人以正确姿态接近目标位置,由远及近逐步调小增量模式的档位,控制工具到合适的工作位置。在例行程序中选中位置数据"p10",点击右下角"修改位置"按钮,如图 8-18 所示。系统会将机器人当前位置数据赋给 p10,这就是示教编程的由来。示教完 P10 点位可以点击"调试"菜单的"PP 移至光标"选项,点击示教器上的单步运行按钮,做到边示教边调试程序。

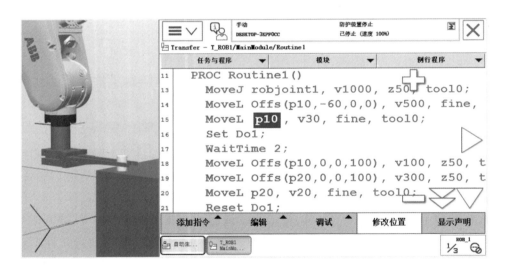

图 8-18　p10 点位示教

B.同样的办法手动操纵机械手将工件夹持至目标点,选中位置数据 p20 进行点位示教,如图 8-19 所示。

C.完成示教后还需对程序进行调试,以检验点位示教和编程是否正确。点击"调试"菜单,将 PP 指针移动到合适位置(一般为程序头)。控制示教器上的启动、步进、步退、停止按钮可进行程序的连续或步进运行,如图 8-20 所示。

第 8 章 ABB 机器人程序编写

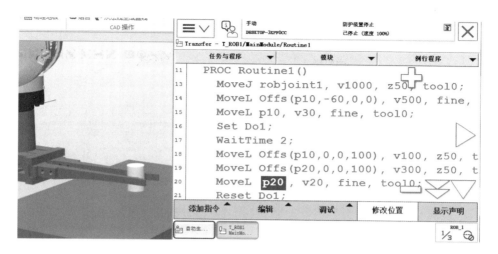

图 8-19　p20 点位示教

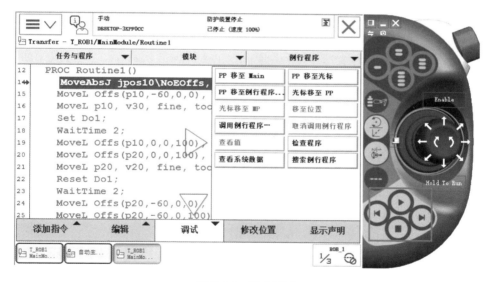

图 8-20　程序调试

8.1.4　码垛程序编写

　　码垛机器人的动作跟搬运动作类似，区别是搬运一般是一个固定点到另一个固定点或者是多个固定点，码垛则是一个或多个固定点到一组相互关联的点，使用循环指令 FOR 或 WHILE 可以有效简化码垛程序的结构。FOR 指令的具体用法如下。

A.新建例行程序 Routine2,添加 FOR 指令如图 8-21 所示。

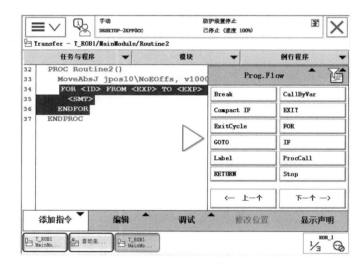

图 8-21 添加 FOR 指令

B.选中 FOR 字段双击进入 FOR 指令结构编辑界面如图 8-22 所示。

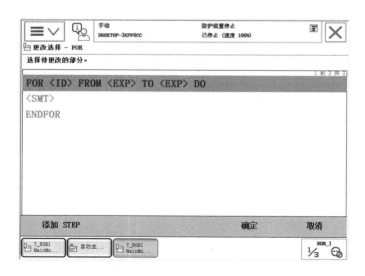

图 8-22 FOR 指令编辑

C.单击"添加 STEP"按钮增加增量参数如图 8-23 所示。

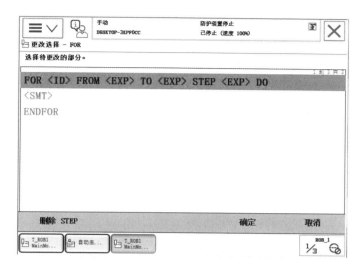

图 8-23　添加 STEP 增量

D.双击"<ID>"字段,打开输入面板并输入计数器变量名(无需提前在程序数据内定义)如图 8-24 所示。

图 8-24　计数器变量输入面板

E.这样在放置位置 Offs 中可以使用计数器 num1 的表达式来表示偏移量,如图 8-25 所示,Z 值偏移量设置为:num1*50+100。

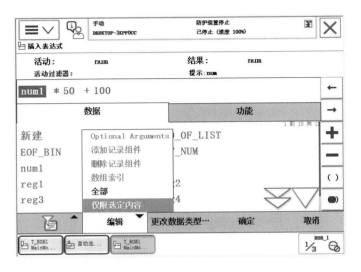

图 8-25 Offs 参数的表达式设置

F.按以上方法继续输入指令,完成 Routine2( )码垛例行程序的编写如下。

PROC Routine2( )

 MoveAbsJ jpos10\NoEOffs, v1000, z50, tool0;

 FOR num1 FROM 0 TO 3 STEP 1 DO

 MoveL Offs(p10,-60,0,0), v500, fine, tool0;

 MoveL p10, v30, fine, tool0;

 Set Do1;

 WaitTime 2;

 MoveL Offs(p10,0,0,100), v100, z50, tool0;

 MoveL Offs(p20,0,0,num1 * 50 + 100), v300, z50, tool0;

 MoveL Offs(p20,0,0,num1 * 50), v20, fine, tool0;

 Reset Do1;

 WaitTime 2;

 MoveL Offs(p20,-60,0,num1 * 50), v100, z50, tool0;

 MoveL Offs(p20,-60,0,num1 * 50 + 100), v300, z50, tool0;

ENDFOR

ERROR
　　MoveAbsJ jpos10\NoEOffs，v1000，z50，tool0；
ENDPROC

8.2 离线编程

　　离线编程可以实现更加复杂的轨迹，而且，不同于示教编程精度完全是靠示教者的目测决定，离线编程拥有很高的理论精度。但是由于机器人在设定工具坐标系和工件坐标系时存在误差，因此离线编程实际运行的轨迹精度与操作者的经验直接相关。提高工具坐标系和工件坐标系的标定精度可以借助工装、外部高精度测量设备或探头等，这样可以进一步降低人为因素对离线编程实际运行轨迹精度的影响。离线编程又可以分为二维离线编程和三维离线编程。

8.2.1 二维离线编程

　　二维离线编程顾名思义是指机器人的 TCP 点轨迹在一个平面内且工具 Z 轴保持不变。由于 Z 轴固定，因此工具姿态变化简单，轨迹容易控制、超程、干涉等情况较少。其操作步骤如下。

　　A.打开 RobotStudio 软件并创建 IRB1200 机器人工作站，加载 myTool 工具并安装到法兰端面。以 RobotStudio 软件自带的模型为例，依次点击菜单"基本—导入模型库—设备—Curve Thing"，如图 8-26 所示。

图 8-26　导入 Curve Thing 模型

B.新导入的模型在机器人基坐标系中需要确定摆放位置,在 Curve Thing 右键菜单中依次选择"位置—设定位置",如图 8-27 所示。

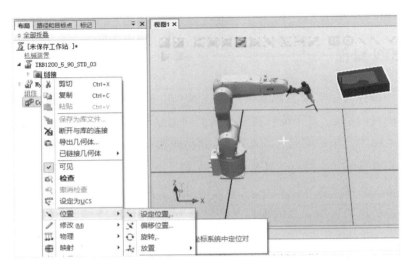

图 8-27 模型位置设定选项

C.在弹出的对话框中设置位置和方向参数,一般情况下参考大地坐标比较容易,移动前后的位置如图 8-28 所示。注意:工件在机器人工作站中的位置最好与实际生产的位置接近,虽然可以通过标定工件坐标系的方式消除这种差别,但是太大的位置差别可能会导致机器人超程或干涉。本案例采用机器人基坐标系为工件坐标系。

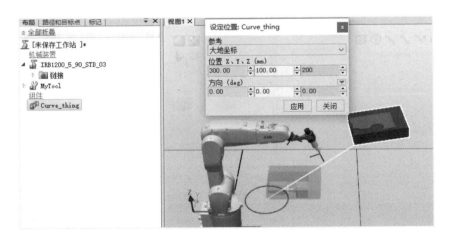

图 8-28 设定位置对话框

D.依次点击菜单"基本—路径—自动路径",弹出由曲线创建路径的对话框,如图 8-29 所示。选择曲线时需注意目标点的 Z 轴方向,图中蓝色轴线代表目标点 Z 向,红色轴线代表 X 向。另外需要注意运动方式(直线、圆弧)、公差(用于直线近似曲线)等选项的设置。

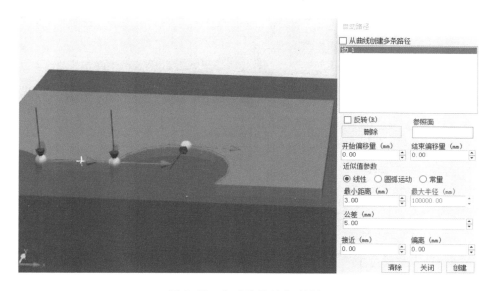

图 8-29 自动路径创建对话框

E.依次选择上表面各条边如图 8-30 所示。

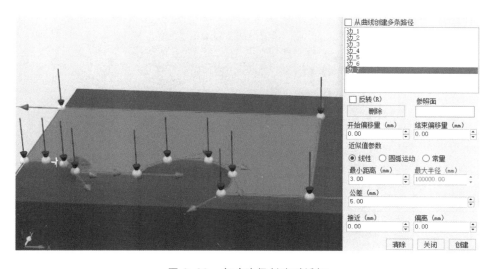

图 8-30 自动路径创建对话框

F.单击"创建"按钮完成 Path_10 路径的创建如图 8-31 所示,可以看到目录树中新生成的目标点和运动指令。

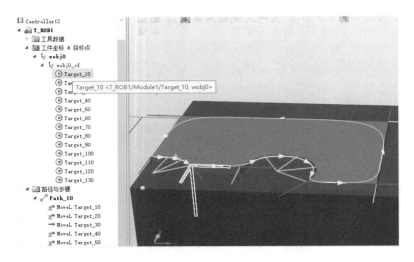

图 8-31　Path_10 创建完成界面

G.选中目标点右键查看机器人目标可以查看该位置的机器人姿态,如图 8-32 所示。各目标点 Z 轴向下,但是 X 轴方向各异,这就很容易导致机器人各轴自相干涉的情况,图中工具尾部的线束就很容易发生干涉。

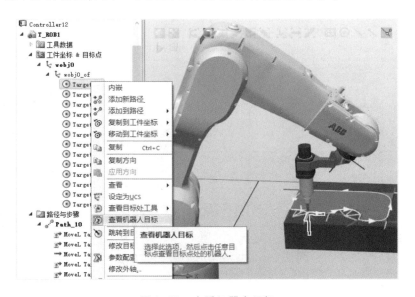

图 8-32　查看机器人目标

H.修改目标点的方向可用旋转工具(右键菜单:修改目标--旋转),如图 8-33 所示,Target_10 目标点绕自身坐标系 Z 轴旋转 180°。

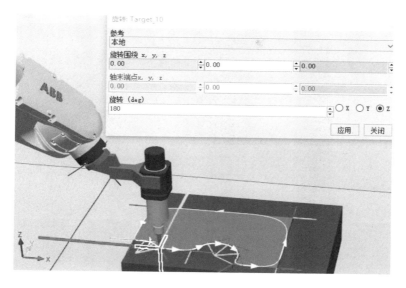

图 8-33　程序调试

Z.复制 Target_10 方向并批量应用到其余目标点,完成方向的批量修改,如图 8-34 所示。

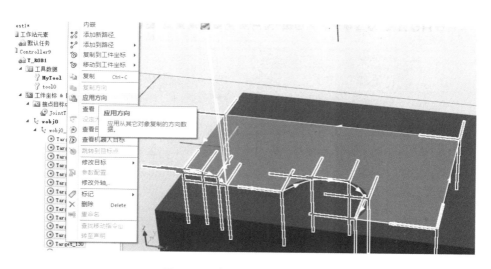

图 8-34　复制 & 应用方向选项

J.程序起始点的添加可采点击菜单"基本—目标点—创建Jointtarget",打开关节数据设置对话框如图8-35所示,单击Values后的按钮弹出输入框进行修改。

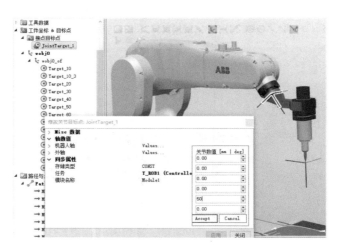

图8-35 关节目标点创建

K.通过右键菜单,可以将程序起点添加到程序路径的开始和结尾处,如图8-36所示。

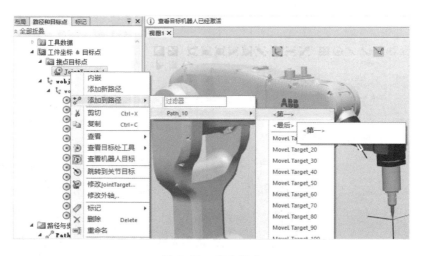

图8-36 程序调试

L.复制并粘贴轨迹的首尾点,修改 Z 值后可以作为轨迹接近点和离开点,如图 8-37 所示。

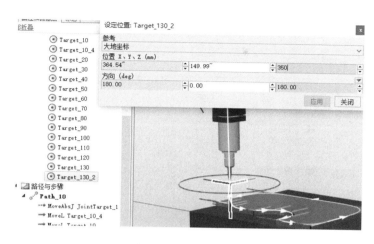

图 8-37　程序调试

M.选中所有目标点,右键依次选择"修改指令—区域—fine"选项,可以批量修改区域参数,实现各区域点的准确到达,如图 8-38 所示。也可以批量修改速度和工具等参数。

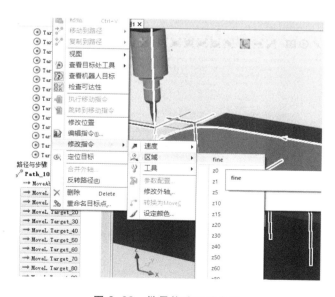

图 8-38　批量修改区域参数

N.路径编辑完成后,可通过选中 Path_10 右键单击"同步到 RAPD"选项,弹出如图 8-39 所示对话框,选中要同步的路径和工具点击确认,即可将路径转化为机器人指令添加到虚拟控制器。

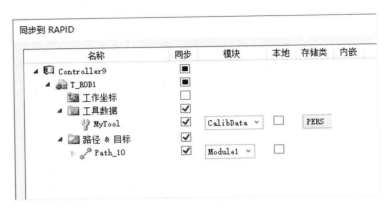

图 8-39　程序调试

O.打开虚拟示教器,可以看到例行程序 Path_10 如图 8-40 所示。

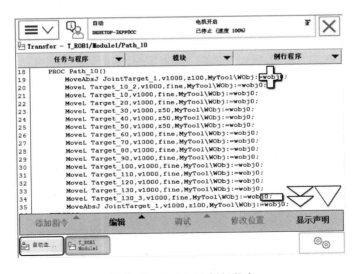

图 8-40　Path_10 例行程序

P.可以根据需求添加 I/O 信号等指令,在虚拟工作站内仿真测试没有问题的程序就可以使用另存模块工具导出,如图 8-41 所示。导出的.MOD 文件路径见图中提示栏,用 U 盘拷贝至真实机器人并导入模块即可完成离线程序的输入。

图 8-41　输出程序文件

8.2.3 三维离线编程

三维离线编程的工具坐标 Z 轴一般随工件形状而变化,如激光切割、激光熔覆等工具要求始终垂直于工件表面,焊接、涂胶等工具要求与工件表面保持一定的角度。如果工件形状复杂就要求工具姿态不断变化,机器人容易出现超程、干涉等情况,有时一条轨迹可能需要分成两个路径来编程,以解决机器人姿态无法连续调整的问题。三维离线编程的方法与二维基本相同,主要操作步骤如下。

A.打开 RobotStudio 软件并创建 IRB1200 机器人工作站,加载 myTool 工具并安装到法兰端面。导入离线编程用的三维模型,选项如图 8-42 所示。

B.设置工件位置坐标如图 8-43 所示,若后续生成的机器人轨迹碰撞和干涉难易消除,可以尝试调整工件位置重新计算轨迹。

C.合理设置工件坐标系可以简化实际操作时工件坐标系的标定工作,图 8-44 所示为在软件内创建坐标系的方法(菜单—基本—框架—创建框架),图中坐标系原点与工件底面圆心重合。

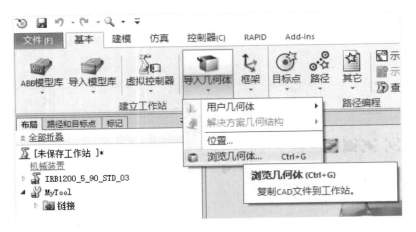

图 8-42　导入外部几何体

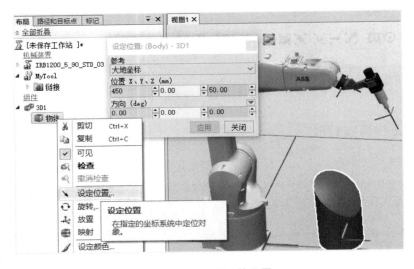

图 8-43　设置工件位置

D.RobotStudio 创建的框架(Frame)实际上就是坐标系,新创建的框架显示在布局视图,如图 8-45 所示,将其转换为工件坐标。

E.新转换的工件坐标系显示在"路径和目标点"视图,如图 8-46 所示,右击设定为激活,同样的操作确保任务 T_ROB1 和工具 MyTool 激活,为下一步编程做好准备。

第 8 章 ABB 机器人程序编写

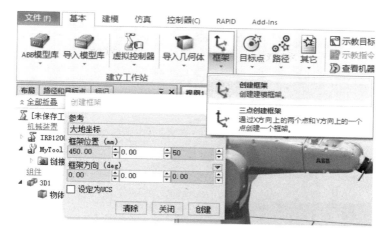

图 8-44 创建框架界面

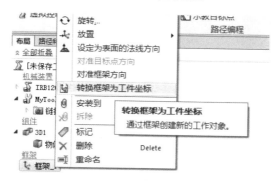

图 8-45 转换框架为工件坐标

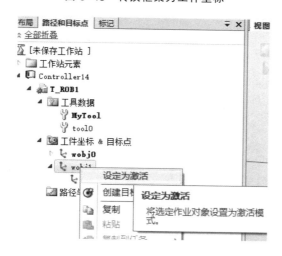

图 8-46 激活工件坐标

181

F.打开"自动路径"工具选择工件上表面边缘如图 8-47 所示,注意设置圆弧运动、公差 0.5mm,选择时注意 Z 轴方向,完成后点击"创建"按钮。

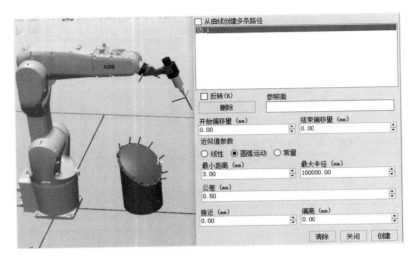

图 8-47 自动创建路径

G.自动创建的目标点其 Z 向垂直于曲面,X 向与轨迹相切,可以选中目标点右键查看机器人姿态如图 8-48 所示。可以看出工具姿态从头至尾绕 Z 轴旋转了 360 度,这很容易造成机器人超程或干涉,图中所示姿态就不太合理。

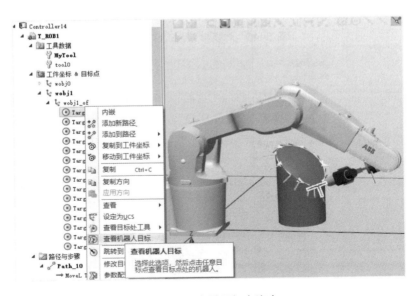

图 8-48 查看目标点姿态

H.切换不同目标点,直至找出机器人姿态比较合理的,如图8-49所示的Target_60目标点。

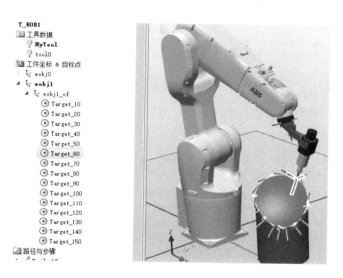

图8-49 程序调试

I.选中其余目标点,右击修改目标,选择"对准目标点方向"选项,如图8-50所示。

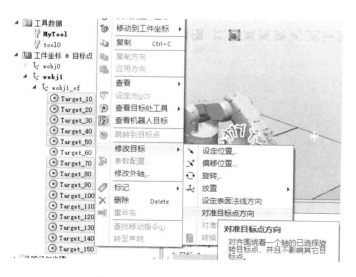

图8-50 对准目标点方向选项

J. 打开"对准目标点方向"对话框如图 8-51 所示,选择参考坐标系为 Target_60,对准轴选择 X 并点击"应用"按钮,对齐效果如图所示。

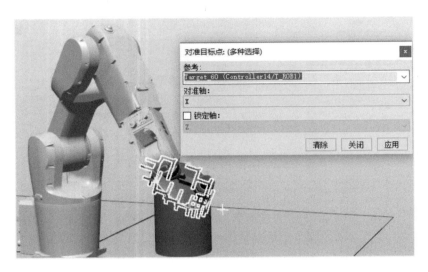

图 8-51　对准目标点对话框

K. 对齐方向之后查看原来有问题的目标点,如图 8-52 所示,可以看到机器人姿态已经没有自相干涉,还可以对当前姿态单独使用旋转工具进行微调。最后在目录中选中整体路径右击"沿着路径运动"进行运动仿真。

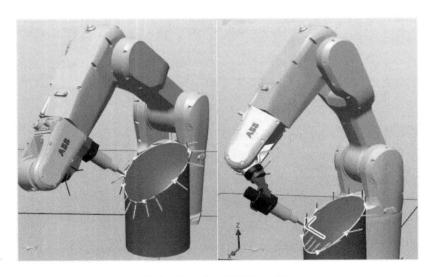

图 8-52　对齐前后姿态对比

I.确保编辑完的轨迹没有问题之后,可以将轨迹同步至虚拟控制器,进一步添加对外设的控制指令,在虚拟控制器内进行运行仿真,确保没有问题之后导出 MOD 文件。

注意:由于实际工件坐标系与软件中设定的工件坐标系相对于机器人基坐标系的位置有一定偏差,因此离线编程的程序在真实机器人上首次运行时一定要进行低速手动操作测试,确保无误后才允许自动运行。

第 9 章 机器人保养与常见故障维修

9.1 ABB 机器人保养

工业机器人由于工作条件恶劣,即使有完善的防护设计标准,仍必须定期进行常规检查和预防性维护。做好机器人维护与保养工作,将有效延长机器人的使用寿命以及降低故障概率,对于生产的稳定运行也有很大的助力,因此对工业机器人的日常保养事项必须了解并严格执行。

9.1.1 机器人保养事项

A.一般维护与保养事项:运行前需检查每个轴电机的刹车,运行每个轴到它最大负载位置,关掉电机检查轴是否停在原来位置,如果刹车良好轴仍然保持原来位置。进行点检,确认机器人的原点位置,不要从电脑或示教器上改变齿轮变速比或其他运动参数,这将影响减速运行功能。

B.工作中使用注意事项:注意机器人的液压、气动、仪表指示、监控信号、保险装置等运行情况,正确使用示教器的使能按钮,按下一半时,系统处于得电状态,松开或全部按下按钮时,系统处于电机关状态。当使能按钮不需要功能编程或调试时,机器人不需要移动时,立即松开使能按钮,当编程人员进入安全区时,必须带上示教器带,避免其他人误操作。进入工作区,控制器的模式开关选择手动,方便操作使能设备断开电脑或遥控操作,注意机械手各臂的旋转轴,另外注意机械手上其他选择部件或其他设备。

C.维护与保养周期:维护周期的时间间隔可根据机器使用的环境、频率、强度和温度而适当调整。

D.本体维护与保养:对于本体维护保养,主要是机械手的清洁和检查、轴制动测试的检查,减速器润滑的检查。

9.1.2 机器人保养标准

9.1.2.1 保养备件

A.润滑油脂。

B.保养备件包:SMB电池、冷却风扇、马达灯、保险丝等。

9.1.2.2 本体标准保养

A.常规检查

- 本体清洁:根据现场工作对机器人本体进行除尘清洁。
- 本体和6轴工具端固定检查:检查本体及工具是否固定良好。
- 各轴限位挡块检查。
- 电缆状态检查:检查机器人信号电缆、动力电缆、用户电缆、本体电缆的使用状况与磨损情况。
- 密封状态检查:检查本体齿轮箱,手腕等是否有漏油、渗油现象。

B.功能测量

- 机械零位测量:检测机器人的当前零位位置与标准标定位置是否一致。
- 电机抱闸状态检查:检测打开电机抱闸电压值,测试各轴电机抱闸功能。

C.保养件更换

- 本体油品更换:机器人齿轮箱、平衡缸或连杆油品更换。
- 机器人SMB板检查及电池更换:检查SMB板的固定连接是否正常,更换电池。

9.1.2.3 控制柜标准保养

A.常规检查

- 控制柜清洁:对机器人控制柜外观清洁,控制柜内部进行除尘。
- 控制柜各部件牢固性检查:检查控制柜内所有部件的紧固状态。
- 示教器清洁:示教器及电缆清洁与整理。
- 电路板指示灯状态:检查控制柜内各电路板的状态灯,确认电路板的状态。
- 控制柜内部电缆检查:控制柜内所有电缆插头连接稳固,电缆整洁。

B.控制柜测量

- 电源电压测量：测量机器人进线电压、驱动电压、电源模块电压，进行整体评估。
- 安全回路检测：检查安全回路(AS,GS,ES)的运行状态是否正常。
- 示教器功能检测：检测所有按键有效性，急停回路是否正常，测试触摸屏和显示屏功能。
- 系统标定补偿值检测：检测机器人标定补偿值参数与出厂配置值是否一致。
- 系统备份和导入检测：检查机器人是否可以正常完成程序备份和重新导入功能。
- 硬盘空间检测：优化机器人控制柜硬盘空间，确保运转空间正常。

C.保养件更换

- 驱动单元冷却风扇更换。
- 控制柜保险丝更换。
- 控制柜操作面板电机上电按钮内指示灯更换。

9.1.2.4 标准保养报告总结内容

A.机器人标准保养后建议。

B.机器人标准保养后建议备件清单。

9.1.2.5 标准保养周期

一年或者运行 3000~4000 小时。

9.1.3 更换 SMB 电池

机器人关掉主电源后，就需要依靠 SMB 电池来保存 6 个轴的数据。电池耗尽会导致机器人零点信息丢失，进而影响机器人精度和正常工作。电池的剩余后备电量(机器人电源关闭)不足 2 个月时，ABB 机器人系统将显示电池低电量警告。

SMB 及电池安装在机器人的本体里面，不同机器人的安装位置会有所区别，具体位置可以在对应机器人型号的产品手册中查询到，IRB1200 机器人的电池如图 9-1 所示。在 SMB 电池即将耗尽之前进行更换电池，可以省去手动找零位的麻烦，因此更换电池的操作规范尤为重要，具体步骤如下：

A.将机器人所有关节回零，完成此步骤的目的是为了方便更换电池后更

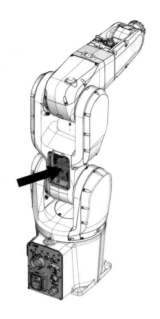

新转数计数器。每个关节均有零位标志。

B.关闭连接到机器人的所有电源、气源和液压源。

C.首先静电放电(ESD),然后卸下下臂连接器盖的连接螺钉并小心打开盖子(开启前需要对盖子清洁,导电杂质可能会引起机器人短路故障),如图9-2所示。[注:ESD(静电放电)是电势不同的两个物体间的静电传导,它可以通过直接接触传导,也可以通过感应电场传导。搬运部件或部件容器时,未接地的人员可能会传递大量的静电荷。这一放电过程可能会损坏敏感的电子设备。所以在有此标识的情况下,要做好静电放电防护]

图 9-1　电池位置示意图

D.断开电池电缆与编码器接口电路板的连接(注意:卡扣需要按下,不要强制拔插头)。用剪刀剪掉扎带,拆下旧电池,如图9-3所示。

E.装回新电池,并用扎带固定,剪下多出来的扎带。将电池电缆与编码器接口电路板相连,用其连接螺钉将连接器盖重新安装到机器人本体上。

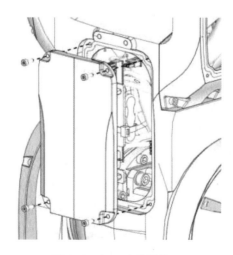

图 9-2　打开连接器盖

第 9 章 机器人保养与常见故障维修

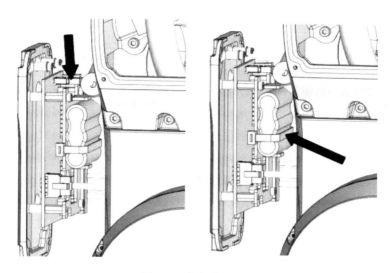

图 9-3 拆卸电池

F.按照 4.4.2 节的讲解更新转数计数器,电池更换完成。

(注意:有些机器人在更换电池时可以不断电,这种情况下更换电池后转数计数器不需要更新,继续使用机器人即可。具体要求需对照机器人查阅机器人使用手册。)

9.1.4 清洁

定期清洁对于机器人的长时间正常运行非常重要,清洁频次取决于机器人的工作环境,不同防护类型可采用不同的清洁方法。清洁之前务必确认机器人的防护类型和说明书规定的清洁注意事项,具体要求如下。

A.务必使用规定的清洁设备,否则其他任何清洁设备都可能缩短机器人的寿命。

B.在清洁之前,务必检查是否已将所有保护罩都装到了机器人上。

C.切勿向连接器、接头、封口或衬垫喷水。

D.不用压缩空气来清洁机器人。

E.切勿用未经 ABB 公司同意的溶剂来清洁机器人。

F.不在 0.4m 以内喷水。

G.在清洁机器人之前,不要拆除任何盖置或其他防护装置。

注意:标准防护等级机器人的清洁方法为采用真空吸尘器和清洁布擦拭,

严谨用水冲洗。IP67 防护等级的机器人可以用水冲洗,但是强烈建议在水中加入防锈剂,并且在清洁后对操纵器进行干燥处理。

9.2 机器人常见故障维修

9.2.1 机器人手动操作错误

表现为无法使用示教器手动操纵机器人,故障排除方法如下。

A.首先确保 FlexPendant 与 Control Module 正确连接。

B.可能是因为在机器人启动过程中,不小心碰到了示教器上的操纵杆(此时报错信息会显示操纵杆无法使用,如图 9-4 所示),点击 ABB 菜单中的"Restart"选项,重新热启动机器人即可。

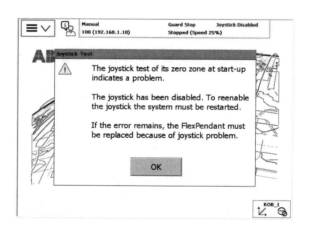

图 9-4　Joystick 错误提示

C.机器人系统可以正常启动,但 FlexPendant 上的控制杆不能工作,可能是控制杆发生故障或发生了偏转,需要确保控制器在手动模式下按下FlexPendant 背面的重置按钮。重置按钮会重置 FlexPendant,但不会重置控制器上的系统。

9.2.2 开机连接错误

如果机器人开机示教器一直显示"Connecting to the robot controller"而无法进入系统,如图 9-5 所示。上述情况是示教器和机器人主控制器之间没有建

第 9 章 机器人保养与常见故障维修

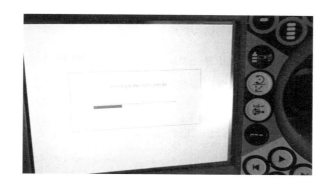

图 9-5 系统连接错误

立通信连接,未建立连接的原因如下。

A.机器人主机故障。

B.机器人主机内置的 CF 卡(SD 卡)故障。

C.示教器到主机之间的网线松动等。

处理方法如下。

A.检查主机是否正常,检查主机内 SD 卡是否正常。

B.检查示教器到主机网线是否连接正常。示教器到主机上的网口均为主机绿色标签网口,如图 9-6 所示。

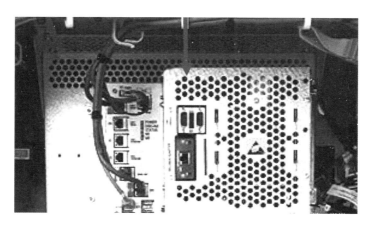

图 9-6 网线连接示意图

9.2.3 SMB 内存数据差异(50296)

如果机器人提示 50296 报警(SMB 内存差异),如图 9-7 所示。故障原因是更换了 SMB 后,SMB 内数据和控制柜内数据不一致。

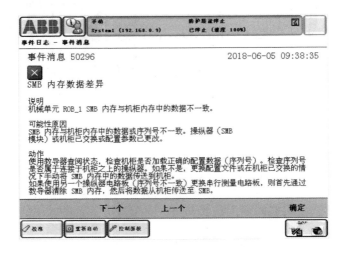

图 9-7　SMB 内存数据差异报警

处理方法如下。

A.点击"ABB"—"校准",点击 SMB 内存,如图 9-8 所示。

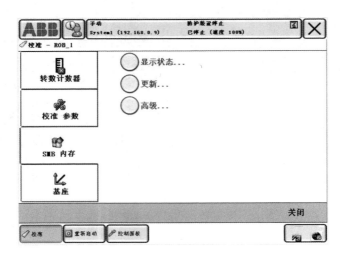

图 9-8　SMB 内存选项

B.选择高级,选择清除 SMB(由于更换了 SMB,如果更换了控制器卡,则选择清除控制柜内存),如图 9-9 所示。

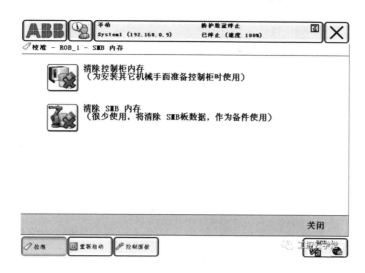

图 9-9　清除 SMB 内存选项

C.点击"关闭"后选择"更新",如图 9-10 所示。

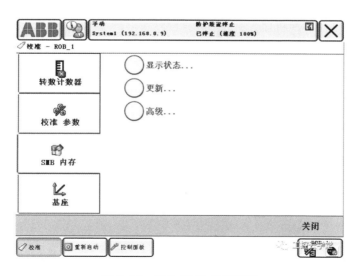

图 9-10　清除 SMB 内存选项

D.选择替换 SMB 电路板,最后重新更新转数计数器即可(详见 4.4.2 节),如图 9-11 所示。

图 9-11　替换 SMB 电路板选项

9.2.4　开机系统故障

机器人在开机时进入了系统故障状态,处理方式如下。

A.重新启动一次机器人。

B.如果不行,在示教器查看是否有更详细的报警提示,并进行处理,完成后重启机器人。

C.如果还不能解除报警则尝试 B 启动,重启控制器后将恢复到上次自动保存的状态。

D.如果还不行,可以尝试 P 启动,重启控制器后将重置 RAPID,删除程序但保留配置数据。

E.如果还不行请尝试 I 启动,这会将机器人恢复到出厂设置状态,删除所有程序和配置数据,需谨慎操作。

9.2.5　控制器性能不佳

如果控制器完全失去响应,导致 FlexPendant 无法操作系统,可能是控制器未连接主电源,或者主保险丝(Q1)可能已经断开,检查并更换即可。

如果是控制器性能低，程序执行迟缓，看上去无法正常执行并且有时停止。可能是计算机系统负荷过高，具体是以下其中一个或多个原因造成。

A.程序仅包含太高程度的逻辑指令，造成程序循环过快，使处理器过载。

B.I/O 更新间隔设置为低值，造成频繁更新和过高的 I/O 负载。

C.内部系统交叉连接和逻辑功能使用太频繁。

D.外部 PLC 或者其他监控计算机对系统寻址太频繁，造成系统过载。

可尝试以下解决方法。

A.检查程序是否包含逻辑指令(或其他"不花时间"执行的指令)，因为此类程序在未满足条件时会造成执行循环。可以通过添加一个或多个 WAIT 指令来进行测试，仅使用较短的 WAIT 时以避免不必要地减慢程序。适合添加 WAIT 指令的位置可以是：在主例行程序中，最好是接近末尾；在 WHILE/FOR/GOTO 循环中，最好是在末尾接近指令 ENDWHILE/ENDFOR 等部分。

B.确保每个 I/O 板的 I/O 更新时间间隔值没有太低，不经常读的 I/O 单元可以进行"状态更改"操作。

C.检查 PLC 和机器人系统之间是否有大量的交叉连接和 I/O 通信，断开不必要的连接和通信。

D.尝试以事件驱动指令而不是使用循环指令编辑 PLC 程序，机器人系统有许多固定的输入和输出可以实现此目的，尝试优化程序。

9.2.6 路径精确性不一致

表现为机器人 TCP 的路径不一致，并且有时会伴有轴承、变速箱或其它位置发出的噪音，精度经常变化进而影响生产的正常进行。可能的原因和解决方法如下。

A.机器人没有正确校准，需检查转数计数器的位置，如有必要重新校准机器人各轴。

B.未正确定义机器人 TCP，可以重定位运动方式或辅助工具检查机器人工具定义是否正确。

C.电机和齿轮之间的机械接头损坏，通常会使出现故障的电机发出噪音，查明噪声来源并检查对应电机。

D.轴承损坏或破损，尤其是耦合路径不一致并且一个或多个轴承滴答声或磨擦噪声，查明噪声来源并检查对应轴承。

E.制动闸未正确松开。

9.3 机器人安全操作规范

工业机器人在运输、安装调试和运行操作时均存在安全风险,因此有必要了解工业机器人的安全风险,掌握工业机器人的通用安全规范,树立正确的安全操作意识。本节主要介绍工业机器人操作的通用安全规范。

A.操作人员必须正确佩戴劳保防护用品。

B.机器人操作人员必须经过专业培训,必须熟识机器人本体和控制柜上的各种安全警示标识,能按照操作要领手动或自动编程控制机器人动作。通电时,禁止未受培训的人员触摸机器人和示教器,以免导致人员伤害或者设备损坏。

C.机器人设备周围必须设置安全隔离带,必须清洁,做到无油、无水及无杂物。

D.开机前要先检查电、气供应是否正常。

E.在调试与运行机器人时,可能会执行一些意外的或不规范的运动,而且所有的运动都会产生很大的力量,会严重伤害个人或损坏机器人工作范围内的任何设备,所以要时刻警惕,与机器人保持足够安全的距离。

F.当进入机器人作业区域时,如果在机器人工作区域内有工作人员,必须手动操作机器人系统。准备好示教器,以便随时控制机器人。靠近机器人工作区域前,确保机器人及运动的工具已经停止。

G.注意夹具并确保夹好工件。如果夹具打开,工件会脱落并导致人员伤害或设备损坏。有的夹具力量很大,如果不按照正确方法操作,也会导致人员伤害。

H.注意液压、气压系统以及带电部件。即使断电,这些电路上的残余电量也很危险。

I.不要戴着手套操作机器人示教盘,如需要手动控制机器人时,应确保机器人动作范围内无任何人员和障碍物,将速度由慢到快逐渐调整,避免速度突变造成人员或设备损害。

J.FlexPendant示教器应小心轻放,切勿跌落、抛掷或重击以免造成破损或故障。在编程和测试过程中,机器人无须移动时必须尽快释放使能器。不使用

FlexPendant时,需将其挂在壁支架上存放,以便确保其不会意外跌落地面或某人不会被其线缆绊倒。

K.自动执行程序前应确保机器人工作区不得有无关的人员、工具或物品,工件夹紧可靠并确认。

L.必须知道所有会影响机器人移动的开关、传感器和控制信号的位置和状态。

M.必须知道机器人控制器和外围控制设备上的紧急停止按钮的位置,随时准备在紧急情况下使用这些按钮。

第 10 章 工业机器人实践应用项目

10.1 ABB 机器人基本操作

要求通过工业机器人实际操作，做到熟练认知 ABB IRB1200 工业机器人的基本结构，熟练掌握 ABB 工业机器人示教器的结构和基本操作方法，做到熟练掌握 ABB 工业机器人单轴运动操作、线性运动操作和重定位运动操作。

具体实训内容如下。

A. 了解实训机器人的主要技术参数。

表 10-1 IRB1200 主要参数

型号	IRB 1200-5
工作范围	901mm
负载	5kg
轴数	6
精度	0.02mm
安装方式	任意角度
防护等级	IP40，IP67
控制器	IRC5 紧凑型
各轴运动范围	
轴 1 旋转	+170°~-170°
轴 2 手臂	+130°~-100°
轴 3 手臂	+70°~-200°
轴 4 手腕	+270°~-270°
轴 5 弯曲	+130°~-130°
轴 6 翻转	+400°~-400°

B.理解机器人系统各组成部分及工作原理。

C.掌握 ABB 工业机器人开机、关机操作方法,以及安全注意事项。

D.认识 FlexPendant 示教器,熟练掌握示教器的握法,掌握各组成部分和使用方法,尤其是以下部分:

- 状态钥匙
- 使能器
- 操纵杆
- 触摸笔

E.熟练掌握示教器各按键的名称和含义,分别如下。

- 预设按键(共 4 个)
- 选择机械单元
- 切换运动模式,重定向或线性
- 切换运动模式,轴 1–3 或轴 4–6
- 切换增量模式
- Step BACKWARD(步退)按钮
- START(启动)按钮
- L Step FORWARD(步进)按钮
- M STOP(停止)按钮

F.掌握 ABB 工业机器人单轴运动操作。

G.掌握 ABB 工业机器人线性运动操作。

H.掌握 ABB 工业机器人重定位运动操作。

10.2 机器人零点转数计数器更新

理解机器人零点概念,掌握机器人机械零点手动回零操作方法和技巧,掌握 ABB 工业机器人零点转数计数器更新操作方法和注意事项。

具体实训内容如下。

A.理解机器人零点的概念,理解必须进行机器人转数计数器更新操作的工况。

B.能够手动操纵机器人快速回到零点。

C.能够正确找到 1–6 轴机械原点位置,如图 10–1 中 A~F 分别是 1~6 轴

第 10 章　工业机器人实践应用项目

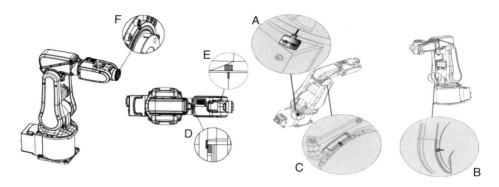

图 10-1　各轴零点位置示意图

的机械零点。

D.能够手动操纵 1—6 轴按顺序正确回零(各轴回零顺序:4→5→6→3→2→1)。

E.了解 ABB IRB1200 机器人电池位置,熟悉电池更换方法。

F.掌握转速计数器更新操作方法。

G.掌握机器人控制系统备份与还原操作要领。

10.3 I/O 信号的配置、监控及快捷键定义

通过该项目的训练,熟练掌握 ABB 工业机器人 I/O 口通信的种类、常用 I/O 板的总线连接和 I/O 信号配置、测试方法。

具体要求如下。

A.在示教器定义一块 DSQC651 的 IO 板并正确配置。

B.在 DSQC651 板上定义 Di1、Di2、Do1、Do2、Gi1、Go1、Ao1 信号。

C.正确连接 Do1 信号端子至图 10-2 所示继电器,当 Do1 输出 1 时,继电器吸合,输出为 0 时,继电器断开。

D.正确连接继电器与二位五通电磁阀,实现 Do1 信号对平行气缸的控制。

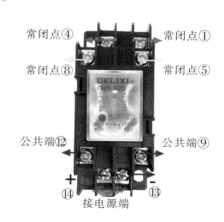

图 10-2　继电器接线图

E.正确连接平行气缸上的磁性开关传感器至 Di1 信号端子,确认 Di1 为 1 是气缸开到位。

F.正确配置示教器快捷键 1、2 对应 Do1,Di1,并进行测试。

G.正确配置 Go1 信号,实现气动吸盘的控制。

H.正确配置 Ao1 端子至 3010 直流风扇(24V)的连接,实现 Ao1 信号对风扇电压的控制,实现调速目的。

10.4 工具坐标系的设定与简单示教编程

通过该项目的训练,要求数量掌握 ABB 工业机器人工具数据 tooldata 的建立方法、操作和使用注意事项,掌握系统默认工具 (tool0) 的工具中心点 (TCP)和坐标轴方向,理解工具坐标 tooldate 的 TCP、质量、重心等参数的意义和设置方法。掌握 ABB 示教编程基本操作方法。

工具中心点(TCP)设定原理如下。

A.首先在机器人工作范围找一个非常精确的固定的尖点做参考点。

B.再在工具上找一个尖点做参考点(最好在工具中心)。

C.操纵工具上的参考点以最少四种不同的姿态尽可能接近固定参考点,一般取 1、2、3、4 四个标定点之间相差 90°且不在同一平面上,可有效提高标定精度,如图 10-3 所示。

D.机器人通过四组解计算,得出 TCP 相对于机器人法兰坐标系的坐标数值。

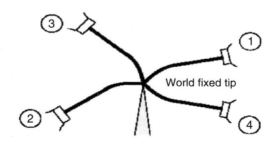

图 10-3　4 点标定法

具体要求如下。

A.新建工具 tool1。

B.合理选择工装夹持一支 0.5mm 水性笔。

C.选择合适位置放置标定用尖点参照物。

D.采用 4 点标定法对 tool1 进行标定,控制误差在 0.5mm 以内。

E.标定完成后手动操作机器人进行重定位运动,工具选择 tool1,观察机器人绕 tool1 的 TCP 点旋转时,笔尖的位置变化。

F.创建例行程序,选择不同练习图形(三角形、四边形、五角星、圆形)放置在操作台上,合理选择 MoveL、MoveJ 等指令编制程序,用笔尖按顺序示教各顶点。

G.添加程序 Home 点(笔竖直朝下),要求程序从 Home 点开始,运行结束后返回 Home 点。

H.添加接近点和离开点,接近点在示教第一点正上方 30mm 处,离开点在示教最后一点上方 30mm 处。

I.移动 PP 指针至程序开始,手动控制机器人完成图形绘制。

10.5 工件坐标系的设定与二维离线编程

通过该项目的训练,需熟练掌握 ABB 工业机器人工件坐标 wobjdata 和有效载荷数据的建立操作。能够操作 RobotStudio 软件进行二维轨迹的离线编程,熟练掌握 RobotStudio 软件程序的导出和 ABB 机器人程序的导入。

机器人可以拥有若干工件坐标系,或者表示不同工件位置,或者表示同一工件在不同位置的若干副本。对机器人进行编程时就是在工件坐标中创建目标和路径。

具体要求如下。

A.选择工具为项目四标定的 tool1。

B.创建工件坐标 wobj1。

C.选择"3 点"法、利用笔尖对长方体型工作台进行标定。

D.验证工件坐标精确度:选定动作模式为线性,工件坐标为 wobj1,线性移动各坐标轴,查看示教器显示屏右上角区域的坐标变化及机器人运动是否符合预期。

E.打开 RobotStudio 软件,定义工件坐标系,要求与步骤 3 标定的工件坐标系点接近,X 方向与基坐标系 X 方向夹角 45°,导入二维图形(镂空字)。

F.定义工具 Pen,TCP 点与机器人工具 tool1 保持一致。

G.进行二维轨迹的离线编程,并添加接近点和离开点。(注意多条互不相连轨迹的图形,需添加多组接近点和离开点)

H.添加程序 Home 点(笔竖直朝下),要求程序从 Home 点开始,运行结束后返回 Home 点。

I.RobotStudio 软件内进行仿真,确认无误后导出程序。

J.使用 U 盘将二维图形的程序导入机器人控制器。

K.移动 PP 指针至程序开始,手动控制机器人完成图形绘制。

10.6 码垛程序设计与示教

通过该项目的训练,需熟练掌握 ABB 工业机器人循环指令、坐标偏移指令的使用方法和点位示教的操作技巧,能够解决工程实际中遇到的各类码垛工况的需求。

如图 10-4 所示单排码垛为例,物料尺寸为 Φ90*8?mm。要求工业机器人使用气动吸盘将位置 P10 处三个一字排开的物料,按顺序移至位置 P20 并码垛为 3 层。具体动作要求如下。

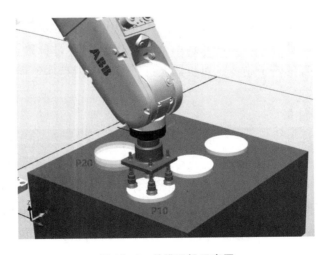

图 10-4 单排码垛示意图

A.首先,机器人的TCP点首先以关节运动的形式从Home点(jpos10)出发前往中间点P10点的正上方20mm处。

B.低速线性运动至P10后,气动吸盘吸取物料。

C.低速线性运动返回至P10上方20mm处,以关节运动形式中速移动到目标位置P20正上方100mm处。

D.低速线性下降到P20点后松开气动吸盘放置物料,快速沿Z向抬升100mm,完成一层搬运。

E.快速运动至第二块物料正上方20mm(物料在Y向间距为120mm),依次类推完成剩余两层码垛。

F.码垛完成后机器人返回Home点(jpos10)。

要求采用主程序调用例行程序的结构,程序示例如下:

PROC main( )　　//主程序

　RInit; //初始化子程序

　SuckerTool; 　//工具拾取子程序,拾取吸盘工具。

　Stack; 　　//工件码垛子程序

　ReleaseTool; 　//工具释放子程序,释放吸盘工具。

　ENDPROC

PROC Stack( )

　MoveAbsJ jops10\NoEoffs,V500,z50,tool1;

　FOR I FROM 0 TO 2 DO

　　MoveJ Offs(p10,0,I*100,20),V500,z50,tool1; //接近点

　　MoveL Offs(p10,0,I*100,0),V100,fine,tool1; 　//抓取点

　　WaitTime 0.5;

　　Set Do1;

　　WaitTime 0.5;

　　MoveL Offs(p10,0,I*120,20),V100,z50,tool1; //低速抬升

　　MoveJ Offs(p10,0,I*120,100),V200,z50,tool1; //中速关节运动

　　MoveJ Offs(p20,0,0,8*I+20),V200,z50,tool1; 　//中速关节运动

　　MoveL Offs(p20,0,0,8*I),V100,z50,tool1; //低速运动至放置点

　　WaitTime 0.5;

　　Reset Do1;

WaitTime 0.5;
MoveL Offs(p20,0,0,8*I+100),V500,z50,tool1; //快速抬升
ENDFOR
MoveAbsJ jops10\NoEoffs,V500,z50,tool1;　初始原点
ENDPROC

工业机器人技术应用

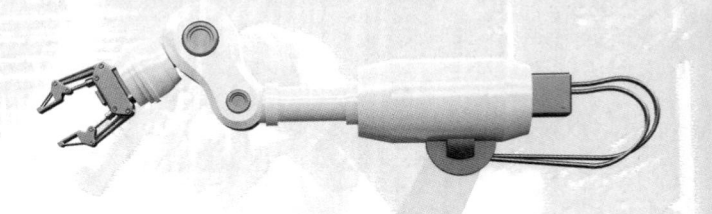

·英文版·

Industrial Robot Technology Application

主　编：王同庆、商丹丹

副主编：周　京、韩志国、马绪鹏、喻　秀

天津出版传媒集团

天津科学技术出版社

Foreword

The robot industry has shown its great vitality since 1959 when the first industrial robot was built in the United States. So far, industrial robots have been widely applied to the manufacturing industry and many other fields in industrially developed countries represented by the United States, Japan and Germany. Industrial robots have become irreplaceable key equipment in the advanced manufacturing industry, and their application has also become an important indicator to measure the manufacturing level and the scientific and technological level of a country. Industrial robots can not only improve production efficiency and product quality, but also replace human operators to perform high-intensity operations in harsh and dangerous environments. Compared with other machinery and equipment, robots are more adaptable and can realize full automation throughout the production process.

With the gradual decrease of the population suitable for working, the average salary of employees in manufacturing enterprises (above designated scale)are increasing, resulting in rapid increase of labor costs. Increased labor costs directly affect the healthy development and profit level of manufacturing enterprises, and the labor costs of labor-intensive production enterprises with low automation level are growing day by day. It is urgent to change the mode of complete with cheap labors in the manufacturing industry. In addition, there is an urgent need to replace labors with automated equipment. In this context, China´s industrial robot market has moved to a rapid rise stage.

Industrial robots have significant positive impact on manufacturing enterprises in the process of replacing human operators with robots. Throughout the industrialization process of developed countries, automated equipment, as a key means to improve the production efficiency, was combined with many industrial manufacturing technologies to play a crucial role in the revolutionary reform of production mode in the traditional equipment manufacturing industry.

Compared with human operators, automated manufacturing equipment has such features as high work efficiency and high manufacturing accuracy. With the increase in labor costs of manufacturing enterprises, intelligent manufacturing equipment can help such enterprises to reduce operating costs and improve profits, while assist them in optimizing production and improving product quality.

At present, intelligent manufacturing is one of the most typical manufacturing modes of advanced manufacturing technologies, the development direction of the manufacturing industry in the new century and the core of the new round of industrial revolution, and also the breakthrough and main target of making China's manufacturing industry go out to the world. The development of intelligent manufacturing, the in−depth integration of informatization and industrialization, and the provision of a new drive power for economic development are of decisive significance for promoting improved product quality, increased production efficiency, better transformation and upgrading of the manufacturing industry, and achieving the strategic target of becoming a great manufacturing power. Furthermore, industrial robots are indispensable in the intelligent manufacturing system.

This book systematically introduces the basic concept, structural composition, installation and commissioning, basic operation, teach programming, communication settings, offline programming and project application of industrial robots according to the skills required for their actual production positions. The applications of some main robots, such as handling robot, stacking robot, welding robot and laser cutting robot, are contained herein. This book is novel in contents and easy to teach and learn, and it focuses on making students master comprehensive knowledge. After studying this book, students can get general understanding and comprehensive knowledge of industrial robots.

With the training contents of this book being close to the actual production conditions of industrial robots, supporting course resources and project cases have been developed in combination with mainstream advanced teaching and practical training equipment of vocational colleges, such as the simulation and practical training room for intelligent and virtual manufacturing, the practical training room for fitting and maintenance of intelligent manufacturing cutting unit, the practical

training room for intelligent manufacturing handling unit, and the practical training room for comprehensive application of intelligent manufacturing. In consideration of China's vocational skill level standards, the EPIP teaching mode is followed and textbook is prepared in both Chinese and English, to meet the needs of making the vocational education related to China's intelligent manufacturing technology go to the world in a better way.

This book is suitable to be used as the textbook of relevant disciplines in higher vocational and technical colleges, such as industrial robot technology, intelligent manufacturing technology, mechatronics technology and electric automation technology, and as a reference for vocational skill training and self-learning of engineering technicians.

Contents

1. Overview of Industrial Robot ·· 1

    1.1 Introduction to robot ·· 1

    1.2 Definition of industrial robot ·· 1

    1.3 Development history of robots ·· 2

    1.4 Development trend of robots ·· 7

2. Basic Composition and Technical Parameters of Industrial Robot ··· 9

    2.1 Basic composition of industrial robot ································ 9

    2.2 Basic components ·· 13

    2.3 Main technical parameters of industrial robot ···················· 16

    2.4 Classification and typical application of industrial robots ········· 20

3. Installation and Commissioning of Industrial Robots ················· 27

    3.1 Unpacking and fixing of robots ······································ 28

    3.2 Commissioning Process of Robots ·································· 32

4. Basic Operation of Industrial Robots ································· 37

    4.1 Knowledge of teach pendant ·· 37

    4.2 Configuration of necessary operating environment ················ 40

    4.3 Manual control of single-axis motion ······························ 50

    4.4 Zero calibration of robots ·· 61

    4.5 Robot Restart ·· 66

    4.6 Emergency stop and recovery of robot ······················· 67

5. RobotStudio Application ·· 69

    5.1 Introduction to RobotStudio ·· 69

    5.2 Installation of RobotStudio ·· 71

    5.3 RobotStudio interface ··· 75

5.4 Creation of virtual workstation ·································· 78

6. I/O Communication of ABB Robot ·························· 85

6.1 I/O communication types of ABB robot ···················· 86

6.2 Introduction to DSQC651 ································· 86

6.3 DSQC651 configuration ·································· 90

6.4 Profibus communication ································· 103

6.5 Programmable key settings of teach pendant ··············· 109

7. Programming Basics of ABB Robot ······················ 113

7.1 Robot system related terminology ························ 113

7.2 Program structure of ABB robot ························· 118

7.3 Program data of ABB robot ···························· 120

7.4 Commonly used instructions for ABB robot ··············· 142

8. ABB Robot Programming ······························ 161

8.1 Teach programming ································· 162

8.2 Offline programming ································· 177

9. Maintenance and Common Fault Repair for Robot ·············· 193

9.1 ABB robot maintenance ······························ 193

9.2 Common fault repair for robot ························· 199

9.3 Safety operation specification for robot ··················· 205

10. Practical Applications of Industrial Robot ··················· 207

10.1 Basic operations of ABB robot ························· 207

10.2 Updating revolution counter for robot zero point ············ 208

10.3 Configuration and monitoring of I/O signals and definitions of

quickset keys ·································· 209

10.4 Setting of tool coordinate system and simple teach programming · 210

10.5 Setting of work object coordinate system and 2D offline

programming ································· 212

10.6 Design and teaching of stacking program ················· 213

1.

Overview of Industrial Robot

1.1 Introduction to robot

What is a robot? A robot is an intelligent machine that can work in a semi−autonomous or full−autonomous way. The term "robot" first appeared in 1920, included in the science−fiction play Rossum's Universal Robots of the Czech writer Karel ?apek. It was derived from the Czech word "Robota"(which means hard labor and slavery)and the Polish word "Robotnik"(which means worker). Therefore, this term represents doing hard and heavy work. In the play, the robot works silently at the commands of its master, without feeling and emotion, and always finishes heavy work in a repeated and unchanged manner. The play foretold the tragic impact of robot development on the human society and attracted wide attention, so it was deemed as the origin of the term "robot". However, "robots" was quoted as a technical term till the middle of the 20th century.

With the unceasing development of modern science and technology, the concept of robot has become a reality. During the development of modern industry, robots have gradually been integrated into many disciplines such as machinery, electronics, kinetics, power, control, sensing detection and computing technology, and have become an extremely important part of modern scientific and technological development.

1.2 Definition of industrial robot

Although decades have passed since the birth of robot, there is still no uniform definition for it. One reason is that many robots are still developing and new

models and functions are emerging. Some countries and international organizations have given the following definitions to robot:

(1)The Robot Institute of America(RIA, established in 1979): A robot is a programmable, multifunctional manipulator designed to move materials, parts, tools, or dedicated devices through variable programmed motions for the performance of a variety of tasks.

(2)Japan Industrial Robot Association(JIRA): An industrial robot is a general-purpose machine with memory devices and end manipulators and can replace human operator to work through automated motions.

(3)Chinese Industrial Robot Research Scholars: A robot is a highly-flexible, automated machine that has some capabilities similar to human or living beings, such as perception, planning, motion and collaboration capabilities.

(4)International Organization for Standardization(ISO): An industrial robot is a bionic, automatically-controlled, reprogrammable manipulator with multiple functions and multiple degrees of freedom.

These definitions are different but have certain commonality, namely, an industrial robot is a machinery consisting of bionic mechanical structure, motor, reducer and control system for performing industrial production and can automatically execute various work commands. It can accept human commands and operate according to preset programs. Modern industrial robots can also act by following the principles and programs formulated in accordance with artificial intelligence technology.

Generally, an industrial robot should have the following four features:

(1)Specific mechanical structure;

(2)Common performance for doing various jobs;

(3)Different degrees of intelligence in perception, learning, computing and decision-making capabilities;

(4)Relative independence.

1.3 Development history of robots

1.3.1 Ancient robots

Long before the word "robot" appeared, human beings had always dreamed of

1. Overview of Industrial Robot

and sought to build a human−like machine to replace human operators for completing relevant jobs. Recorded examples include: a wooden bird, which was designed and made by Lu Ban in the Spring and Autumn Period in ancient China, could fly in the air; Muniuliuma, which imitated the walking of ox and were fabricated by Zhuge Liang in the Three Kingdoms Period, was a kind of ancient wooden transportation vehicle used to transport food and grass for the army; in 1662, a Japanese, Takeda Omi, invented an automatic robot doll that could give performance by using the clock technology; in the 18th century, a French genius technician invented a robotic duck. All these were early and long−lasting explorations of human beings from dream to reality in terms of robot.

1.3.2 Recent robots

In the late 1940s, the research and invention of robots got more concern or attention. After the 1950s, the Oak Ridge National Laboratory (a laboratory in the United States) started to research on a remotely controlled manipulator capable of handling nuclear raw materials, as shown in Fig. 1−1. This was a master−slave control system, to which force feedback was added, so that the operator could know the magnitude of the applied force. The master and slave manipulators were

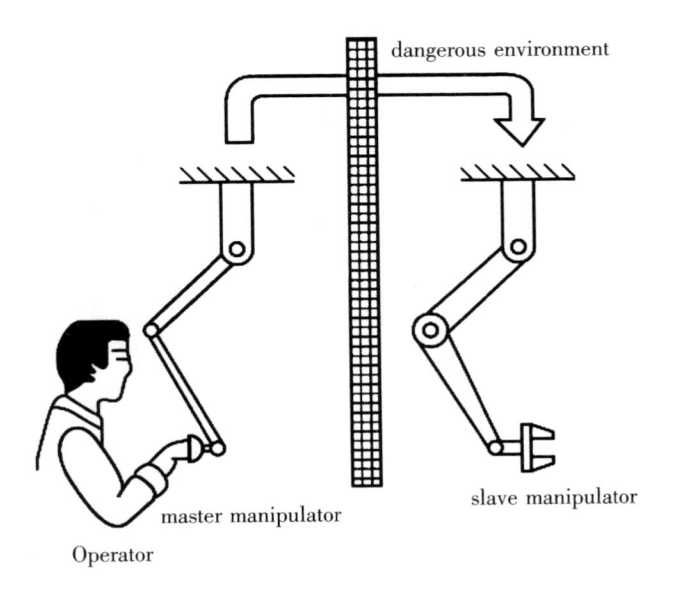

Fig. 1−1　Remote Control Manipulator

3

separated by a protective wall, and the operator could effectively monitor the slave manipulator through an observation window or a closed–circuit television. The appearance of the master–slave manipulator system has paved the way for the emergence of robot and the design and manufacture of modern robots.

The "master manipulator" acted under the guidance of the user, and the "slave manipulator" imitated the action of the master manipulator as accurate as possible. Later, force feedback was added to the "mechanically coupled master–slave manipulator" action, so that the operator could feel the force exerted by an object in the environment on the manipulator. In the mid–1950s, mechanical coupling devices in these manipulators were replaced by hydraulic ones. Examples include the robot "Mister Handy" of General Electric Corporation and the Type I robot "Monster" of General Manufacturing Plant.

1.3.3 Modern robots

Robot was born under the guidance of control and information theories and by integrating the achievements of mechanics, microelectronics, computer technology, sensing technology and other disciplines. Therefore, with the development of these disciplines, especially computer technology, the emergence of modern robot has become a matter of course.

In 1954, G. C. Devol, an American, built the first programmable robot in the world. In that year, he proposed a "general–purpose robot for repeated operations", which was patented in 1961. In 1958, Joseph F. Engel Berger, known as the "Father of Industrial Robots", founded the world's first robotics company — Unimation (meaning Universal Automation). In 1959, G. C. Devol and Joseph F. Engel Berger (also an American inventor)jointly build the world's first industrial robot — Unimate, as shown in Fig. 1–2. This was a five–axis hydraulically–driven robot for die casting. The arm of the robot was controlled by a computer. This robot incorporated separate solid CNC components and was equipped with a magnetic drum for storing information, which could memorize 180 working steps.

At the same time, another American company, AMF, started its work in developing an industrial robot, i.e. Versatran(meaning Versatile Transfer), as shown in Fig. 1–3. This robot was hydraulically driven and was mainly used for transporting materials between machines. Its arm could rotate around the base, rise and fall in the vertical direction, and extend and contract in the radius direction. In

1. Overview of Industrial Robot

Fig. 1–2　Unimate Robot

Fig. 1–3　Versatran Robot

general, Unimate and Versatran are considered to be the world's oldest industrial robots.

In 1978, Reis Robotics, a company in Germany, developed the first six–axis robot RE15 with an independent control system, as shown in Fig. 1–4. In 1979, Unimation launched its PUMA series industrial robot, which was a technological-ly–advanced robot with all–electric drive, joint structure, multi–CPU two–level microcomputer control, VAL specific language, and configurable, visual and tac-tile force sensing receptors. In the same year, the University of Yamanashi devel-oped a SCARA robot with planar joints.

5

Industrial Robot Technology Application

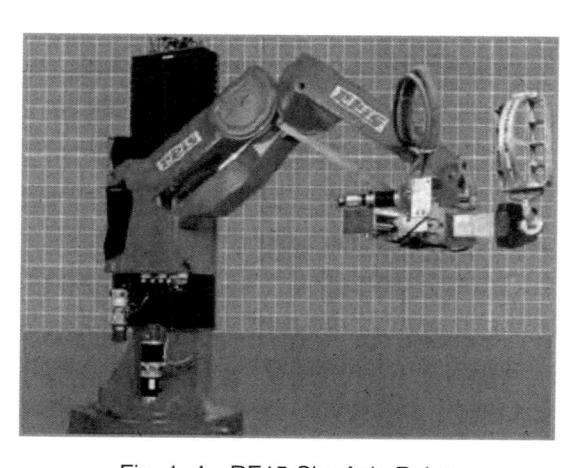

Fig. 1-4   RE15 Six-Axis Robot

In the 1970s, more robotic commodities emerged and were gradually popularized and applied to industrial production. With the development of computer science and technology, control technology and artificial intelligence, rapid advances were made in the research and development of robots both in level and scale. According to foreign statistics, by 1980, more than 20,000 robots were introduced for industrial applications in the world.

In the past decade, the advantages of the world's robots, both in terms of technological level and quantity used, have been concentrated in a few developed industrialized countries and regions represented by Japan, Europe and the United States. In 2019 and 2020, 2.75 million and 3.15 million industrial robots were installed for use in the world respectively, with a rapid growth rate over 10% per year. The sales of the four major families of industrial robots, namely ABB, KUKA, FANUC and YASKAWA, accounted for more than 75% of the global robot market share. At present, six-axis robot is the most widely-used industrial robot. This industrial robot with six joints is very similar to a human arm, and the six degrees of freedom are suitable for working at almost any track or angle. The second widely-used industrial robots are three-axis, four-axis and five-axis double-arm types. The selection of specific axis number usually depends on the requirements for flexibility, economic cost and speed in the actual application.

1.4 Development trend of robots

1.4.1 Industrial robots are developing towards intelligentization, modularization and bionics.

(1)Intelligentization

Intelligentization of industrial robots means that they can sense and perceive, quickly and accurately perform detection and judge various types of complex information. With the technological progress in execution and control, autonomous learning and intelligent development, the development of robots will gradually transition from pre-programming, teaching and reproduction control, direct control, remote operation and other manipulated operation modes to autonomous learning and autonomous operation.

Artificial intelligence is about the intelligent behaviors of artificial objects, including perception, reasoning, learning, communication and other behaviors in a complex environment. The long-term goal of artificial intelligence is to invent machines that can better perform these behaviors than human beings.

(2)Modularization

There will be the trend of manufacturing industrial robots by assembling standardized modules. At present, various countries are researching and developing combined robots, which will be assembled from standardized industrial robot components such as servo motor, sensor, arm, wrist, and robot body.

The research on new robot structures will be the future development trend. The new micro-motion robot structure can improve the operation accuracy and environment of an industrial robot. The development of new industrial robot structures will be required to adapt to high-intensity operations in complex environments.

(3)Bionics

Until recently, most robots were deemed as one of biological categories. A tool type robot retains the basic elements of a robot, such as claw-shaped manipulator, gripper and wheel. However, it looks like a machine in any perspective. In contrast, humanoid robots are mostly similar to the humans who created them, with their own hands on moving arms, real feet connected to their lower limbs, faces like humans, and, most importantly, their movements like humans. Some humanoid

robots are not only human–like in appearance, but also "thinking" themselves. Due to their thoughts and intelligence, they can replace human beings to perform more complex tasks.

2. Overall development trends of industrial robots

(1)Technological development trend

In terms of technological development, industrial robots are developing towards lightweight, intelligent, modular and systematic ones. The main development trends in the future include:

1)Modularization and reconfiguration of robot structure;

2)High performance and networking of control technology;

3)Openness and advanced language of control software architecture;

4)High incorporation and integration of servo drive technology;

5)Integration and intelligence of multi–sensor fusion technology;

6)Simplification and collaboration of human–machine interface.

(2)Application trend

The automobile industry has been the main venue for the application of industrial robots since their advent. In its annual report of 2014, the Robotic Industries Association of North America pointed out that by the end of 2013, the automobile industry was still the largest application market for robots in North America, but the delivery of robots to electronic/electrical industry, metal processing industry, chemical industry, food and other industries grew rapidly. Therefore, the future application of industrial robots will rely on the automobile industry and will be rapidly extended to many other industries. This is a quite positive signal for the robotics industry.

(3)Industrial development trend

According to the World Robotics Report 2020 released by the International Federation of Robotics(IFR), the top five markets for annual installation of industrial robots in 2019 throughout the world were China, Japan, the United States, South Korea and Germany. Currently, China is the world's largest and fastest growing market for robots. It is believed that the era of industrial robots will come soon in the near future, and will initiate a revolution in the field of intelligent manufacturing.

2.

Basic Composition and Technical Parameters of Industrial Robot

2.1 Basic composition of industrial robot

An industrial robot usually consists of four parts, i.e., actuator, drive system, control system and sensing system, as shown in Fig. 2–1.

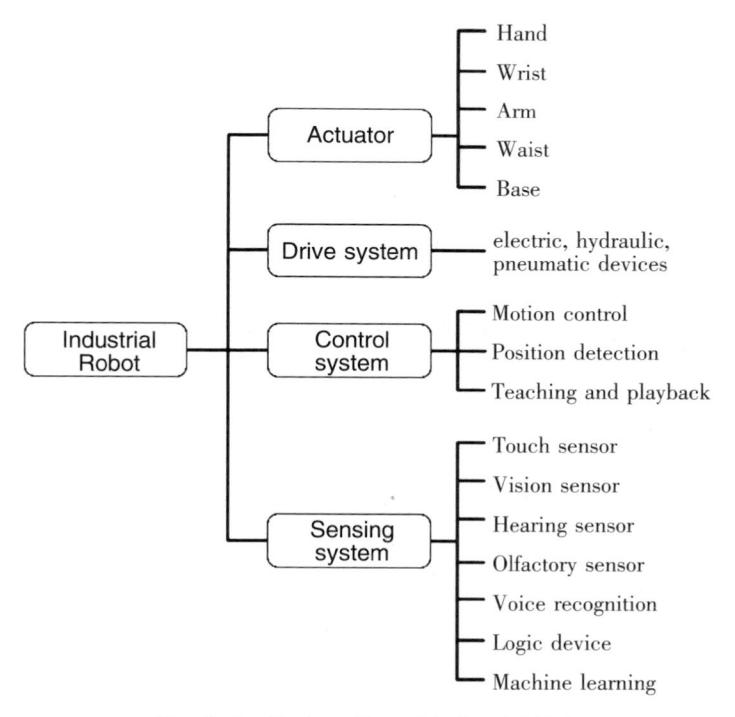

Fig. 2–1　Composition of Industrial Robot

Industrial Robot Technology Application

2.1.1 Actuator

An actuator is the entity through which a robot completes tasks and it is usually composed of a series of connecting rods, joints or kinematic pairs in other forms. An actuator can be divided into hand, wrist, arm, waist and base by function. The connection of each part, commonly known as "joint", can be driven by a servo motor to rotate, as shown in Fig. 2–2.

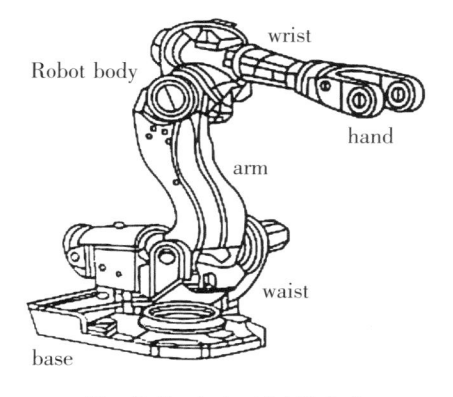

Fig. 2–2　Industrial Robot

2.1.2 Drive system

The drive system of an industrial robot is a device to provide power for each part of the actuator system. This system consists of a drive device and a transmission mechanism, which are usually integrated with the actuator. Usually, drive devices include electric, hydraulic, pneumatic devices and an integrated system that combine them for application. Commonly–used transmission mechanisms include harmonic, screw, chain, belt transmission mechanisms and various gear transmission mechanisms.

(1)Pneumatic drive: A pneumatic drive system usually consists of air cylinder, air valve, air reservoir and air compressor, and the actuator is driven by compressed air. The advantages of this system include convenient air source, rapid action, simple structure, low cost, easy maintenance, fire and explosion prevention, and no leakage impact on the environment, while its disadvantages include small operating force, large volume, and difficult speed control, slow response, unstable action and impact due to the high compression rate of air. Since the air

2. Basic Composition and Technical Parameters of Industrial Robot

source pressure is only about 60 MPa in general, pneumatically driven robots are suitable for applications requiring small pick-up forces.

(2)Hydraulic drive: A hydraulic drive system usually consists of hydraulic motor (including various oil cylinders and oil motors), servo valve, oil pump and oil tank, and the actuator is driven by compressor oil. This system is characterized by large operating force, small volume, stable transmission, sensitive action, shock resistance, vibration resistance and good explosion resistance. Compared with pneumatically driven robots, hydraulically driven ones have much greater pick-up capability, which is up to hundreds of kilograms or more. However, the hydraulic drive system has stricter requirements for sealing and is not suitable to work at high or low temperatures.

(3)Electric drive: The electric drive system directly drives the robot, or indirectly drives the robot via a reduction gear, with the force or the moment of force generated by the motor, to get the desired position, speed and acceleration. Electric drive is the most widely-used drive method for robots at present, due to its advantages such as easy access to power supply, no environmental pollution, fast response, large driving force, convenient signal detection, transmission and processing, various flexible control options, high movement accuracy, low cost and high drive efficiency. Generally, the drive motor used is a stepping motor, a DC servo motor or an AC servo motor.

2.1.3 Control system

There are two position control modes for industrial robots: point position control mode and continuous path control mode. To be specific, the point position control mode only concerns the start and end points of the robot's end effector, but not the motion track between these two points, and this control mode can complete spot welding, loading & unloading, handling and other operations under barrier-free conditions. In contrast, the continuous path control mode not only requires the robot to reach the target point with certain accuracy, but also has certain accuracy requirements for the motion track, and robots applied to painting and arc welding operations are examples of this control mode. Actually, the continuous path control mode is based on the point position control mode, and the position track interpolation algorithm meeting the accuracy requirements is used between every two points to achieve the continuity of tracks.

Industrial Robot Technology Application

2.1.4 Sensing system

Sensing system is an important part of a robot. Sensors forming this system are generally divided into internal and external sensors according to the location of the information collected. Internal sensors are necessary sensors for robot motion control, such as position and speed sensors, and they are used to collect the internal information of the robot and are indispensable and basic components of the robot. External sensors detect the environment of the robot, the state of an external object, or the relationship between the robot and the external object. Commonly used external sensors include force sensor, tactile sensor, proximity sensor and vision sensor. The classification of robot sensors is shown in Table 2-1.

Table 2-1　Classification of Robot Sensors

Sensor classification	Purpose	For precise control of robot
Internal sensors	Information sensed	Position, angle, speed, acceleration, pose, orientation, etc.
	Sensors used	Microswitch, photoelectric switch, differential transformer, encoder, potentiometer, resolver, tachogenerator, accelerometer, gyroscope, tilt angle sensor, force(or torque)sensor, etc.
External sensors	Purpose	For knowing the work object, the environment or the state of robot in the environment, and flexibly and effectively operating the work object
	Information detected	Work object and environment: shape, position, range, mass, pose, motion, speed, etc.
		Robot and environment: position, speed, acceleration, pose, etc.
		Operations of work object: non-contact operations (detection of spacing, position, pose, etc.), contact operations (obstacle detection, collision detection, etc.), touch operations (tactile sense, sense of contact force, slip sense), clamping force, etc.
	Sensors used	Vision sensor, optical ranging sensor, ultrasonic ranging sensor, tactile sensor, capacitive sensor, electromagnetic induction sensor, limit sensor, voltage-sensitive conductive rubber, elastomer and strain gauge, etc.

2. Basic Composition and Technical Parameters of Industrial Robot

Internal sensors are only applied to traditional industrial robots for precise control of their motion, position and pose. By using external sensors, robots can adapt to the external environment to some extent, thus showing a certain degree of intelligence.

Generally, other control components, such as drive devices and controllers, are placed in the same cabinet, which is called "control cabinet". The operating unit for man–machine interaction is the teach pendant, which consists of hardware and software and is a complete computer. Therefore, an industrial robot can be simply understood as consisting of three major parts shown in Fig. 2–3: robot body A, control cabinet B (also called the "control system", including main computer control module, axis computer board, axis servo drive, SMB measuring board connecting servo axis encoder, and I/O board), and teach pendant C (a hand–held operator device).

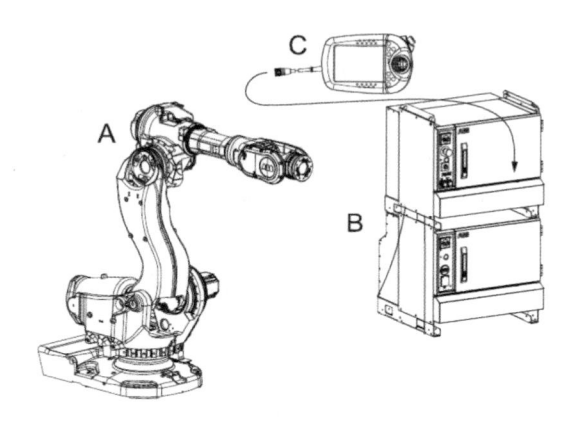

Fig. 2–3   Basic Composition of Industrial Robot

2.2 Basic components

2.2.1 Robot body

It is the main mechanical body of an industrial robot, which is also the actuator for various operations. In general, it includes:

- Manipulator;
- Drive and transmission devices;
- Various internal and external sensors.

Industrial Robot Technology Application

During operation, the end fixture, also known as the "end effector", is used to clamp and move the working target of the robot.

(1)Manipulator: The manipulator of an articulated robot is the aggregation of several connected mechanical joints, and it mainly consists of the following parts.

1) Base: It is the supporting part of the robot, and the actuator and drive device of the robot are installed inside it.

2)Waist: It is the middle supporting part connecting the base and the upper arm of the robot. During operation, the waist can rotate on the base through joint 1.

3)Arm: The arm of a 6-joint robot is generally composed of an upper arm and a forearm. The upper arm is connected to the waist through joint 2, while the forearm is connected to the upper arm through joint 3. During operation, the upper arm and the forearm are driven by corresponding joint motors to move or rotate.

4)Wrist: It is the part connecting the forearm and the end effector and mainly used to change the spatial pose of the end effector. It can combine all joints of the robot to achieve the expected movements and states of the robot.

(2)Body drive and transmission devices: When the robot is in motion, the movement of each joint must be done by a drive device and a transmission mechanism. Fig. 2-4 shows a typical driving joint consisting of an AC servo motor and a precision reducer.

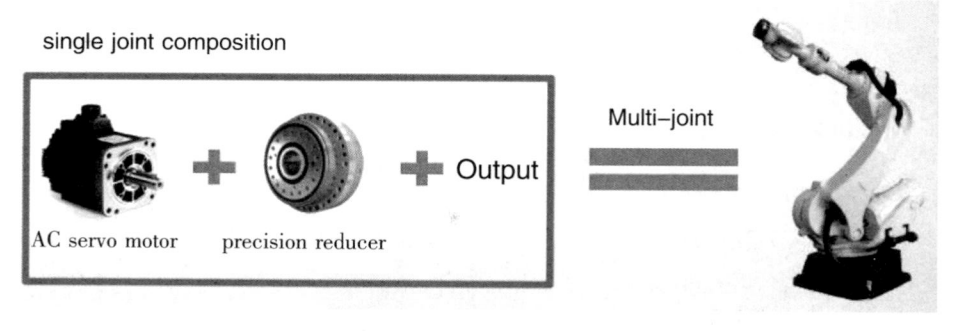

Fig. 2-4   Composition of Typical Driving or Moving Joint

2. Basic Composition and Technical Parameters of Industrial Robot

The drive device is a device that provides power and motion to each manipulator. Power sources and transmission modes of the drive system vary for different types of robots. The drive system mainly applies four transmission modes: hydraulic, pneumatic, electrical and mechanical transmission modes.

(3)Sensors used for robot body: The sensors used for the robot body are internal sensors, including microswitch, photoelectric switch, differential transformer, encoder, potentiometer, resolver, accelerometer, gyroscope, tilt angle sensor, and torque sensor.

2.2.2 Control system

Control system is the neural center of an industrial robot, and it consists of main computer unit, drive unit, axis computer unit, I/O interfaces and some dedicated circuits. During operation, the control system controls the robot body to complete certain actions or paths according to written commands and sensing information. This system is mainly used to process all information needed for operation of the robot. Its front internal structure is shown in Fig. 2–5.

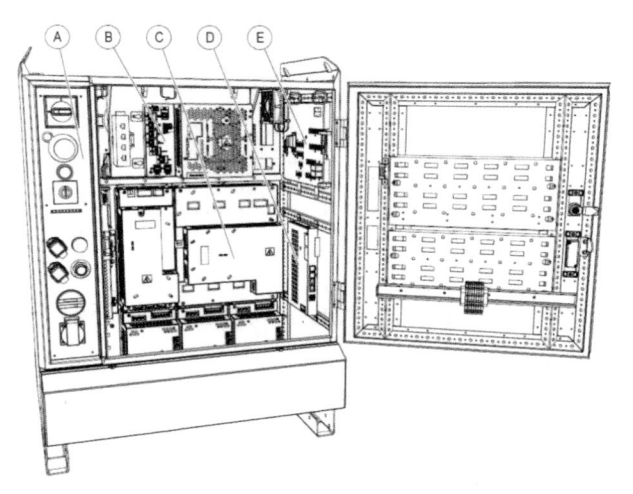

A. Control panel B. Main computer unit C. Drive unit D. Axis computer unit E. Security panel

Fig. 2–5　Robot Control Cabinet

2.2.3 Teach pendant

Teach pendant is an interface for man–machine interaction, also known as "teach box" or "teach programmer". It mainly consists of an LCD screen and

Industrial Robot Technology Application

touchable keys, as shown in Fig. 2-6. During operation, the operator should hold the teach pendant and press corresponding keys to input all the control information into the memory inside the control cabinet via the cable connected to the controller, so that the robot could be controlled. The teach pendant is an important part of the robot control system. Through the teach pendant, the operator can carry out manual teaching for the robot in order to control it to reach different positions/poses and record the coordinates of these positions/poses, or carry out on-line programming with the robot language in order to realize program playback and make the robot complete the designated trajectory motion in accordance with the written program.

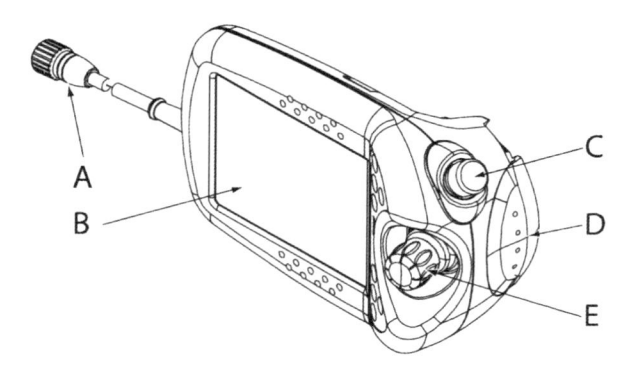

A. Connector; B. Touch screen; C. Emergency stop button; D. Actuator; E. Joystick

Fig. 2-6   Teach Pendant

2.3 Main technical parameters of industrial robot

The technical parameters of an industrial robot are as follows:
- Degree of freedom(DOF)
- Resolution
- Positioning accuracy
- Repeated positioning accuracy
- Operating range
- Movement speed
- Load-carrying capacity

These technical parameters are detailed as follows:

16

2. Basic Composition and Technical Parameters of Industrial Robot

(1)Degree of freedom(DOF)

DOF refers to the number of independent coordinate axis motions of the robot, excluding the opening and closing DOF of the end effector. Generally, one DOF of the robot corresponds to one joint, so DOF is equal to joint in terms of concept. DOF is a parameter indicating the flexibility of robot action. The higher the DOF, the more flexible the robot, with more complex structure and greater control difficulty. For this reason, the DOF of a robot should be designed according to its intended use, generally ranging from 3 to 6. Fig. 2−7 shows a 6−DOF robot, with the rotation direction of each axis indicated by an arrow.

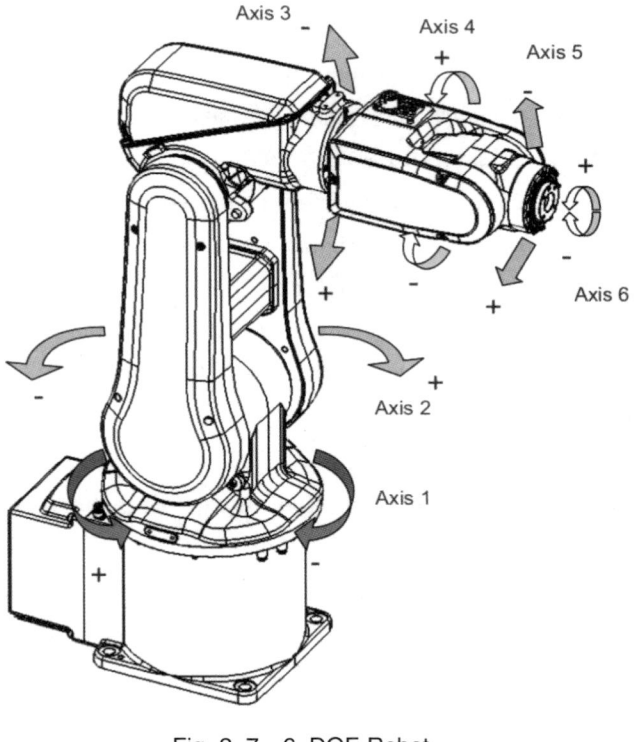

Fig. 2−7 6−DOF Robot

(2)Resolution

Resolution refers to the minimum moving distance or the minimum rotation angle that can be achieved by each joint of the robot. Resolution of an industrial robot is divided into programming resolution and control resolution.

17

Industrial Robot Technology Application

• Programming resolution: It refers to the minimum distance that can be set in the control program, also known as reference resolution. The reference resolution of a robot is 0.01 mm when a joint motor of the robot rotates by 0.1° and the joint end point of the robot moves for 0.01 mm in terms of linear distance.

• Control resolution: It refers to the minimum displacement detected by the system position feedback circuit, i.e. the rotation angle of the joint motor when a single pulse is generated by the coding disc installed coaxial with the robot's joint motor.

(3) Positioning accuracy

Positioning accuracy refers to the deviation between the actual and target positions of the robot's end effector, consisting of mechanical error, control algorithm and system resolution. In general, the positioning accuracy of typical industrial robots is in the range of ±0.02 mm to ±5 mm.

(4) Repeated positioning accuracy

Repeated positioning accuracy refers to the position dispersion of the robot when it repeats its movement several times under the same environment, condition, target action and command, and it is the statistical data about accuracy. Since the repeated positioning accuracy is not affected by the changes of operating load, it is usually used as an important indicator to measure the level of teaching-playback industrial robots.

(5) Operating range

Operating range refers to the set of position points that can be reached by the arm end or wrist center when the robot is moving. Fig. 2-8 shows the operating range of ABB IRB120 robot. The operating range is related not only to the size of each connecting rod of the robot, but also to the overall structural form of the robot. The shape and size of the operating range are very important, because a robot may be unable to complete its tasks due to the blind area that its hand cannot reach when performing a certain operation. Therefore, a robot meeting the requirements for the current operating range must be selected for performing its tasks.

Fig. 2-9 shows the schematic diagram of the operating range of a cartesian robot. In terms of structure, the robot has such advantages as easy control through computer, easy access and high accuracy, while its disadvantages include covering

2. Basic Composition and Technical Parameters of Industrial Robot

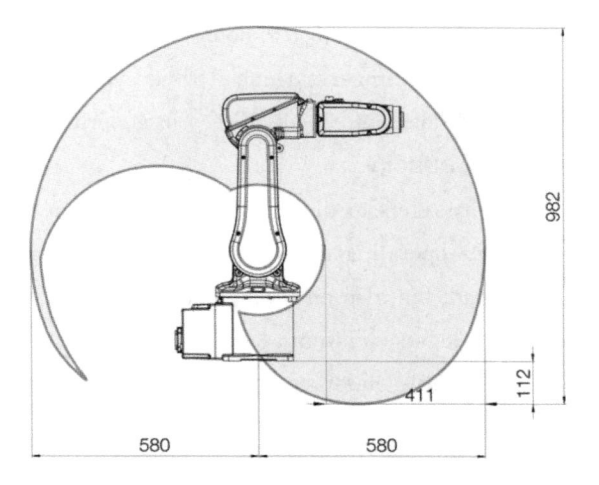

Fig. 2–8   Operating Range of ABB IRB120 Robot

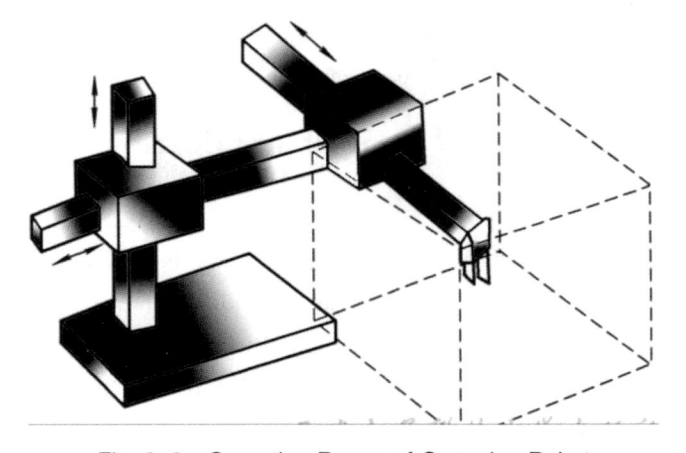

Fig. 2–9   Operating Range of Cartesian Robot

a large floor area, hindering work, moving at low speed and giving poor sealing performance.

(6)Movement speed

Movement speed may affect the working efficiency and motion period of the robot, and it is closely related to the pick–up weight and position accuracy of the robot. If the movement speed is high, the dynamic load borne by the robot would increase, which inevitably makes the robot bear a greater inertia force during acceleration and deceleration, thus affecting the operating stability and position ac-

19

curacy of the robot. As far as the current technology is concerned, for most general-purpose robots, the maximum linear movement speed is below 1,000 mm/s and the maximum slewing speed does not exceed 120°/s in general.

(7) Load-carrying Capacity

Load-carrying capacity refers to the maximum mass that a robot can bear in any position/pose within its operating range. Load-carrying capacity depends not only on the mass of the load, but also on the operating speed of the robot and the magnitude and direction of its acceleration.

Industrial robots are roughly classified into the following by load-carrying capacity:

① Mini-robot, with a load-carrying capacity less than 1 N;

② Small robot, with a load-carrying capacity not more than 105 N;

③ Medium robot, with a load-carrying capacity of 105 ~ 106 N;

④ Large robot, with a load-carrying capacity of 106 ~ 107 N;

⑤ Heavy-duty robot, with a load-carrying capacity more than 107 N.

2.4 Classification and typical application of industrial robots

Many kinds of industrial robots have been developed, and their functions, characteristics, driving modes, applications and other parameters are not exactly the same. At present, there is no unified classification standard for robots in the world. Generally, industrial robots can be classified by structural feature, control mode, drive mode and application field.

(1) Classification by structural feature

• Cartesian robot

• Cylindrical robot

• Polar robot

• Articulated robot

• Parallel robot

①Cartesian robot

Cartesian robot is a kind of multi-purpose robot with three DOFs that can be automatically controlled, repeatedly programmable and spatially perpendicular to each other in industrial applications, as shown in Fig. 2-10. This robot has three

2. Basic Composition and Technical Parameters of Industrial Robot

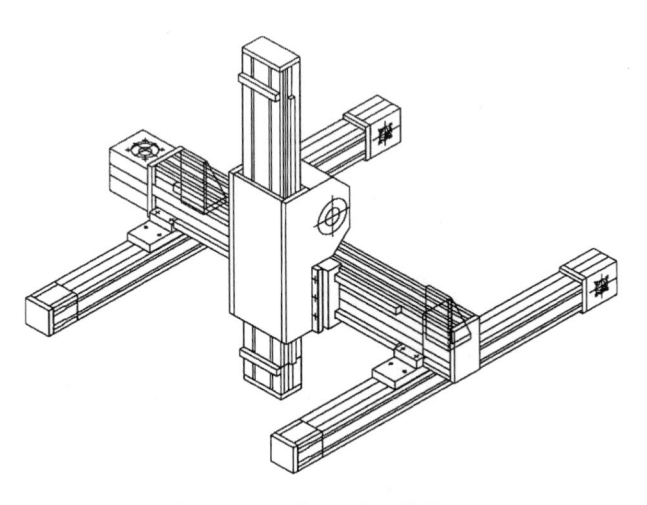

Fig. 2–10　Cartesian Robot

perpendicular moving joints X/Y/Z in the space coordinate system, and each joint can move in an independent direction. It is characterized by linear motion and simple control. However, its disadvantages include poor flexibility and large space occupation.

Cartesian robot can be conveniently applied to various automatic production lines and can complete a series of tasks, such as welding, handling, loading & unloading, packing, stacking, detection, flaw detection, classification, assembly, labeling, code spraying, code printing, painting, target following and explosion clearing.

②Cylindrical robot

Cylindrical robot is a kind of robot whose axis can form a cylindrical coordinate system. In terms of structure, it mainly consists of a rotating joint formed by a rotating base and two moving joints that move vertically and horizontally, as shown in Fig. 2–11. The pose of the end effector of a cylindrical robot is determined by(x, R, θ).

The cylindrical robot has such advantages as small spatial structure, large operating range, high speed of end effector, simple con-

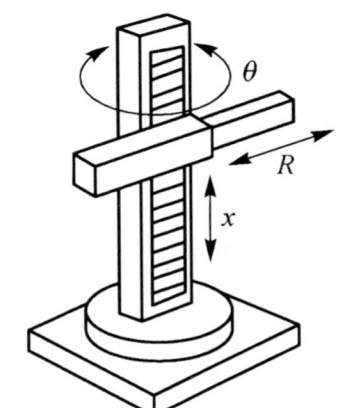

Fig. 2–11　Cylindrical Robot

21

Industrial Robot Technology Application

trol and flexible movement. However, its disadvantages include a moving space a-
long axis r required for moving forward and backward during operation and a low
space utilization rate. At present, cylindrical robots are mainly used for loading &
unloading, handling and other operations of heavy objects. Versatran, a famous
robot, is a typical cylindrical robot.

③Polar robot

A polar robot generally consists of two
slewing joints and a moving joint. Its axes
are configured in accordance with polar
coordinates. R is the moving coordinate, β
is the arm's swing angle in the vertical
plane, and θ is the rotation angle around
the axis perpendicular to the arm support
base. The track surface formed by the
movement of this robot is a hemispherical
surface, so this robot is also called spheri-
cal coordinate robot, as shown in Fig. 2-12.

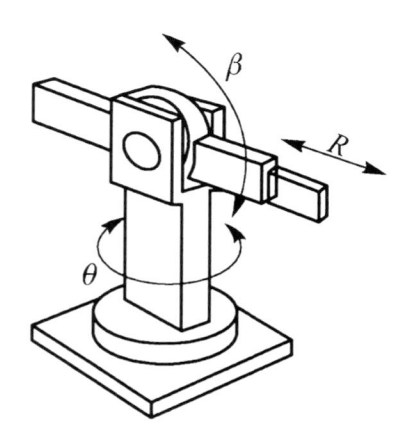

Fig. 2-12　Motion of Polar Robot

The polar robot also has the advantages of small space occupation, flexible
operation and large operating range, but its kinematic model is complicated and
difficult to control.

④Articulated robot

Articulated robot, also known as articulated arm robot or articulated robot
arm, is the most widely-used industrial robot type. Articulated robots are further
divided into vertically articulated robot (as shown in Fig. 2-13)and horizontally
articulated robot by joint configuration.

The articulated robot also features less space occupation, flexible operation
and large operating range, but its kinematic model is complicated and difficult to
control. This robot consists of multiple rotating and swinging joints, with compact
structure, large operating space and close action to human beings. During operation, it can bypass some obstacles around the base. It well adapts to various operations such as assembly, painting and welding, and is suitable for motor drive.
Since its joints are sealed, it is easy to be dustproof. At present, some companies,
including ABB in Switzerland, KUKA in Germany, YASKAWA in Japan and oth-

22

2. Basic Composition and Technical Parameters of Industrial Robot

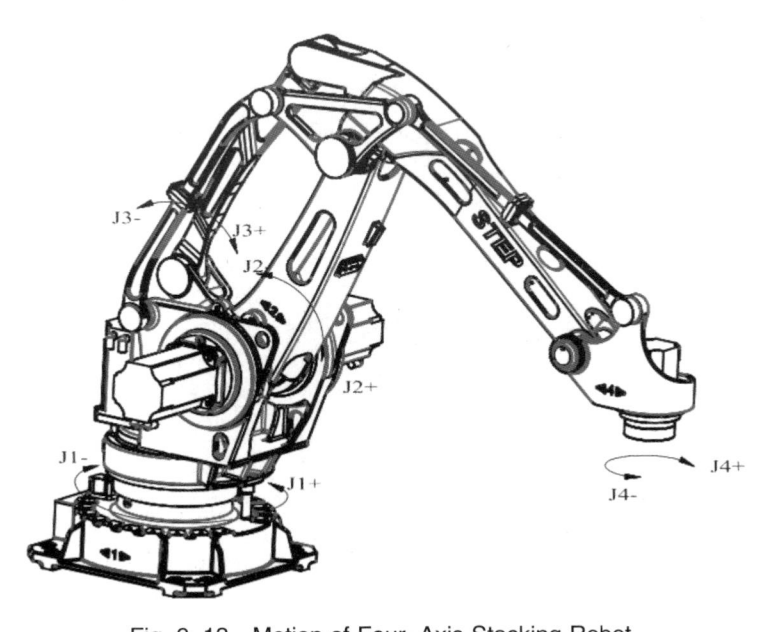

Fig. 2–13   Motion of Four–Axis Stacking Robot

ers in China, are launching such products.

Horizontally articulated robot, also known as SCARA (Selective Compliance Assembly Robot Arm), has four axes and four DOFs of motion, namely, rotation around axes A, B and C and translation in direction Z, enabling the translational DOF in directions X, Y and Z and the rotational DOF around axis Z, as shown in Fig. 2–14.

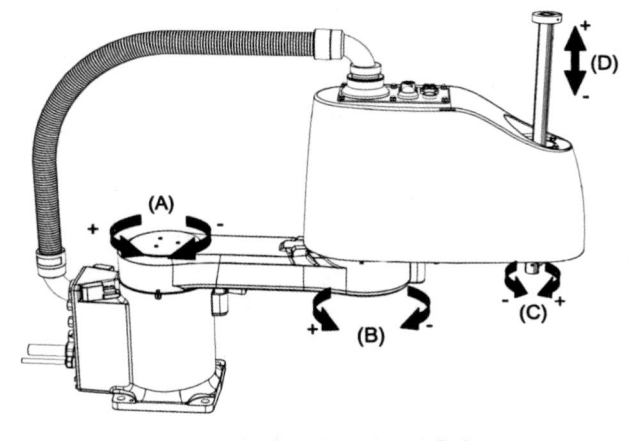

Fig. 2–14   Horizontal Articulated Robot

Industrial Robot Technology Application

⑤Parallel robot

Parallel robot is a new type of robot developed in recent years, which is formed by connecting a fixed base and an end effector with several DOFs with at least two independent kinematic chains. Fig. 2-15 shows a Delta parallel robot.

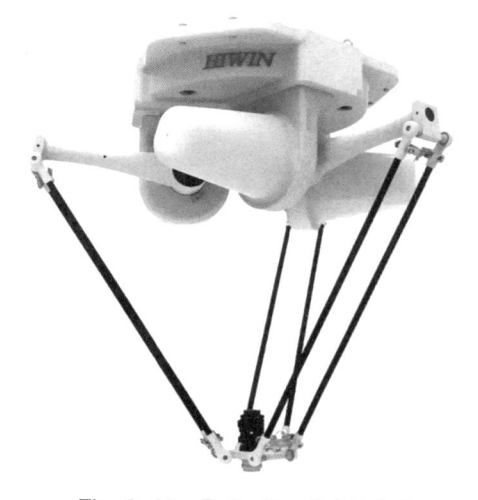

Fig. 2-15　Delta Parallel Robot

A parallel robot has the following characteristics:

• No cumulative error, with high accuracy;

• Drive device placed either on or near the fixed platform, and light weight, high speed and quick dynamic response of moving part;

• Compact structure, high rigidity and large load-carrying capacity;

• Good isotropy;

• Small workspace.

(2)Classification by control mode, drive mode and application field

①Classification by control mode:

• Servo controlled robot

• Non-servo controlled robot

②Classification by drive mode:

• Hydraulically-driven robot

• Pneumatically-driven robot

• Electrically-driven robot

2. Basic Composition and Technical Parameters of Industrial Robot

• Neo−driven robot

③Classification by structural feature or application field:

• Welding robot(as shown in Fig. 2−16)

• Handling robot(as shown in Fig. 2−17)

• Assembly robot(as shown in Fig. 2−18)

• Painting robot(as shown in Fig. 2−19)

Fig. 2−16　Welding Robot

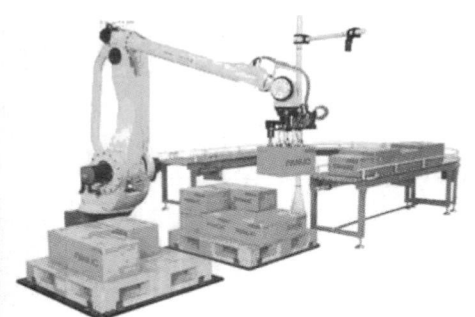

Fig. 2−17　Handling Robot

Fig. 2−18　Assembly Robot

Fig. 2−19　Assembly Robot

3.

Installation and Commissioning of Industrial Robots

Industrial robots are precision electromechanical equipment with special requirements for transportation and installation. Each brand of industrial robots has its own instruction manual for installation and connection, and these manuals are similar. The general installation process of industrial robots is shown in Fig. 3–1.

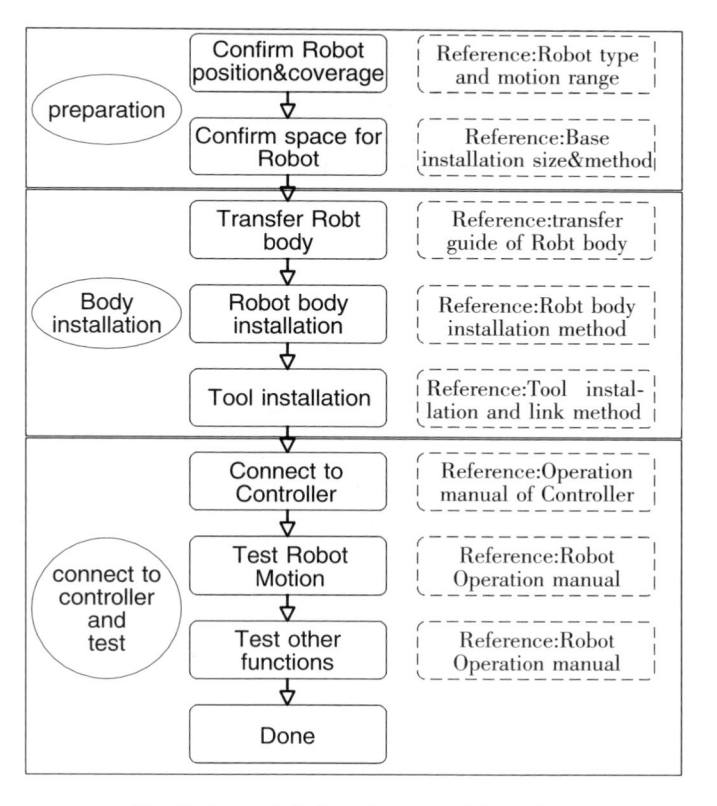

Fig. 3–1   Installation Process of Robots

Industrial Robot Technology Application

Please refer to the relevant parts of the instruction manual carefully for each step.

3.1 Unpacking and fixing of robots

Industrial robots are packed in accordance with the standard process and then delivered to the customer's site. The specific work of unpacking and installation are as follows:

(1)Place the robots with the arrow up after arriving at the site, as shown in Fig. 3-2. Check the appearance for any damage, water inflow and other abnormalities firstly. If there is any problem, contact the manufacturer or supplier immediately.

Fig. 3-2   Arrival Status of Robots

(2)First, remove the top cover of the packing box with tools, and then take out the accessory boxes in turn, as shown in Figure 3-3, which avoid damaging the accessories when removing the side plate of the wooden case.

(3)Count the standard packing items of industrial robots and take ABB robot IRB120 as an example, which includes 4 main items, that is, robot body, teach pendant, cable accessories and control

Fig. 3-3   Schematic Diagram of Robot Unpacking

3. Installation and Commissioning of Industrial Robots

cabinet. In practice, the packing items should be counted according to the packing list of the manufacturer. The attached documents also include: SMB battery safety instructions, ex–factory list, basic operation instructions and.

(4)Remove the screws (4 in total) fixing the robot to the base with a wrench, as shown in Fig. 3–4.

Fig. 3–4　Base of Robot

(5)As shown in Fig. 3–6, When transporting the robot body to the installation position, the special lifting accessories provided by the manufacturer shall be installed and used, otherwise the working parts of the robot may be damaged; It is strictly prohibited to change the standard posture of the robot body when it leaves the factory during the transfer process or when the robot base is not fastened, because moving the arm will shift the center of gravity of the robot and may cause the robot to overturn.

Industrial Robot Technology Application

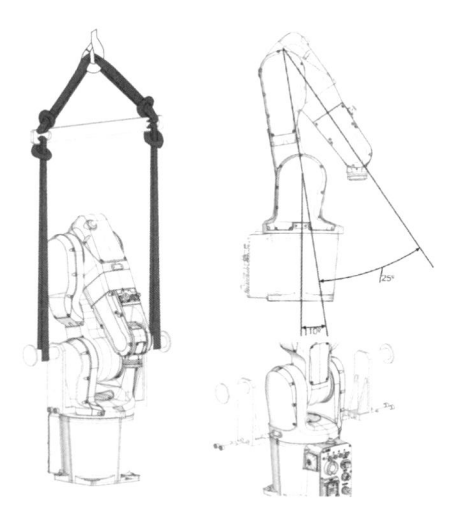

Fig. 3-5　Standard transfer posture of robot body

(6)Prepare the power cable, SMB cable and teaching pendant cable, and read the wiring diagram of ABB IRC5 Compact control cabinet carefully, as shown in Fig. 3-6.

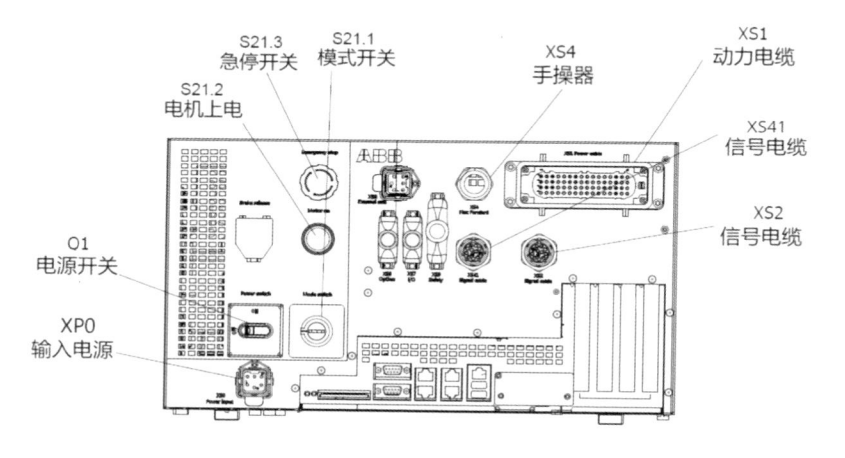

Fig. 3-6　Wiring diagram of ABB IRC5 Compact

(7)Connect the power cable plug marked with XP1 to the XS1 interface of the control cabinet, and connect the plug marked with R1.MP of the power cable to the interface on the base of the robot body.

30

3. Installation and Commissioning of Industrial Robots

(8)Connect the SMB cable (straight end) plug to the XS2 interface of the control cabinet, and connect the SMB cable (elbow) plug to the SMB interface on the base of the robot body.

(9)Insert the connector of teach pendant cables (red) into the XS4 port of control cabinet.

(10)According to the power supply parameters of ABB IRB1200, prepare the power supply line and make the connector end for the control cabinet. Definition of power connector end for the control cabinet is shown in Figure 3–7.

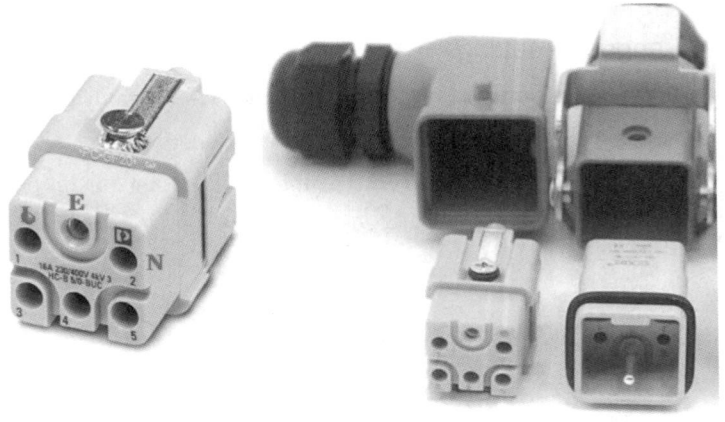

Fig. 3–7　power connector of control cabinet

(11)After confirming that all the above connections are correct again, insert the power connector into the XP0 interface of the control cabinet and lock it. So far, the basic connection between the robot body and the controller has been completed. The finished status is shown in Figure 3–8.

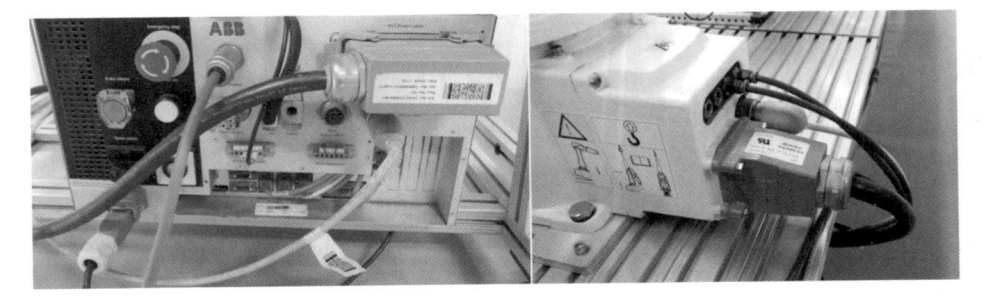

Fig. 3–8　Final status of robot completed wiring

31

Industrial Robot Technology Application

（12）Install the teach pendant bracket to a proper position (easy access to the teach pendant and not easy to collide with the operator), and then place it properly.

（13）After confirming that all connections are correct, turn on the power switch for trial operation. Turn the main power knob on the control cabinet from [OFF] to [ON] to start the machine, and turn the knob counterclockwise from [ON] to [OFF] to cut off the system power supply.

3.2 Commissioning Process of Robots

3.2.1 Language setting

When the robot is powered on for the first time, the default language is English, which needs to be changed to Chinese for easy operation.

（1）Tap the "ABB" menu icon in the upper left corner of the teach pendant screen–"Control Panel" option, as shown in Fig. 3–9.

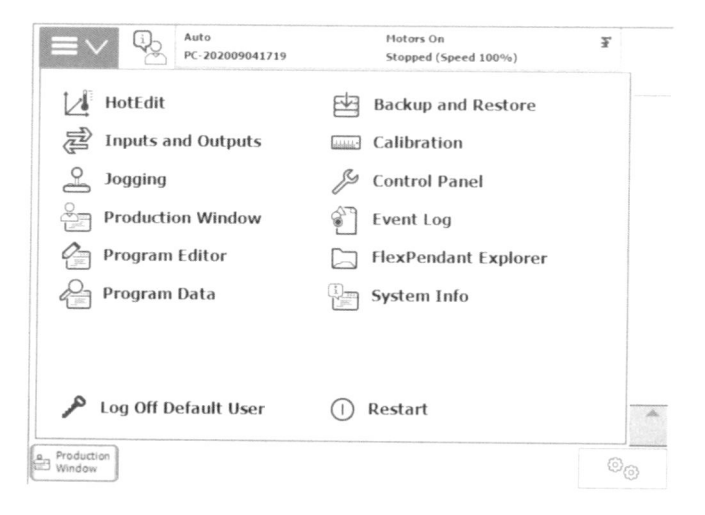

Fig. 3–9　"Control Panel" Option

（2）Tap the "language" option, select "Chinese" and tap "OK" for confirmation, as shown in Fig. 3–10.

32

3. Installation and Commissioning of Industrial Robots

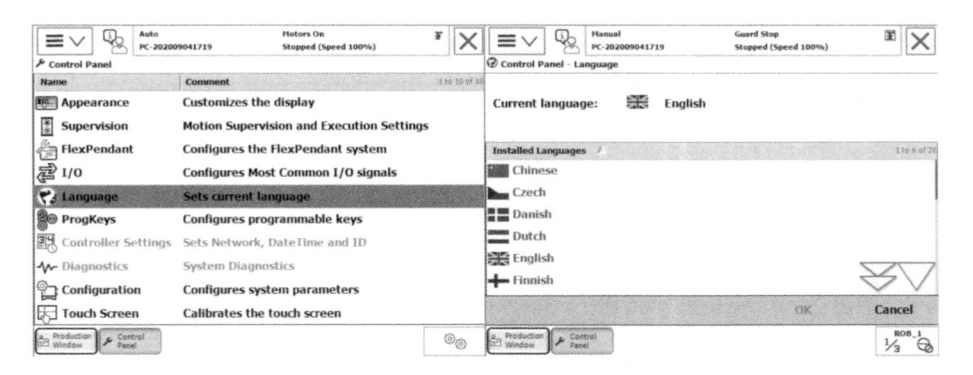

Fig. 3–10   Language Setting

(3)Tap the "Yes" button in the displayed prompt dialog box to restart the robot and complete the language setting, as shown in Fig. 3–11.

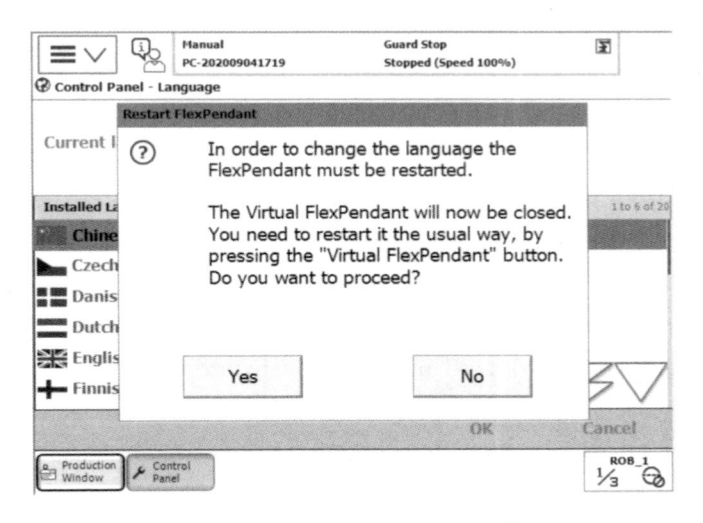

Fig. 3–11   SMB Cable(Straight Head)Connector

3.2.2 Backup and restore

Regular backup for the data of the robot is a good habit to ensure its normal operation. The backup file can be stored on the internal memory of the robot or a USB flash disk.

The file contains contents such as operating procedures and system configuration parameters. In case of an error in the robot system, the status before back-

Industrial Robot Technology Application

up can be restored quickly by using the backup file. Always make a backup before changing the program. It should be noted that the backup recovery data is unique and cannot be used for another robot.

3.2.3 Zero calibration

Each joint axis of the ABB robot has a mechanical origin, which has been calibrated before delivery, and there are markings on six joint shafts of the robot body. This zero is used as the reference for the movement of each joint axis. Therefore, zero return operation should be carried out for the robot after installation to check whether the zero is lost. The new robot can be used directly without calibration as long as the zero is not lost.

3.2.4 System I/O configuration and wiring

The ABB robot is provided with many I/O communication interfaces, realizing easy communication with surrounding equipment. The following takes the most commonly used standard I/O board DSQC651 of ABB robots as an example for detailed explanation. The common signal processing provided by the standard I/O board of ABB robots includes digital input DI, digital output DO, analog input AO, and conveying chain tracking.

3.2.5 Signal check

Tap the "ABB" icon to enter the system menu, and tap the "Input/Output" to monitor I/O signals. 0 indicates no signal and 1 indicates that there is a signal. Check whether the configured signal corresponds to the actual signal. For example, doGripperA and doGripperB represent two fixture cylinders of the robot respectively. Tap one of them and then tap 0 or 1 to change the fixture status, release and close the fixture forcibly, and check whether the solenoid valve wiring is wrong.

3.2.6 Import procedure

Tap the "ABB" icon, select "Program Editor", tap "Module", select "Loading module" in the module interface, and load the program module required to be loaded from the path where the program module is stored. The module is usually stored in the PROGMOD folder and can be opened with a notepad.

3.2.7 Setting of work object coordinate system

(1)Coordinate system of robot

Setting of work object coordinates is the premise of teaching, and all teaching

points must be established in the corresponding work object coordinates. If teaching points are set up on wobj0, all points must be taught again after handling of the robot. If the teaching is carried out on the left side of the corresponding work object, you can modify the work object coordinates without re-teaching all points.

(2)Necessity of correctly setting work object coordinates

It is difficult for the robot to move in the X/Y direction of the work object object due to inaccurate work object coordinates.

(3)Coordinate setting

Create a wobj1 project in the teach pendant, define work object coordinates and verify their accuracy.

3.2.8 Reference point

Tap the "ABB" icon to enter the main system interface, tap "Program data" and then "robtarget", and select the station to be modified. pPick1 and pPkck2 correspond to the basic positions of stations 1 and 2 respectively. On the drop-down menu after tapping the station, select "Edit", and "Modify position" in the Edit option. After the modification, the robot will automatically record the new position.

3.2.9 Parameter adjustment

Accurately adjust the operation program parameters of the robot. For example, the stacking robot needs to adjust slightly the length, width and height of the carton to be handled, number of cartons, grasping position and stacking placement location to ensure that all movements of the robot are stable and reliable.

3.2.10 Manual commissioning

The commissioned program must pass the manual operation test before automatic operation. In the process of the manual operation commissioning, the robot will stop immediately after the controller is released in case of any problem.

3.2.11 Automatic operation

The automatic operation can only be carried out after the manual commissioning is completed without error, and a lower speed must be set in the early stage. In the process of automatic operation, press the pause key to stop the robot. At this time, the motor is still on. Press the start key to continue the operation. Precautions during commissioning:

(1)Make sure that the power supply wiring is correct and firm and grounded effectively before power on;

(2)The teach pendant shall be connected and disconnected after power off;

(3)Be sure to wait for about 1 minute after complete shutdown and then restart the robot to prevent data loss;

(4)Re-check the procedure contents after modification;

(5)After modification of program parameters, be sure to manually operate the program at a low speed before automatic operation;

(6)Make a backup in time before modification.

)Make a backup in time before modification.

4.

Basic Operation of Industrial Robots

4.1 Knowledge of teach pendant

The robot teach pendant is an important peripheral device of industrial robots. It is a hand−held output device for our operator to "talk" with industrial robots. Manual control of the robot, as well as manual programming and debugging of the industrial robot, and modification of robot system parameters can be carried out through the robot teach pendant. Therefore, the teach pendant plays an important role in the industrial robots. The ABB IRB120 teach pendant is shown in Fig. 4−1, and the parts in the figure include the following:

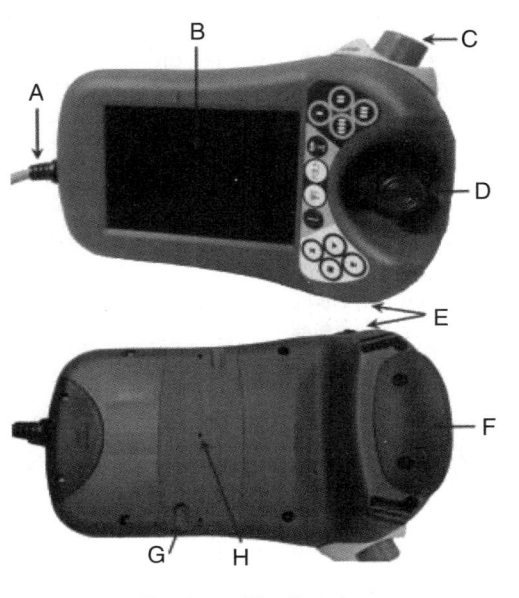

Fig. 4−1   FlexPendant

Industrial Robot Technology Application

- A Connecting cable
- B Touch screen
- C Emergency stop switch
- D Manual operating rocker
- E USB port
- F Enabler key
- G Touchscreen pen
- H Reset button of teach pendant

In order to use the teach pendant correctly, hold the teaching pen with the right hand to operate the screen and buttons, and place the left hand on the enabling key, so that the teach pendant can be placed on the left hand comfortably, and the screen and buttons can be operated with the right hand. This teach pendant of the ABB robot is designed ergonomically and suitable for left-handed operation(as shown in Fig. 4-2). It can be adapted to operation habits of left-handed people just by switching between the screens.

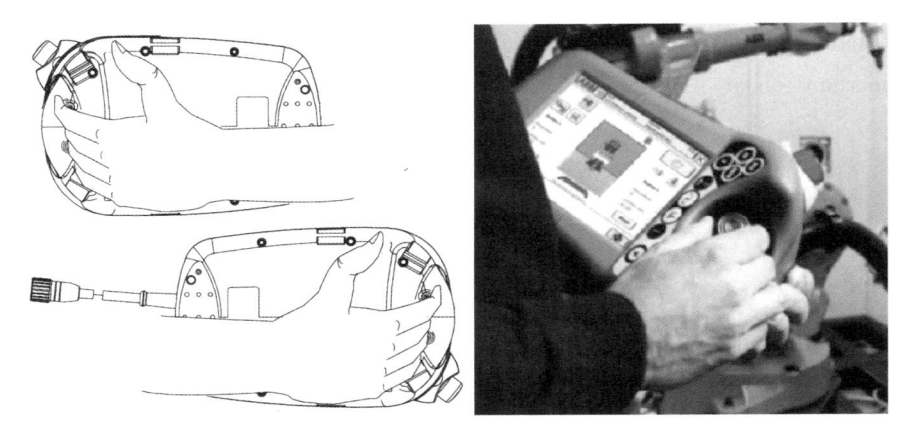

Fig. 4-2   Schematic Diagram of Teach Pendant Grip Method

After the system is started, the interface of the teach pendant is shown in Fig. 4-3. The description of each area marked in the drawing is as follows:

- 1--Main menu: display the main menu interface of various functions of the robot.
- 2--Operator window: The interaction interface between the robot and operator displays the current status information.

4. Basic Operation of Industrial Robots

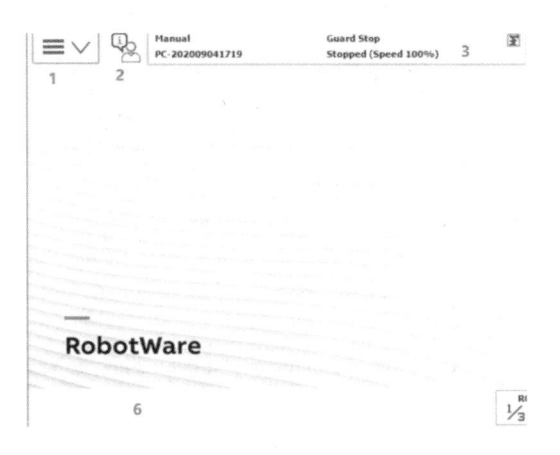

Fig. 4-3   Startup Interface of Teach Pendant

• 3--Status bar: display the current status of the robot, including operating mode, motor status and alarm information.

• 4--Close button: close the current window.

• 5--Quick setting menu: set the robot function interface quickly, including speed, operation mode and increment.

• 6--Task bar: task list of the currently opened interface. Up to 6 interfaces can be opened simultaneously.

The red "ABB" button in the upper left corner of the teach pendant screen is the main menu. After tapping the button, the menu is expanded as shown in Fig. 4-4. For the function description of various options, please refer to Table 4-1.

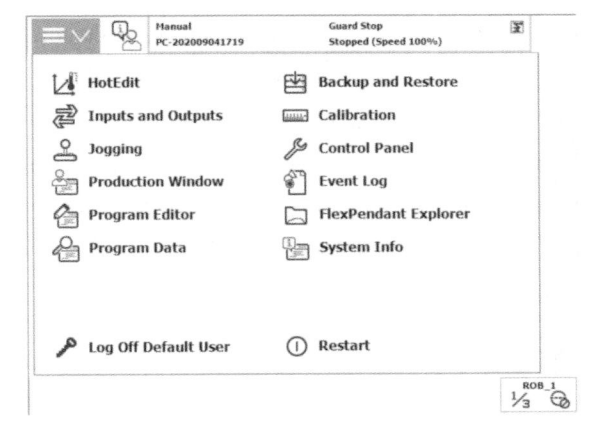

Fig. 4-4   Main Menu of Teach Pendant

39

Industrial Robot Technology Application

Table 4-1   Functions of Main Menu Options

Name of Options	Function
HotEdit	Used for compensation setting of track point position under the program module
Input/output	Used for setting and viewing I/O signals
Manual Control	Used for viewing and configuring manual control properties
Automatic Production Window	Used for debugging the program directly and running it in automatic mode
Program Editor	Used for programming and debugging of the robot
Program Data	Used for viewing and configuring variable data
Backup and Restore	Used for backup and restore of system data
Calibration	Used for calibrating mechanical zero of the robot
Control Panel	Used for configuring system parameters
Event Log	Used for viewing various prompts appearing in the system
Flex Pendant Explorer	Used for managing system resources and backup files
System Information	Used for viewing system controller properties and hardware and software information
Logout	Used for exiting the current user right
Restart	Used for restarting the teach pendant

4.2 Configuration of necessary operating environment

4.2.1 Time setting of robot system

The time of the robot system should be set as the time of the local time zone before various operations in order to facilitate file management and fault inspection and management, and the specific operations are as follows:

Tap the ABB menu button at the upper left corner of the teach pendant screen, and select the "Control panel"−"Date and time" options in turn to enter the date and time setting interface, as shown in Fig. 4−5. After modification of the date and time, tap "OK" to complete the setting.

4.2.2 Correct use of enabler key

The enabler key is set for the industrial robot to ensure the personal safety of the operator. Manual control and program debugging can only be carried out for the robot by pressing the enabler key and keeping it in the "click−on" state. If any

40

4. Basic Operation of Industrial Robots

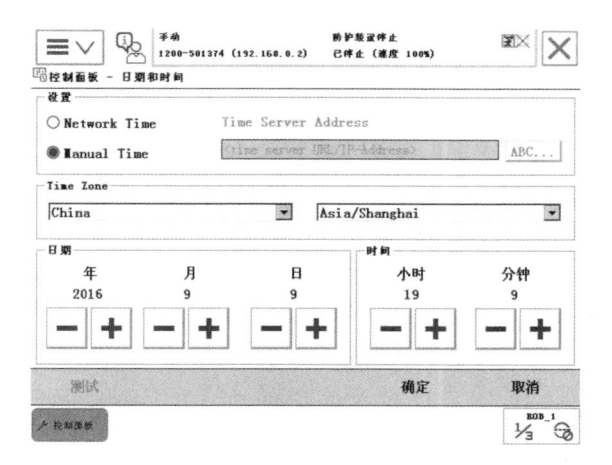

Fig. 4-5　Date and Time Setting Interface

danger occurs, people will instinctively release or press the enabler key, and the robot will stop immediately to ensure safety.

(1)The enabler key is located on the right side of the manual operating rocker of the teach pendant. The operator shall operate the key with four fingers of the left hand, as shown in Fig. 4-6.

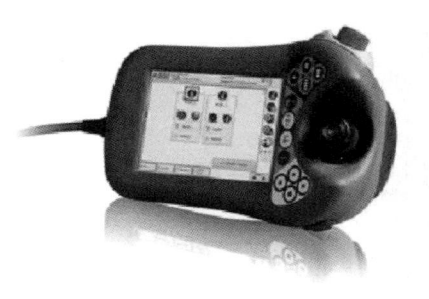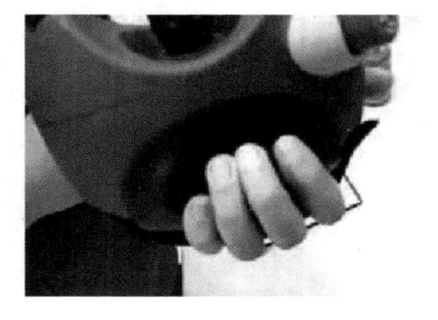

Fig. 4-6　Schematic Diagram of Enabler Key Operation

(2)The ABB robot enabler is a special type of device, also known as three-position actuator. The enabler can be activated only when its key is pressed by half. It is impossible to operate the robot in the fully pressed and fully released positions. In the manual state, lightly press the enabler to the middle gear, and the robot will be in the motor-on state. After complete pressing of the enabler key, the robot will be in the protective device stop state, as shown in Fig. 4-7.

41

Industrial Robot Technology Application

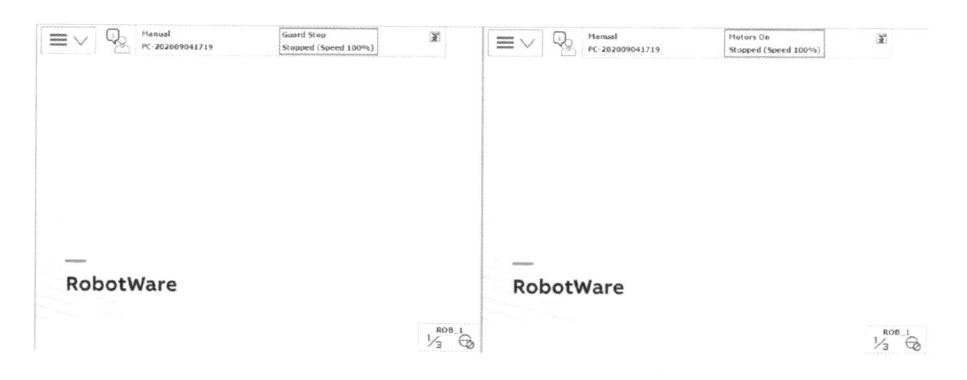

Fig. 4-7　Robot State Change Diagram after Enabler Is Pressed

4.2.3 Viewing for Common Information and Event Log of ABB Robot

The operator can view the common information and event log of the ABB robot through the status bar on the screen of the teach pendant. The status bar contains the following information, as shown in the mark 3 of Fig. 4-3:

(1)Robot state(manual, full-speed manual and automatic).

(2)System information of the robot.

(3)Motor state of the robot.

(4)Program running state of the robot.

(5)Current usage state of the robot or outer shaft.

Tap the status bar in the screen to view the event log of the robot, as shown in Fig. 4-8.

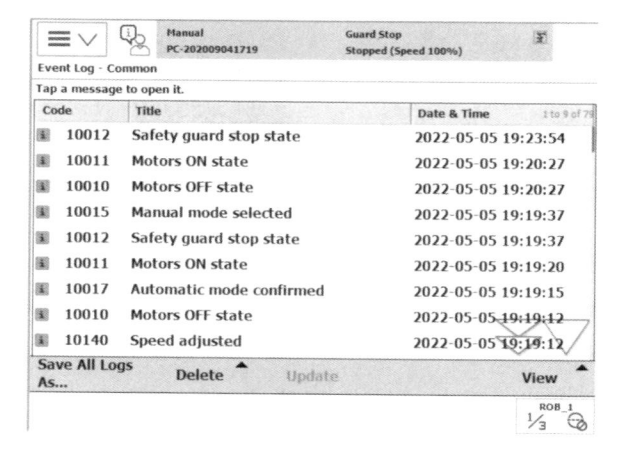

Fig. 4-8　Status Bar of Teach Pendant Screen

4. Basic Operation of Industrial Robots

4.2.4 Backup and restore of ABB robot data

Regular backup for ABB robot data is a good habit to ensure its normal operation. The objects of ABB robot data backup are all RAPID programs and system parameters running in system memory. When the robot system is out of order or a new system is reinstalled, the robot can be quickly restored to the backup state through backup data.

4.2.4.1 Backup of ABB robot data

(1)Tap the ABB menu button in the upper left corner of the teach pendant screen and select the "Backup and Restore" option, as shown in Fig. 4–9.

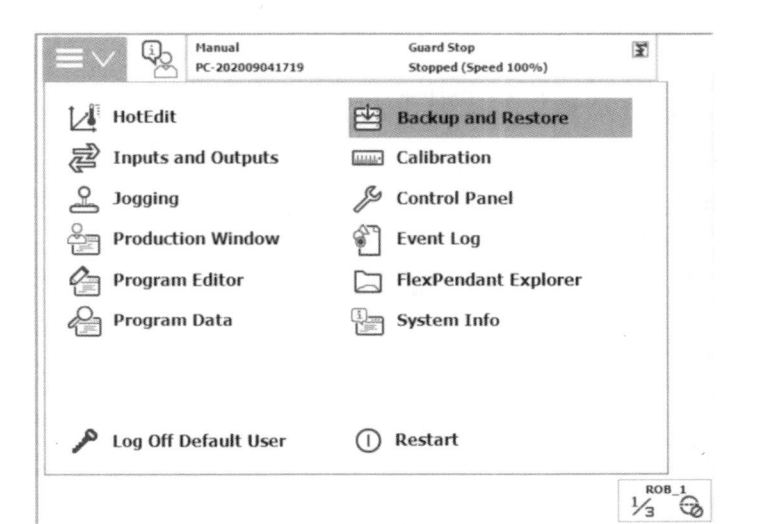

Fig. 4–9　Backup and Restore Option

(2)Tap the "Backup current system..." option, as shown in Fig. 4–10.

(3)Enter the backup path setting interface, as shown in Fig. 4–11, and tap "ABC..." button to set the name of the backup data directory; tap "..." button to select the backup storage location (robot hard disk or USB storage device); finally tap "Backup" button for backup operation.

Industrial Robot Technology Application

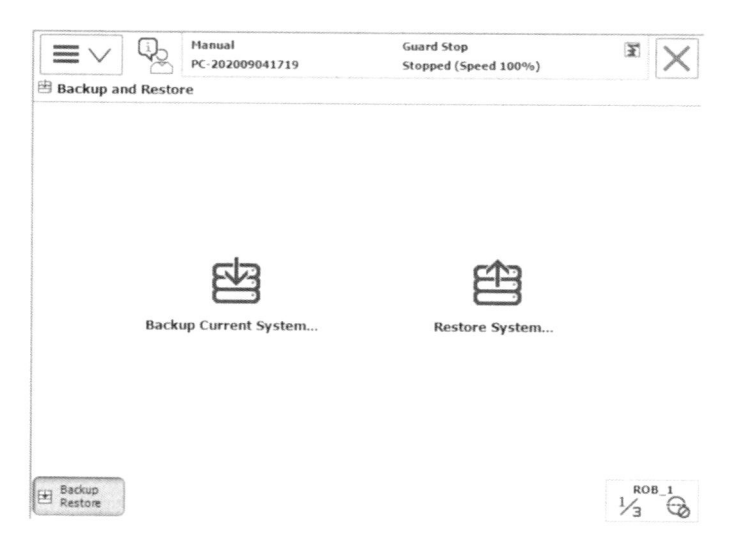

Fig. 4–10　Backup Current System Option

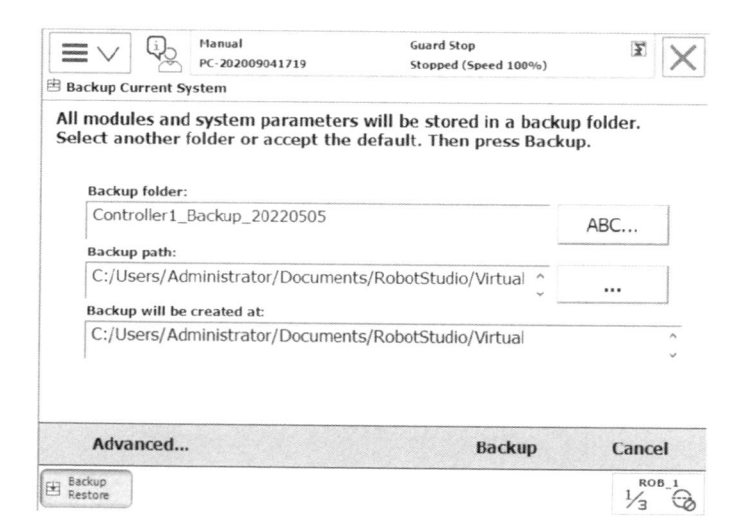

Fig. 4–11　Setting of Backup Path

4.2.4.2 Recovery of ABB robot data

(1)Tap the "Recover system..." button in the interface shown in Fig. 4–10 to enter the recovery path selection interface shown in Fig. 4–12; tap "..." button to select the backup storage directory, and finally tap "Recover".

44

4. Basic Operation of Industrial Robots

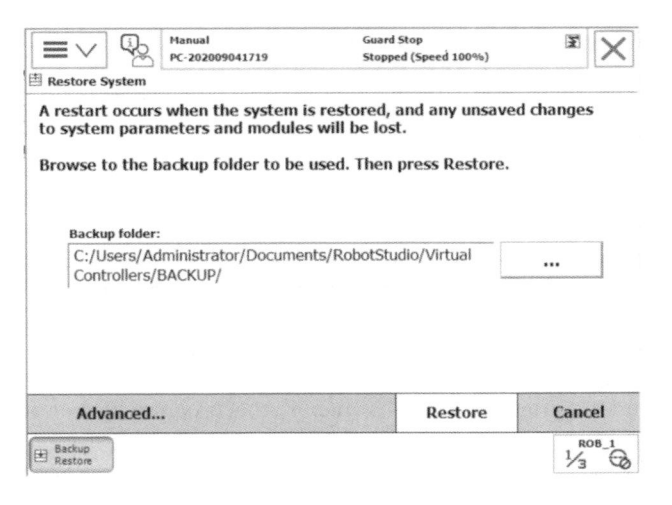

Fig. 4–12   Selection of Recovery Path

(2)The system pops up the recovery confirmation dialog box, as shown in Fig. 4–13. Before tapping "Yes", reconfirm whether the recovery folder is correct. During recovery, it should be noted that the backup recovery data is unique and cannot be used for another robot. Otherwise, system failure may occur. However, the definitions of programs and I/Os are often common to facilitate series production. At this time, the actual needs can be solved by importing the program and EIO files separately.

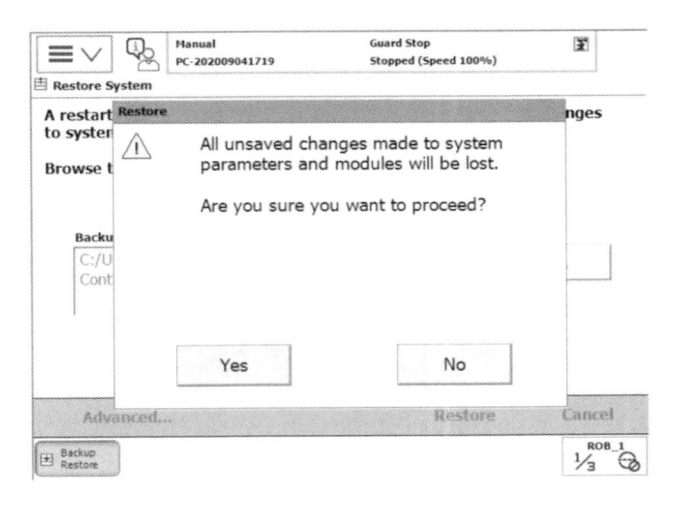

Fig. 4–13   Recovery Confirmation Dialog Box

45

Industrial Robot Technology Application

4.2.4.3 When do you need to make a backup for the robot?

(1)After power-on of a new machine for the first time;

(2)Before any modification;

(3)After completion of the modification;

(4)Regular backup(e.g. once a week)if the robot is important.

(5)It is better to make a backup on USB flash disk as well;

(6)Regular deletion of old backups to free up hard disk space. .

4.2.5 Import procedure

The import program can quickly import the programs generated by off-line programming or the programs of other robots into the current robot, saving manual programming time and improving the utilization rate of industrial robots.

(1)Tap the ABB menu button in the upper left corner of the teach pendant screen and select the "Program Editor" option, as shown in Fig. 4-14.

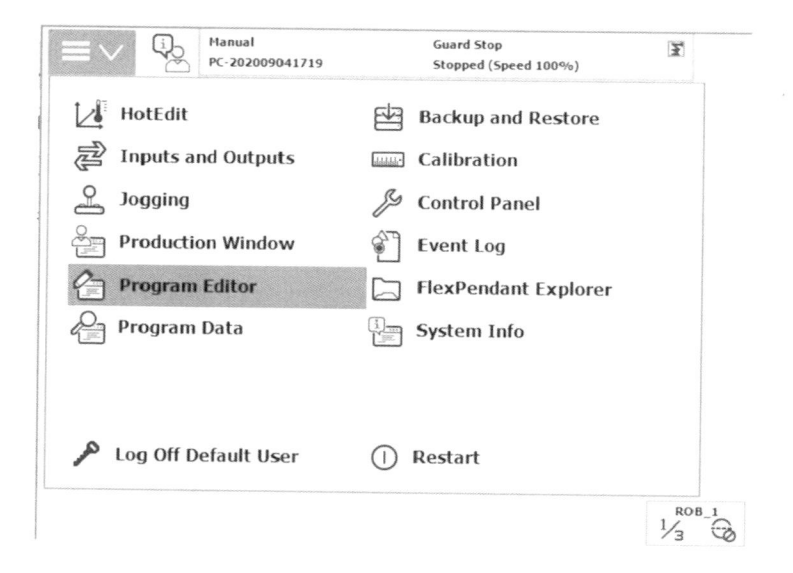

Fig. 4-14　Program Editor Option

4. Basic Operation of Industrial Robots

(2)Tap the "Module" tab, as shown in Fig. 4-15.

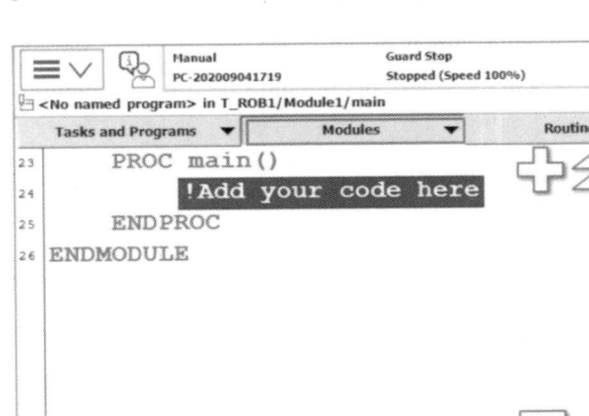

Fig. 4-15　Program Editor Module Option

(3)Open the "File" menu, tap "Loading module ..." and load the required program module from the "Backup directory/RAPID" path, as shown in Fig. 4-16.

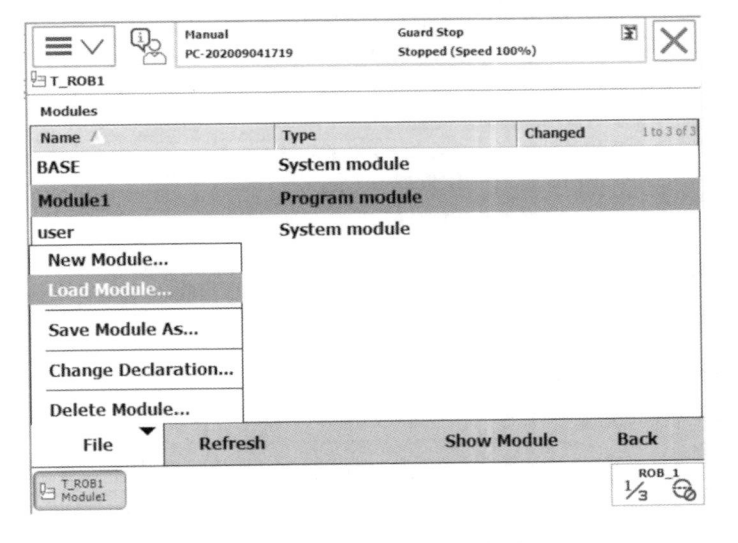

Fig. 4-16　Program Editor Option

47

Industrial Robot Technology Application

4.2.6 Importing EIO file

(1)Tap the ABB menu button in the upper left corner of the teach pendant screen, and select the "Control panel" option and then the "Configuration" option, as shown in Fig. 4–17.

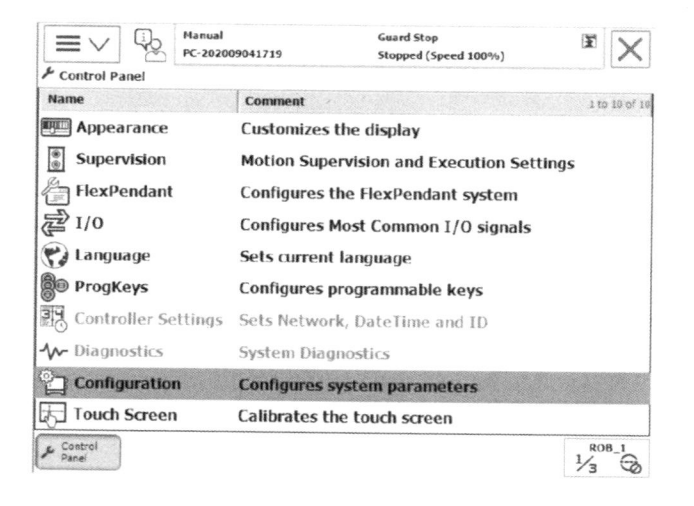

Fig. 4–17　Configuration System Parameter Option

(2)Open the "File" menu and tap "Load parameters", as shown in Fig. 4–18.

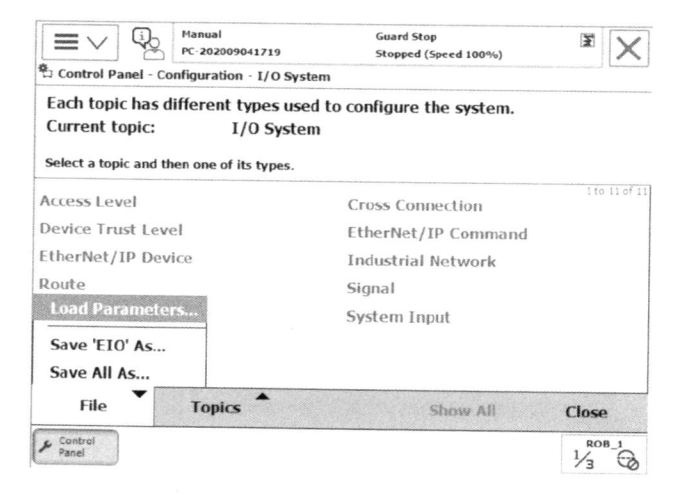

Fig. 4–18　Loading Parameter Option

4. Basic Operation of Industrial Robots

(3)Select "Load after deleting existing parameters" and tap "Loading...", as shown in Fig. 4−19.

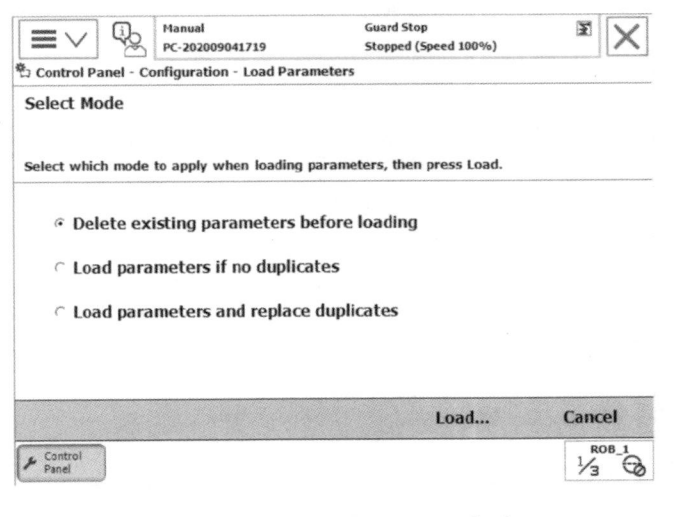

Fig. 4−19　Loading Parameter Option

(4)Find the EIO.cfg file in the "Backup directory/SYSPAR" path, and tap "OK", as shown in Fig. 4−20.

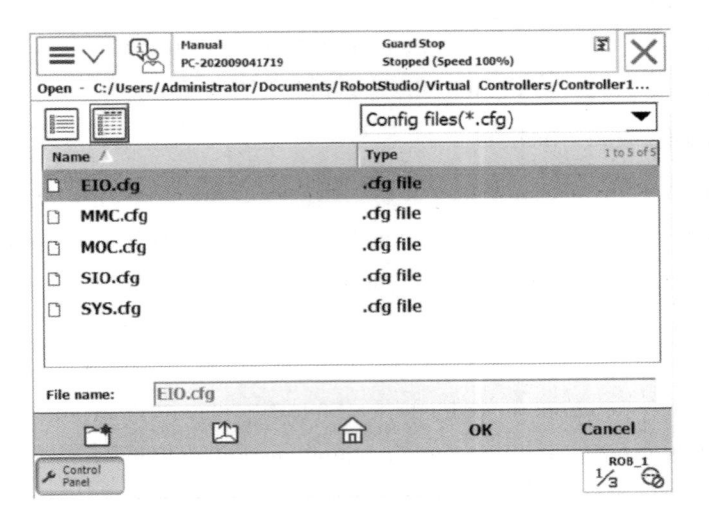

Fig. 4−20　Selection of EIO File

49

Industrial Robot Technology Application

(5)After the system pops up the restart screen as shown in Fig. 4–21, tap "Yes" to restart the robot. After the restart, the new imported EIO takes effect.

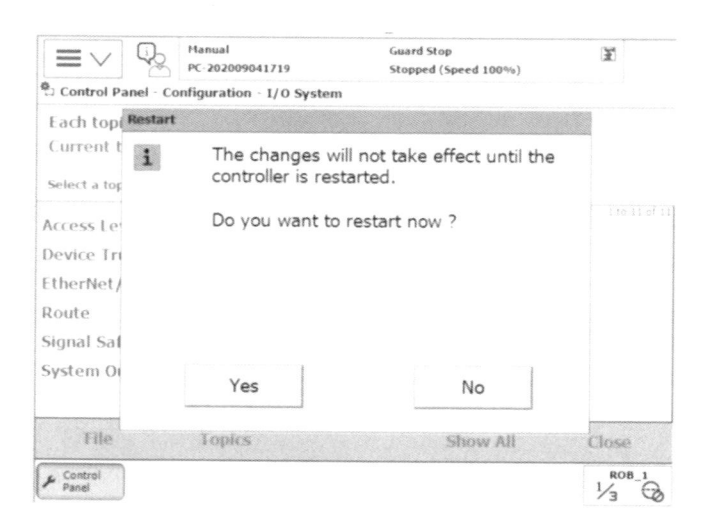

Fig. 4–21   EIO Import Completion Screen

4.3 Manual control of single–axis motion

There are three modes of manually operated robot motion, including single–axis motion, linear motion and reorientation. The following describes how to manually control the robot to perform three motions.

4.3.1 Single–axis motion

In general, the six joint axes of the ABB robot are driven by six servo motors respectively (Fig. 3–9), so each manual control of one joint axis motion is called single–axis motion. The following is the method for manual control of single–axis motion.

The time of the robot system should be set as the time of the local time zone before various operations in order to facilitate file management and fault inspection and management, and the specific operations are as follows:

(1)Select "Manual control", as shown in Fig. 4–22.

(2)Tap "Action mode", as shown in Fig. 4–23.

50

4. Basic Operation of Industrial Robots

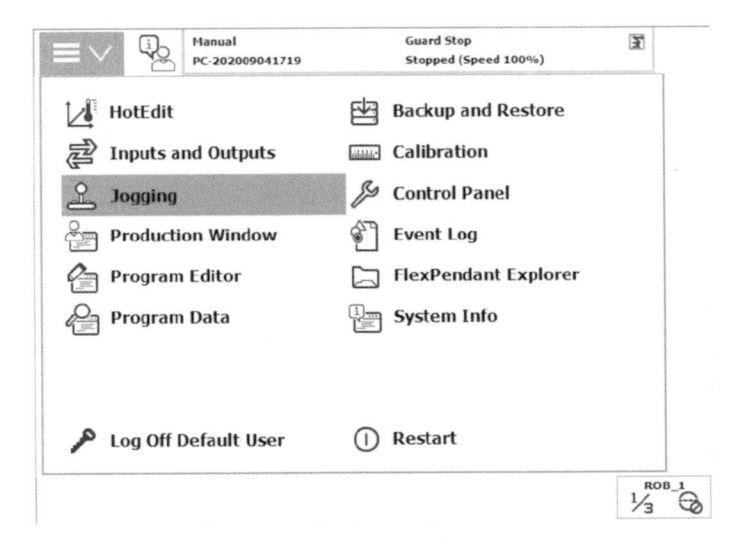

Fig. 4–22 Manual Control Option

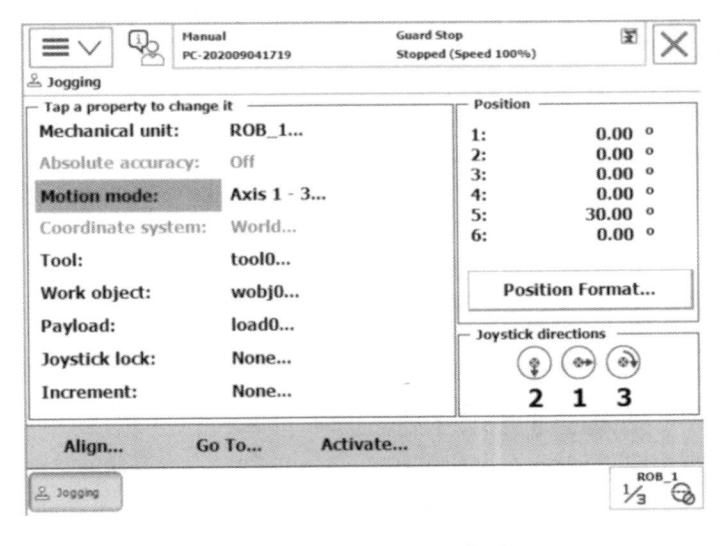

Fig. 4–23 Action Mode Option

(3)Enter the action mode selection interface, and tap "Action mode", as shown in Fig. 4–24. Select "Axis 1–3" option to control axes 1~3 and "Axis 4–6" option to control axes 4–6.

51

Industrial Robot Technology Application

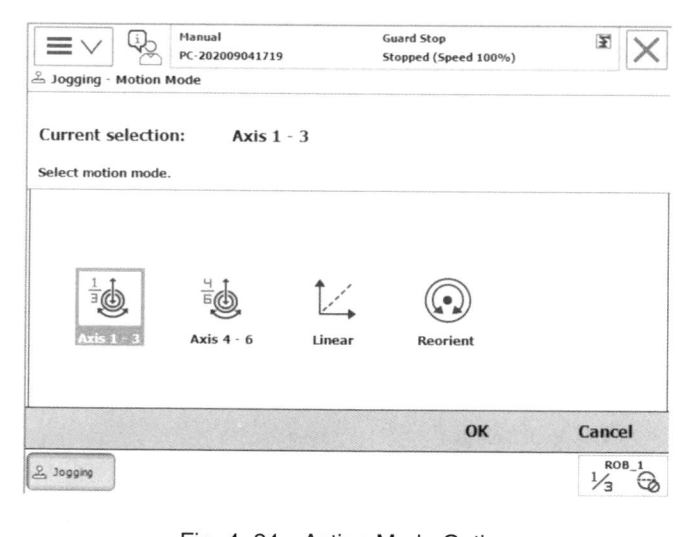

Fig. 4-24　Action Mode Option

(4)Return to the manual control interface, and press the enabler key to enter the "motor-on" state, as shown in Fig. 4-25. The robot axes 1-3 can be controlled for rotation by using the joystick in the current state. The up and down, left and right, and rotation of the joystick correspond to axis 2, axis 1 and axis 3 respectively, and the arrows in the lower right corner represent the positive motion direction of each axis.

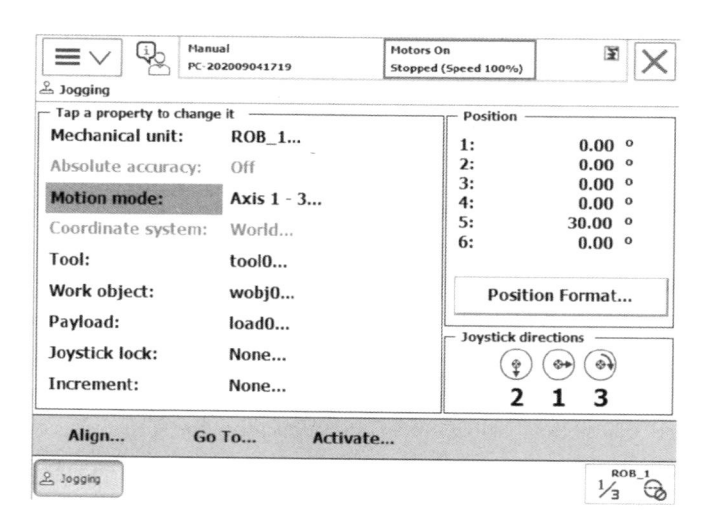

Fig. 4-25　Schematic Diagram of Single-axis Motion Direction

52

4. Basic Operation of Industrial Robots

(5)Operation tips for joystick:

The joystick of the robot can be regarded as a vehicle's throttle, and its control range is related to the motion speed of the robot. If the control range is small, the robot will move slowly. If the control range is large, the robot will move fast. Therefore, try to move the robot slowly with a small control range to start manual control learning during operation.

4.3.2 Linear motion

The linear motion of the robot means that the TCP of the tool installed on the sixth axis flange of the robot moves linearly in space. The following are the manual control steps:

(1)Tap the ABB menu button in the upper left corner of the teach pendant screen, tap "Manual control"−"Action mode" in turn, and select "Linear" option, as shown in Fig. 4−26:

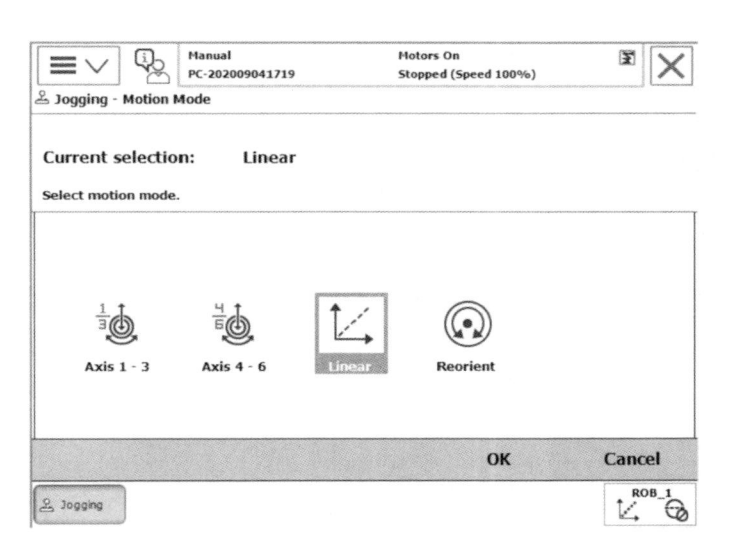

Fig. 4−26　Linear Motion Mode Option

(2)Tap the "Tool coordinate" option, as shown in Fig. 4−27.

(3)Select the corresponding "tool 1" and tap the "Tool coordinate" option, as shown in Fig. 4−28. The default tool coordinate of the system is the robot flange coordinate, and the tool coordinate established by the user is based on the flange coordinate.

53

Industrial Robot Technology Application

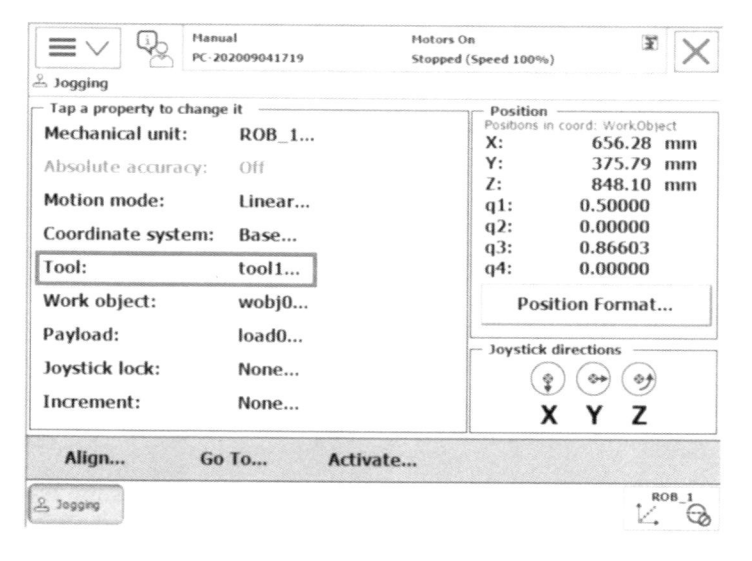

Fig. 4-27   Tool Coordinate Option

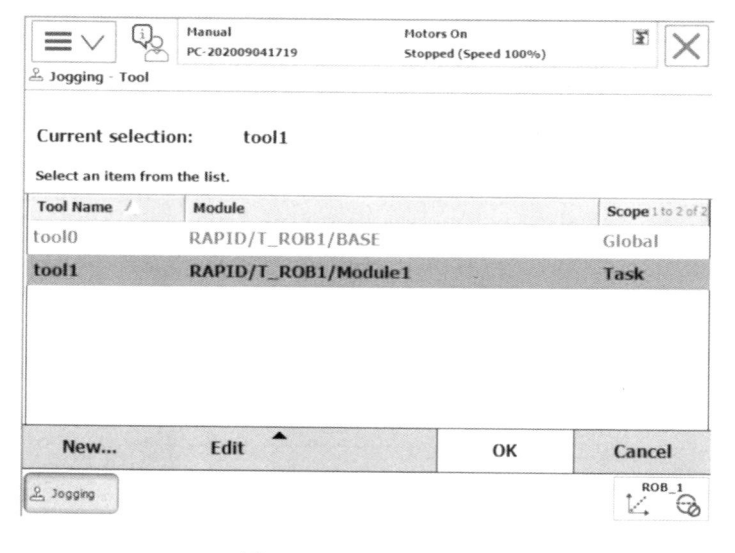

Fig. 4-28   Tool Option

(4)Return to the manual control interface, and press the enabler key to enter the "motor-on" state, as shown in Fig. 4-29. The up and down, left and right, and rotation of the joystick correspond to X, Y and Z respectively, and the arrows in the lower right corner represent the positive motion direction of each axis.

54

4. Basic Operation of Industrial Robots

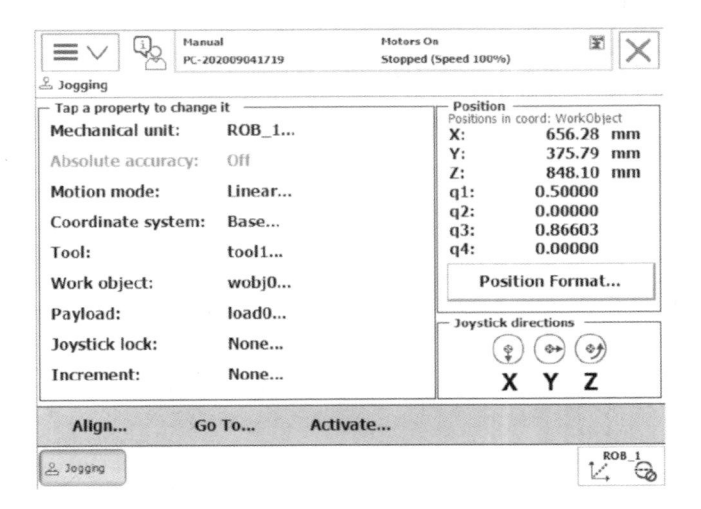

Fig. 4-29　Schematic Diagram of Linear Motion Direction

(5)Operate the joystick on the teach pendant in the current state, so that the TCP point of the robot tool moves linearly in the robot reference coordinate system, as shown in Fig. 4-30.

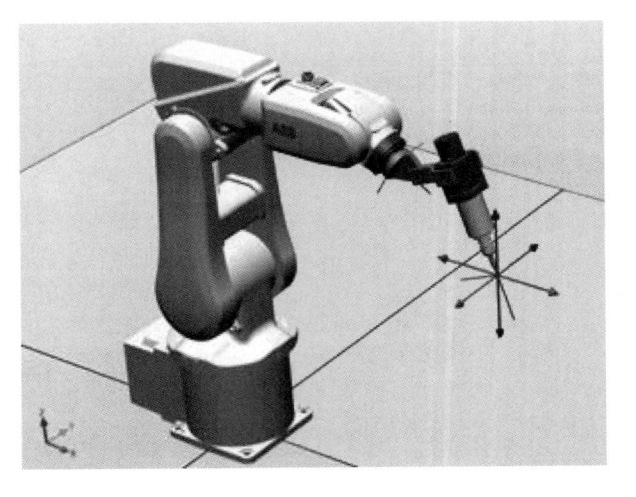

Fig. 4-30　Schematic Diagram of Linear Motion of TCP Point

(6)Use of incremental mode: If not familiar with using the joystick to control the motion speed of robot through displacement amplitude, you can use the "incremental" mode to control the robot motion and select the "Incremental option" in the "Manual operation" interface, as shown in Fig. 4-31.

55

Industrial Robot Technology Application

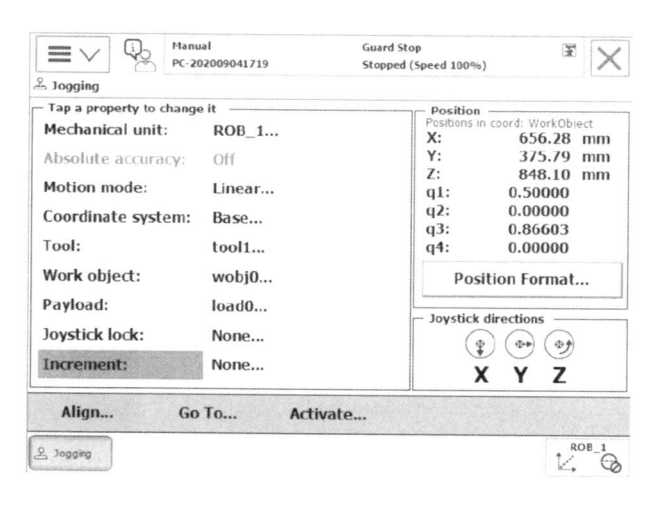

Fig. 4-31　Incremental Mode Option

The robot moves one step every time the joystick moves in the incremental mode. If the joystick lasts for one second or several seconds, the robot will continue to move at a rate of 10 steps/s. The movement amount of each step is selected in the incremental mode, as shown in Fig. 4-32.

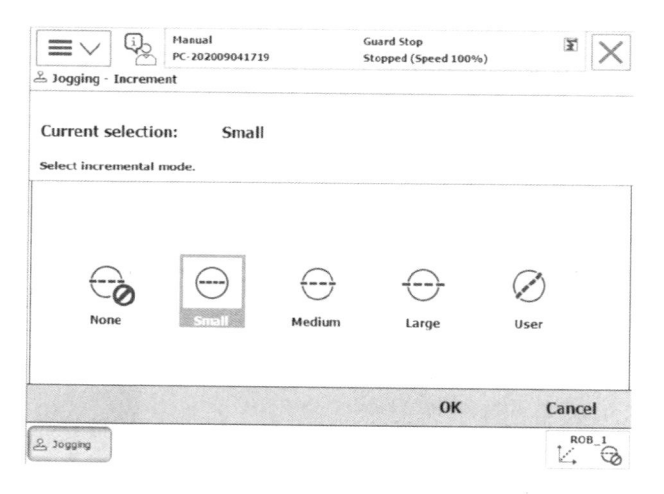

Fig. 4-32　Incremental Size Option

4.3.3 Reorientation

The reorientation of the robot refers to that the tool TCP point on the sixth axis flange of the robot rotates around the coordinate axis in space, and can also

4. Basic Operation of Industrial Robots

be understood that the robot adjusts its pose around the tool TCP point. The following is the method for manual control of reorientation:

(1)Tap the ABB menu button in the upper left corner of the teach pendant screen, tap "Manual control"–"Action mode" in turn, and select "Relocation" option, as shown in Fig. 4–33:

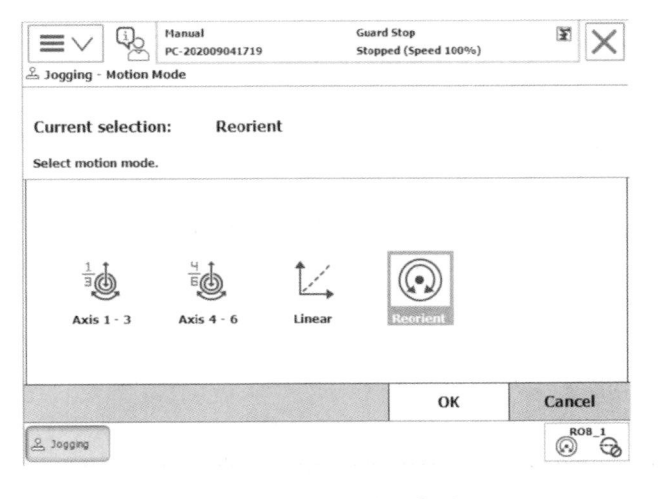

Fig. 4–33　Relocation Option

(2)Return to the manual control interface and select the "Coordinate system" option, as shown in Fig. 4–34.

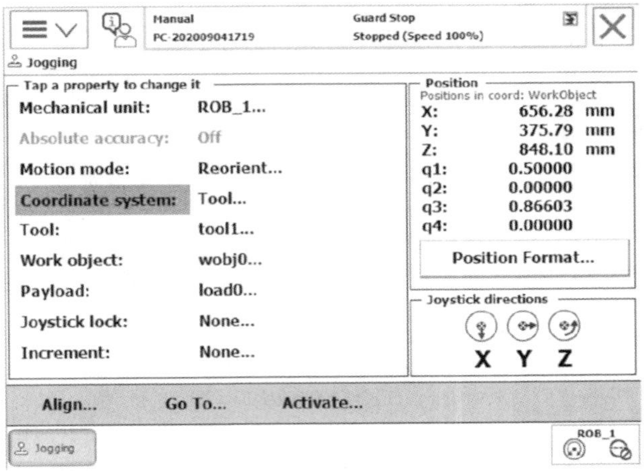

Fig. 4–34　Coordinate System Option

Industrial Robot Technology Application

(3)Select the "Tool" option and confirm it, as shown in Fig. 4–35.(The "Base Coordinate" is the coordinate of robot body, and the geodetic coordinate is a fixed reference coordinate in the workshop)

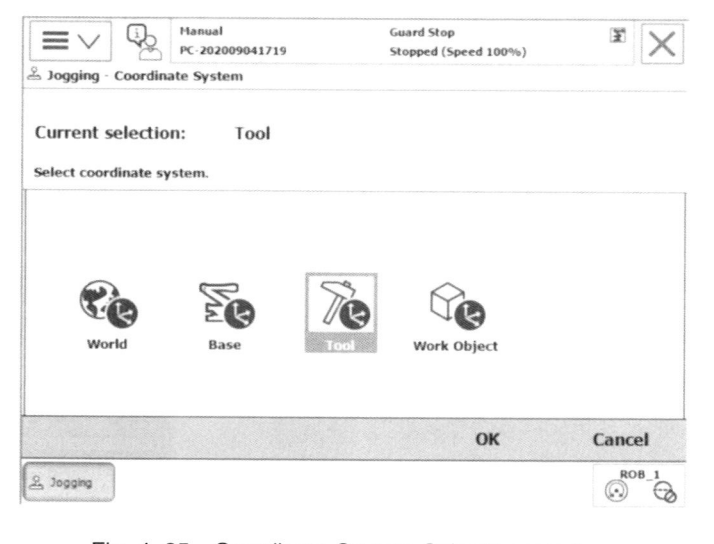

Fig. 4–35　Coordinate System Selection Interface

(4)After confirming that the current tool coordinate is that of the target TCP point, press the enabler key to enter the "motor–on" state, as shown in Fig. 4–36.

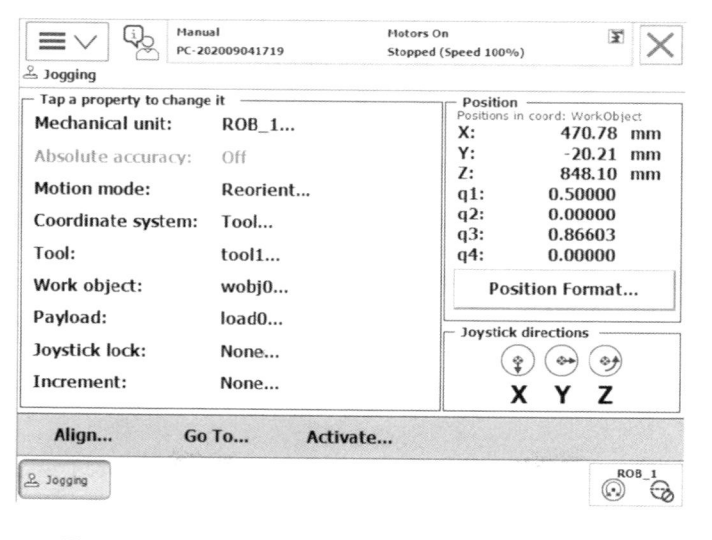

Fig. 4–36　Schematic Diagram of Relocation Direction

4. Basic Operation of Industrial Robots

(5)Operate the joystick on the teach pendant in the current state, so that the robot adjusts its pose around the tool TCP point, as shown in Fig. 4–37.

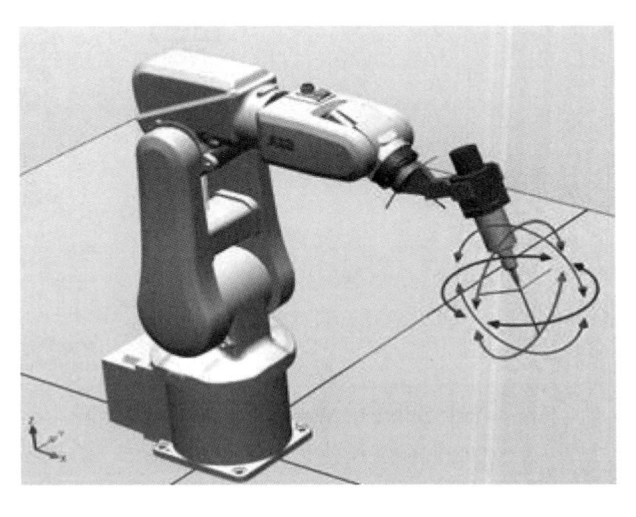

Fig. 4–37　Schematic Diagram of Linear Motion of TCP Point

4.3.4 Manual control shortcut

There are 4 shortcut keys and a shortcut menu on the ABB teach pendant to facilitate quick setting for manual control.

(1)There are 4 shortcut keys on the teach pendant, as shown in Fig. 4–38. These keys represent the following functions respectively:

1)A. Switching of robot/outer shaft

2)B. Switching of linear motion/reorientation

3)C. Switching of joint motion axes 1–3/axes 4–6

4)D. Incremental On/Off

(2)The shortcut menu button is in the lower right corner of the teach pendant screen. After tapping the button, the shortcut menu will pop up. Tap the first item "Manual control" to expand the menu, as shown in Fig. 4–39.

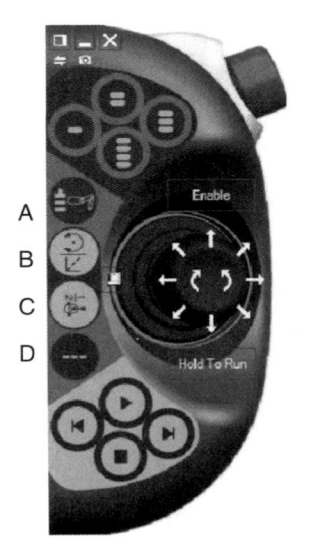

Fig. 4–38　Schematic Diagram of Linear Motion of TCP Point

59

Industrial Robot Technology Application

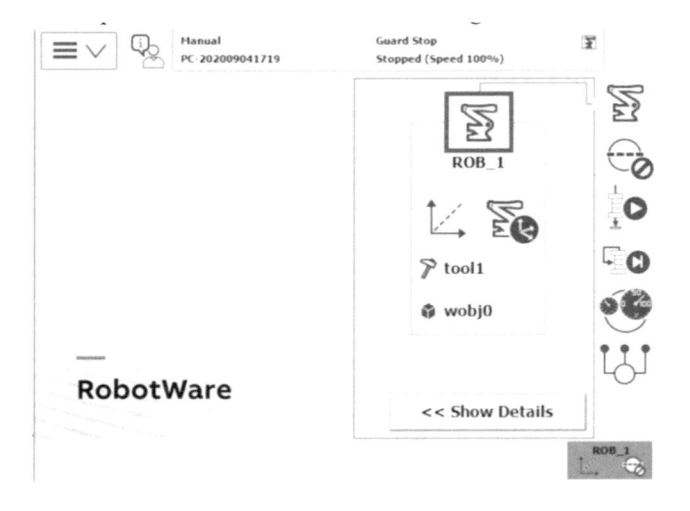

Fig. 4-39　Shortcut Menu of Teach Pendant

Fig. 4-40　Manual Control Shortcut Menu Option

(3)Tap "Display details" to display all menu options, as shown in Fig. 4-40. The functions are as follows:

1)A. Selection of currently used tool data

2)B. Selection of the currently used work object coordinates

3)C. Joystick rate

4)D. Incremental On/Off

4. Basic Operation of Industrial Robots

5)E. Selection of coordinate system

6)F. Selection of action mode

(4)Tap the second item "Incremental mode" button of the shortcut menu to display the incremental option, as shown in Fig. 4-41. Method to customize incremental value: select the "User module" and tap the "Displayed value" to customize incremental value.

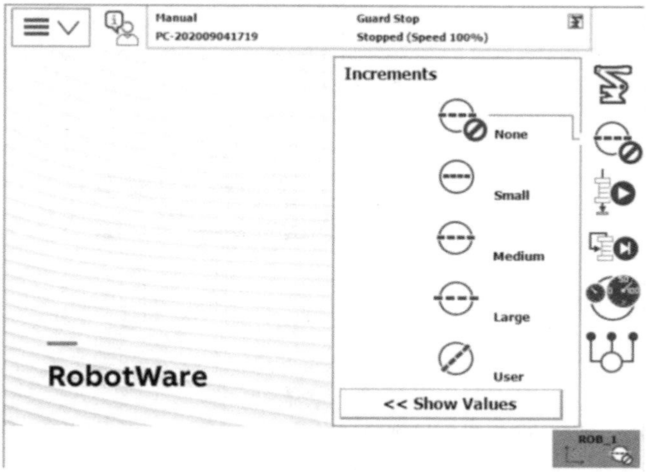

Fig. 4-41   Incremental Mode Shortcut Option

4.4 Zero calibration of robots

The robot zero information refers to the reading of the motor encoder of each axis when the robot axis is at mechanical zero. This information is stored on the serial measuring board of the body. Data can be saved only after power-on, and will be lost after power-off. In case of the following circumstances, it is necessary to update the revolution counter at the position of mechanical origin:

(1)After the battery of the servomotor revolution counter is updated;

(2)When the revolution counter fails and is repaired;

(3)After the revolution counter is disconnected from the measuring board;

(4)After power-off, the robot joint axis moves;

(5)When the system alarm prompts "10036 revolution counter is not updated".

Industrial Robot Technology Application

4.4.1 Robot zero position

Robots of different brands and models have different zero mark positions. The pose of ABB IRB1200 industrial robot is shown in Fig. 4–42 after correct zero return. The enlarged views 4, 5 and 6 show the zero mark positions of axes 4, 5 and 6 respectively.

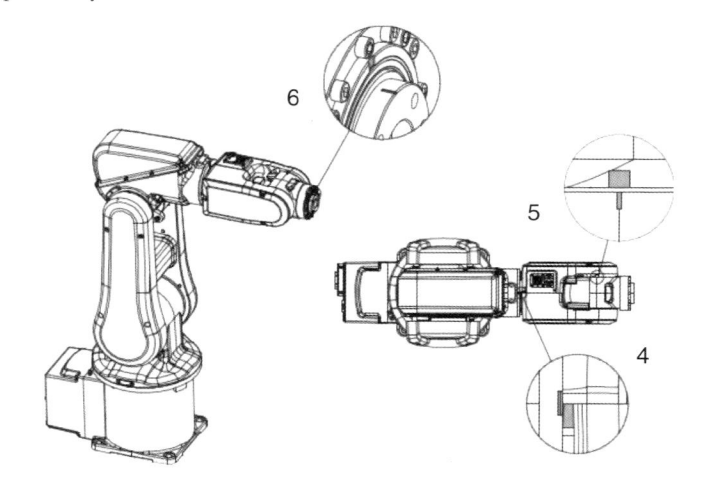

Fig. 4–42   Schematic Diagram of Zero Position of IRB1200 Robot

Fig. 4–43 shows the zero mark positions for axes 1–3.

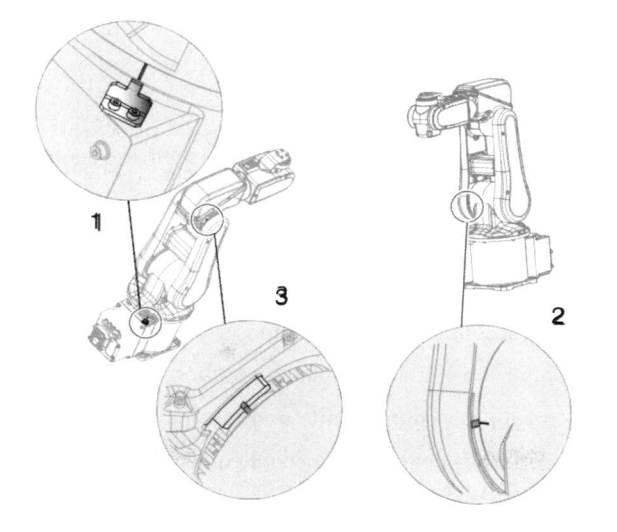

Fig. 4–43   Schematic Diagram of Zero Mark Positions

62

4. Basic Operation of Industrial Robots

4.4.2 Update of revolution counter

It is necessary to adopt manual control to update the revolution counter, so that each joint axis of the robot moves to the position of the synchronization marks. The general sequence is 4–5–6–1–2–3. Robots of different models have different positions of synchronization marks. Please refer to the Operation Instruction of the robot for comparison with the robot body. After the zero mark position of each axis of the robot is adjusted, the FlexPendant can be operated to update the revolution counter. The detailed steps are as follows:

(1)Tap the ABB menu button in the upper left corner of the teach pendant screen, and tap "Calibration" option to enter the selection interface of calibration mechanical unit, as shown in Fig. 4–44.

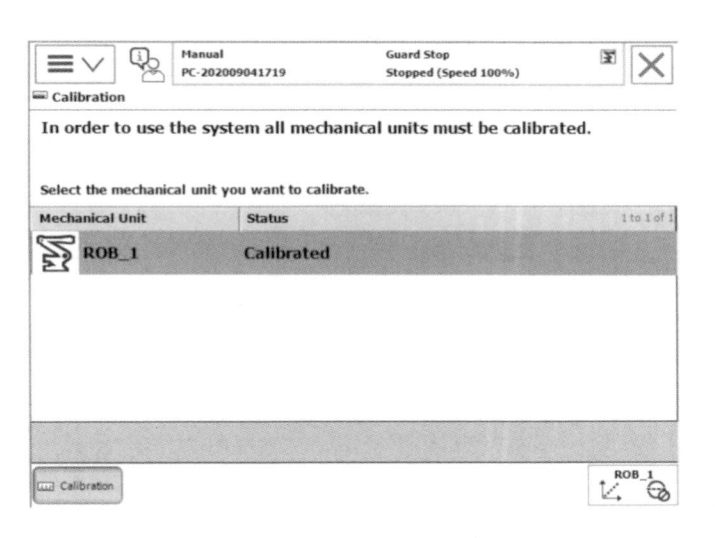

Fig. 4–44　Selection Interface of Calibration Unit

(2)Tap "ROB_1" to enter the calibration design interface and select the "Calibration parameters" option, as shown in Fig. 4–45.

(3)Tap the "Edit electrode calibration offset…" option to enter the setting interface of motor offset value, as shown in Fig. 4–46. There is a nameplate recording the motor calibration offset value on the side of the robot body. Carefully check the value displayed in the teach pendant, modify it to be consistent with the label value on the body, and tap the "OK" button.

63

Industrial Robot Technology Application

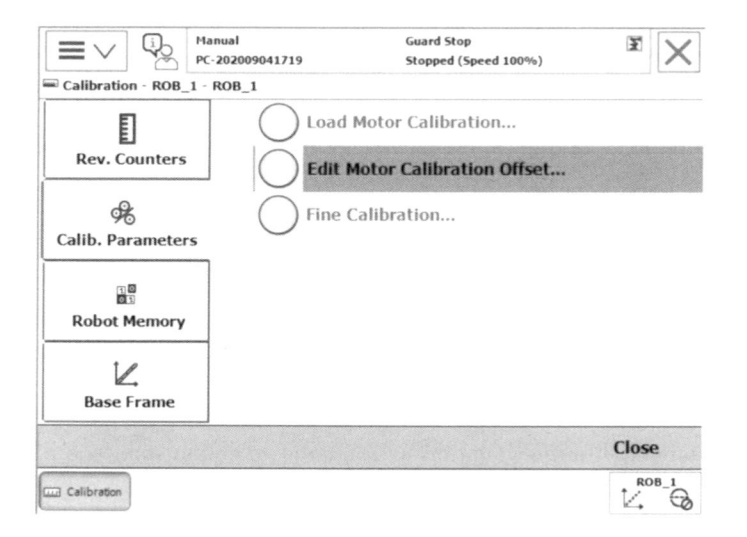

Fig. 4-45　Selection Interface of Calibration Unit

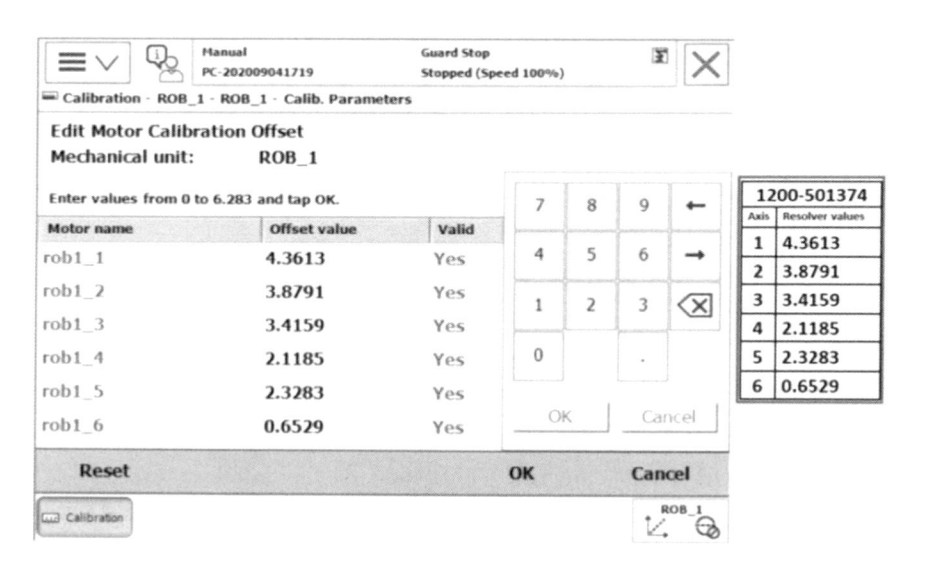

Fig. 4-46　Setting Interface of Motor Offset Value

4. Basic Operation of Industrial Robots

(4)When the system prompts to restart the robot controller, tap "Yes" to restart, as shown in Fig. 4–47.

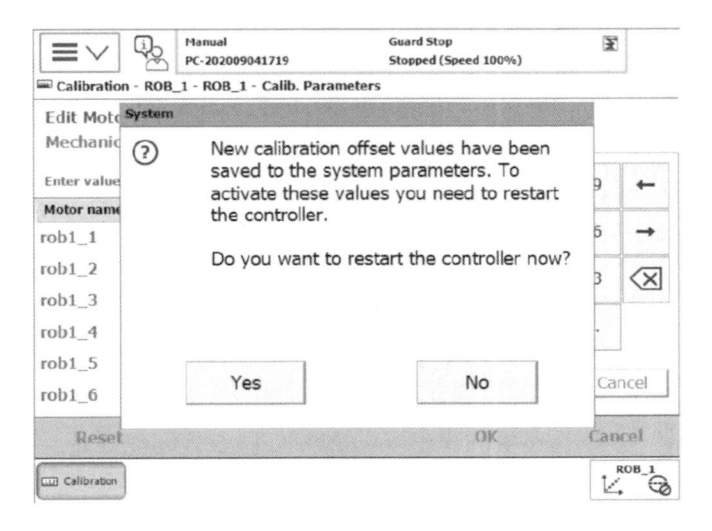

Fig. 4–47　Restart Confirmation Dialog Box

(5)After restarting, re–enter the "Revolution counter" option on the "Calibration" setting interface, and tap the "Update revolution counter..." option, as shown in Fig. 4–48.

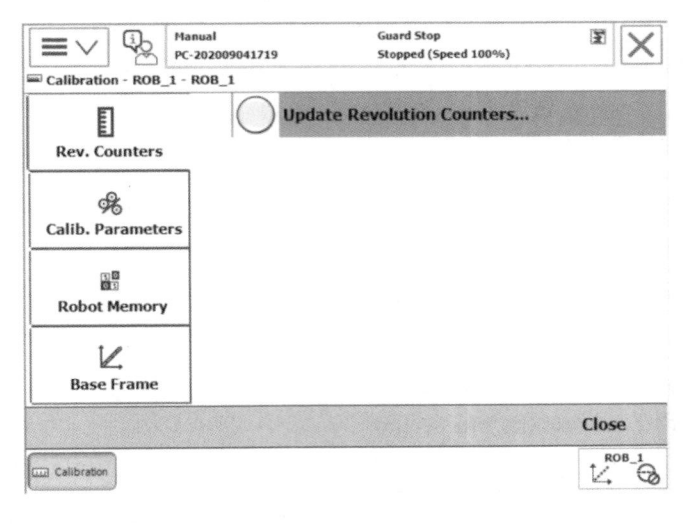

Fig. 4–48　Update Revolution Counter Option

65

Industrial Robot Technology Application

(6)Tap "Yes" button in the pop–up confirmation dialog box to enter the "Mechanical unit" confirmation interface, and tap "OK" as prompted to enter the "Axis" selection interface as shown in Fig. 4–49. Select all six axes and tap the "Update" button. After the update, each axis will display "Revolution counter updated", indicating that the revolution counter has been updated.(Note: If six axes of the robot cannot reach the position of synchronization marks simultaneously due to the installation position, the corresponding joint axis can be selected one by one to update the revolution counter.)

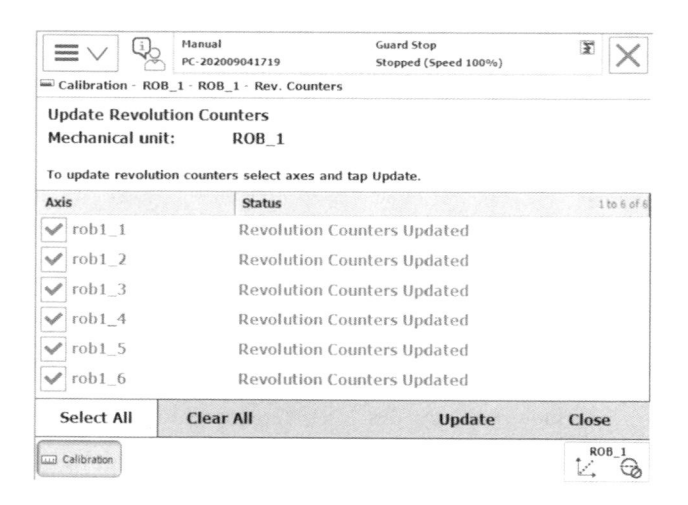

Fig. 4–49   Update Interface of Revolution Counter

4.5 Robot Restart

The ABB robot system can work for a long time without periodic restart. However, the system needs to be restarted when:

(1)A new hardware is installed;

(2)The robot system configuration parameters are changed;

(3)A system failure(SYSFAIL)occurs;

(4)RAPID program failure occurs.

Tap the ABB menu button in the upper left corner of the teach pendant screen, and then the "Restart" option to restart the option interface, as shown in Fig. 4–50.

66

4. Basic Operation of Industrial Robots

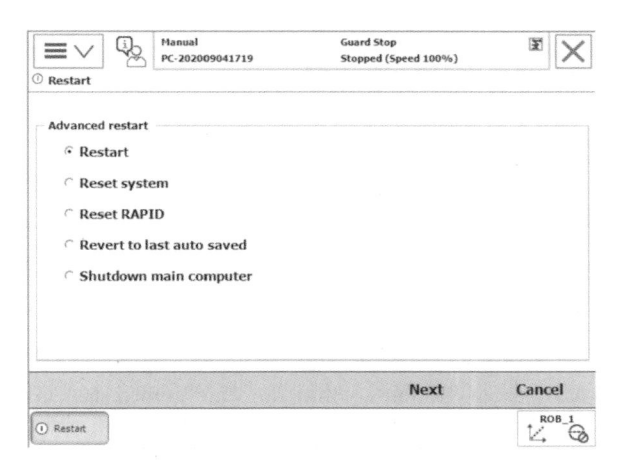

Fig. 4-50   Update Interface of Revolution Counter

Types of restart include restarting, system resetting, RAPID resetting, restoring to the last automatically saved state, and shutting down the main computer. Details of each option are as follows:

• Restarting: restart the current system with the current settings.

• System resetting: restart the system, discard current system parameter settings and RAPID programs, and use original system installation settings.

• RAPID resetting: restart the system and discard current RAPID programs and data, but retain system parameter settings.

• Restoring to the last automatically saved state: restart the system and attempt to restore to the last automatically saved state. This option is generally used when recovering from a system crash.

• Shutting down the main computer: switch off the robot control system, which shall be used in case of controller UPS failure.

4.6 Emergency stop and recovery of robot

During the manual control of the robot, when the operator is prone to collision or other emergencies due to unskilled operation, the emergency stop button (one on the control cabinet and one on the teach pendant)can be pressed to start the safety protection mechanism of the robot and stop it. The emergency stop precedes over any other control operations of ABB robot. This operation will discon-

67

nect the drive power of ABB robot motor, stop all moving parts, and cut off power supply of potentially hazardous functional parts controlled by the ABB robot system.

In the emergency stop state, all power supplies in the ABB robot (except for the manual brake release circuit)are disconnected, and the robot cannot be operated by using the teach pendant. Before operating the robot, it is necessary to perform recovery steps, that is, resetting the emergency stop button and pressing the "motor–on" button to return to normal operation. Corresponding configuration can be made in the ABB robot system so that the emergency stop is in the following state:

(1)Uncontrolled stop: disconnect the power supply of ABB robot motor and stop ABB robot immediately.

(2)Controlled stop: stop ABB robot, but do not disconnect the power supply of ABB robot motor to keep the ABB robot path. After the operation, the power supply is disconnected.

The system default setting is an uncontrolled stop. However, a controlled stop can minimize additional, unnecessary wear and tear on ABB robot and operations necessary to return ABB robot to production status. An emergency stop should not be used as a normal program stop as this would cause additional, unnecessary wear and tear on ABB robot.

5.

RobotStudio Application

5.1 Introduction to RobotStudio

RobotStudio is a computer simulation software developed and produced by ABB Group. Its main function is to help users simulate and build a robot system on the computer, and debug its programs and functions, so as to find the most suitable design and work plan and ensure that users can complete the realistic robot system deployment smoothly. This software is applicable to each cycle stage of the robot system. It is a must-have software for designing and deploying robots, with characteristic functions as follows:

5.1.1 CAD import

Data in various mainstream CAD formats (including IGES, STEP, VRML, VDAFS, ACIS and CATIA)can be easily imported by using RobotStudio. Based on these precise data, robot programmers can create more precise robot programs, thereby improving product quality.

5.1.2 AutoPath

It is one of the most time-saving functions in RobotStudio. By using the CAD model of the part to be machined, the robot position (path)required to track the machining curve can be automatically generated within only several minutes. However, this task used to take several hours or even days.

5.1.3 Program editor

ProgramMaker can generate robot programs, enabling users to develop or maintain these programs offline in Windows environment, which can shorten pro-

gramming time and improve program structure significantly.

5.1.4 Path optimization

If the program contains robot motions close to the singularity, RobotStudio can automatically detect it and make an alarm, thus preventing this phenomenon from happening in actual operation of the robot. The simulation monitor is a visual tool for robot motion optimization, with red lines showing what can be improved to make the robot operate in the most efficient manner. It can optimize the TCP speed, acceleration, singularity or axis and shorten the cycle time.

5.1.5 Autoreach

Autoreach can automatically analyze accessibility and is convenient for use. Through this function, users can move the robot or work object at will until all positions are reachable, and verification and optimization of work unit layout can be completed within several minutes.

5.1.6 Virtual teach pendant

It is the graphic display of the actual teach pendant, with VirtualRobot as its core technology. Essentially, all the work that can be done on the actual teach pendant can be completed on the virtual teach pendant(QuickTeach?). Therefore, it is an excellent teaching and training tool.

5.1.7 Event sheet

It is an ideal tool for verifying the structure and logic of programs. This tool allows direct observation of the I/O status of the work unit in the process of program execution. I/O can be connected to simulation events to realize simulation of robots and all equipment in the station. It is an ideal debugging tool.

5.1.8 Collision detection

Using collision detection function can avoid serious loss caused by equipment collision. After selection of objects to be detected, RobotStudio can conduct automatic monitoring and display whether they will collide during program execution.

5.1.9 Visual Basic for Applications

VBA can be used to improve and expand the functions of RobotStudio, and develop powerful add−ins, macros, or customize user interfaces as per specific needs of users.

5.1.10 PowerPac's

ABB collaborates with partners to develop a series of RobotStudio-based applications by using VBA, so that RobotStudio can better suit for arc welding, management of plate bending machine, spot welding, CalibWare(absolute accuracy), blade grinding, and BendWizard(management of plate bending machine)and other applications.

5.1.11 Upload and download

The entire robot program can be downloaded directly to the actual robot system without any conversion.

5.2 Installation of RobotStudio

(1)Access to RobotStudio:

1)Log in www.robotstudio.com, find the content as shown in Fig. 5-1, click to enter the personal information filling page, fill in the personal email address carefully and then submit it, and start downloading after receiving and opening the email link. The installation trial license allows a 30-day trial of RobotStudio.

Download RobotStudio® for free

Test the software including RobotStudio and RobotStudio PowerPacs for 30 days.

DOWNLOAD IT NOW

Fig. 5-1   Download Button of RobotStudio

2)Log in https://new.abb.com/products/robotics/robotstudio/downloads. The page contains the latest version information of RobotStudio, as shown in Fig. 5-2. The operation after clicking is the same as that in a).

RobotStudio

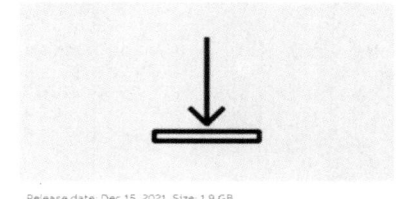

Release date: Dec 15, 2021, Size: 1.9 GB
RobotStudio 2021.4
RobotWare can be installed from RobotApps within RobotStudio.
→ Register to download it now

Fig. 5-2   Download Button of RobotStudio

Industrial Robot Technology Application

（2）Decompress the installation package and double-click setup.exe to start installation, as shown in Fig. 5-3.

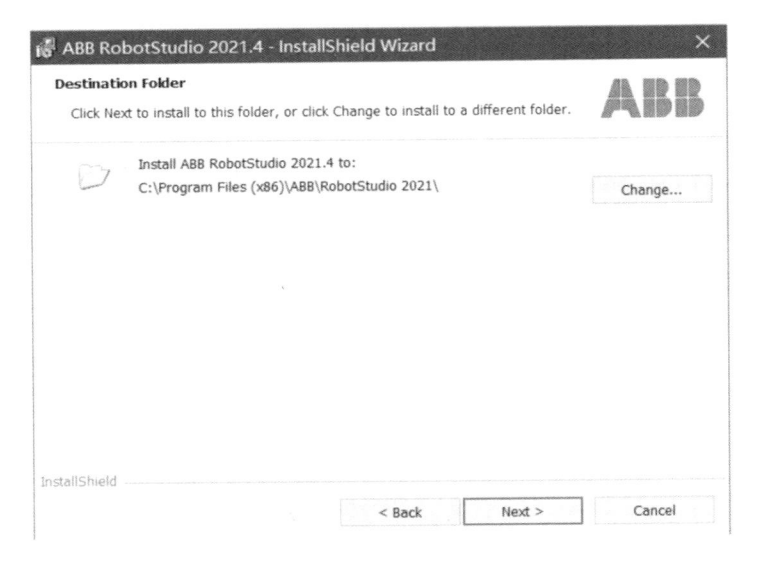

Fig. 5-3　Installation Files

（3）Tick to accept the license agreement as prompted and click "Next" to confirm the software installation path, as shown in Fig. 5-4.

Fig. 5-4　Setup of Installation Path

5. RobotStudio Application

(4)The selection of installation type is shown in Fig. 5–5. Complete installation: install all functions required for complete operation, and use all functions of basic version and advanced version; Custom installation: install user–defined functions, and choose not to install unnecessary robot library files and CAD converters.

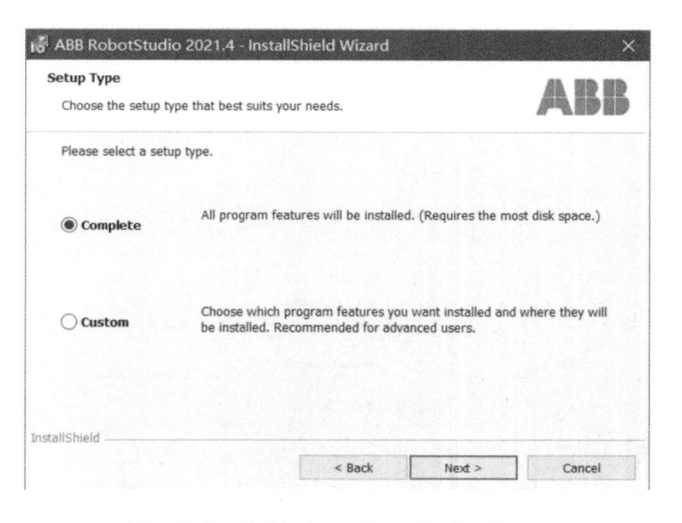

Fig. 5–5　Selection of Installation Type

(5)Click the "Next" to start installation, as shown in Fig. 5–6, and restart the computer as prompted.

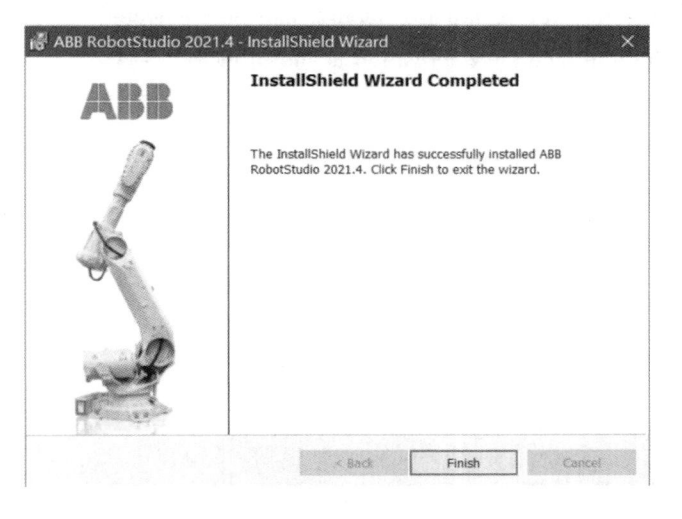

Fig. 5–6　Installation Completion Interface

73

Industrial Robot Technology Application

(6)When you open the software for the first time, an activation dialog box will pop up automatically, as shown in Fig. 5–7. Otherwise, the activation interface should be opened manually by clicking the menu "File–Help–Manage Authorization–Activation Wizard" in turn. For the option "I want to activate the standalone license key", it is necessary to enter the license of genuine software; For the option "I want to request a trial license", you can apply for a 30–day trial period (Note: Each trial license can only be installed once on one computer).

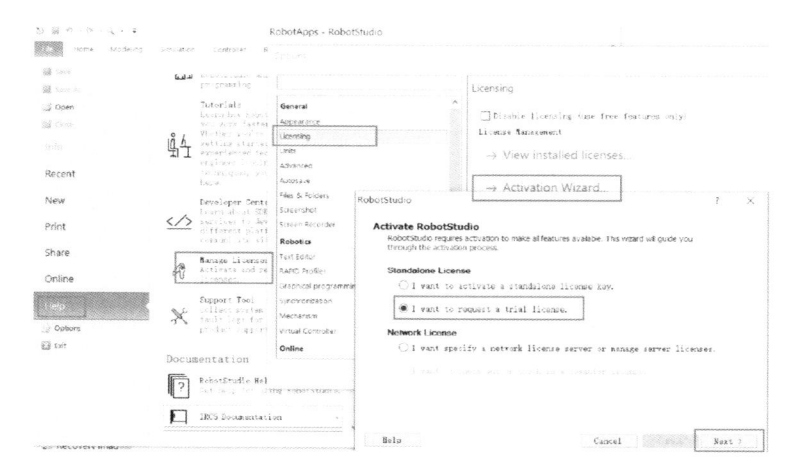

Fig. 5–7   Application for Trial License

(7)Ensure that the installed computer can be connected normally to the Internet, and the interface after activation is shown in Fig. 5–8.

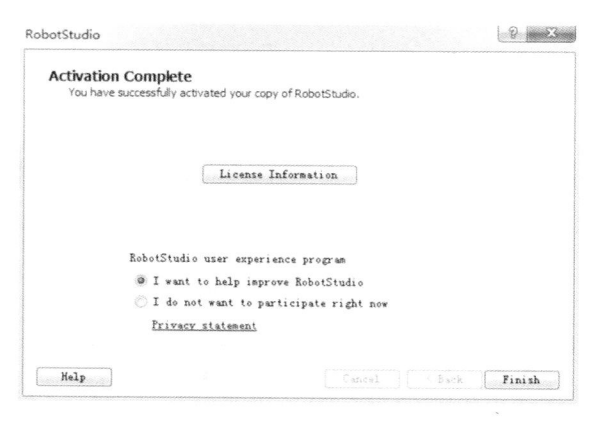

Fig. 5–8   Activation Completion Interface

74

(8)After installation of the trial version, the output column will prompt that the license will expire in 30 days, as shown in Fig. 5-9. After expiration, only the basic functions of the software can be used.

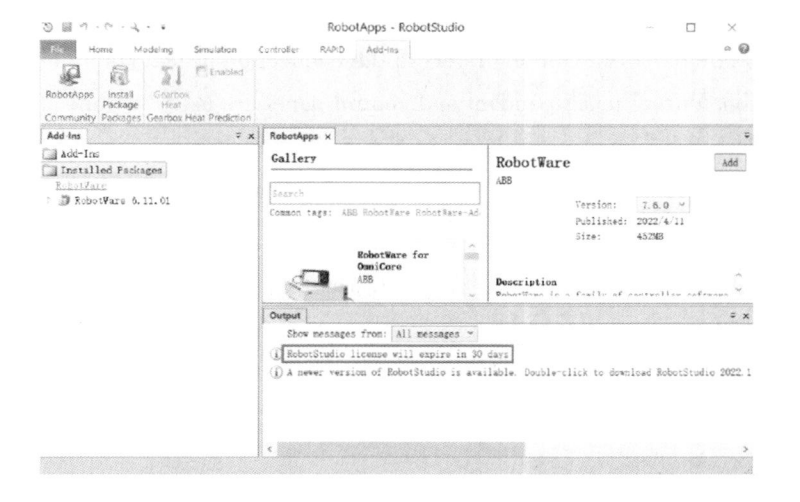

Fig. 5-9   Trial License Period

5.3 RobotStudio interface

The RobotStudio interface is opened for the first time, as shown in Fig. 5-10. The seven menus include: File, Basic, Modeling, Simulation, Controller, RAPID and Add-Ins.

Fig. 5-10   RobotStudio Interface

Industrial Robot Technology Application

Their main functions are as follows:

(1)After clicking the "File" menu, you can open the RobotStudio background view, and save (save as), open, create, view, close and exit the workstation, and share the workstation after packing, and view the message and manual documents through the help. Note: "Online" refers to the connection of real controllers through RobotStudio. Virtual teach pendant and control panel are used "Offline".

(2)The "Basic" menu is shown in Fig. 5–11, which can be used for workstation creation, system building, path programming, basic setting and basic control of the robot.

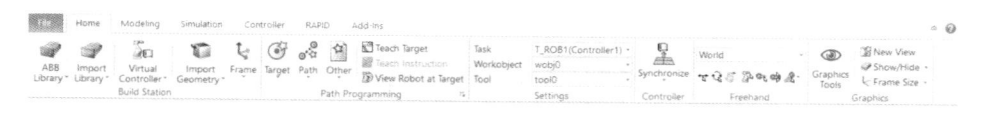

Fig. 5–11 "Basic" Menu

(3)The "Modeling" menu is shown in Fig. 5–12, which can be used for creation, component grouping, model import, creation of some simple components, measurement and control of 2D and 3D models.

Fig. 5–12 "Modeling" Menu

(4)The "Simulation" menu is shown in Fig. 5–13, mainly including functions such as creation, configuration, simulation control, monitoring, signal analysis and recording simulation.

Fig. 5–13 "Simulation" Menu

5. RobotStudio Application

(5)The "Controller" menu is shown in Fig. 5–14, and mainly used for the functions of online and offline controllers. It is used for synchronization, configuration, task assignment and control measure of controllers.

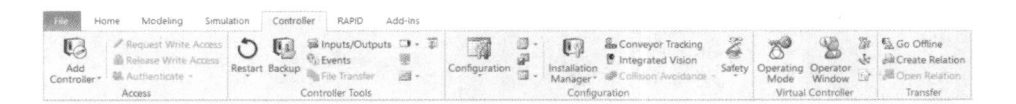

Fig. 5–14   "Controller" Menu

(6)The "RAPID" menu is shown in Fig. 5–15, mainly corresponding to some functions related to RAPID programs.(e.g. program creation, editing and management)

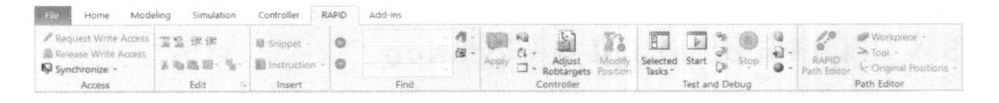

Fig. 5–15   "RAPID" Menu

(7)The "Add–ins" menu is shown in Fig. 5–16, and mainly used for relevant functions of secondary development. Other versions of RobotWare are added online in the case of networking. RobotWare is the robot system software. Virtual controller cannot be created without it. RobotWare can be added online, or offline by downloading the installation package "ABB.RobotWare–6.13.0164.rspak" and clicking the "Installation package" button.

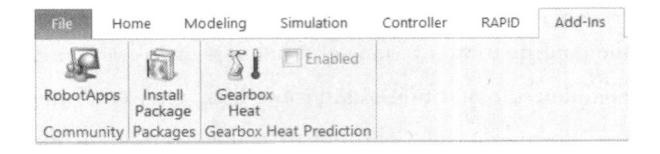

Fig. 5–16   "Add–ins" Menu

For a more detailed introduction to the RobotStudio interface, you can also refer to the HTML help document provided with the system. The document can be opened by clicking "File–Help–RobotStudio Help" in turn, as shown in Fig. 5–17.

77

Industrial Robot Technology Application

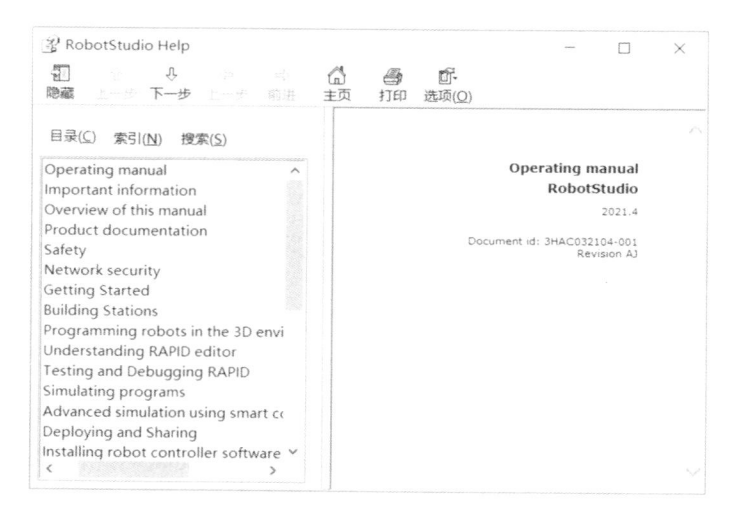

Fig. 5-17 HTML Help Document

5.4 Creation of virtual workstation

RobotStudio is a special software developed for ABB robots. It can not only be used to develop new robot programs, but also create a virtual robot in the computer to help users for offline programming and simulation test. This software covers all ABB robots. The simulated teach pendant has the same operation and function as the real one, and is very convenient for operation and programming learning of ABB robots. The steps for creating a robot virtual workstation are as follows:

(1)Click "File-New-Empty workstation-Create" in turn to create a new workstation, as shown in Fig. 5-18.

(2)Click "Basic-ABB Model Library-IRB 1200" in turn to select the robot model to be added, as shown in Fig. 5-19.

(3)Enter the confirmation interface of robot version, as shown in Fig. 5-20, and click "OK" to complete the addition of the robot body after confirming parameters.

(4)Click "Basic-Virtual Controller-From Layout..." in turn, as shown in Fig. 5-21.

(5)Enter the selection interface of RobotWare robot control system, as shown in Fig. 5-22. If the system fails to install RobotWare normally, the drop-down option is empty and the virtual controller cannot be added.

5. RobotStudio Application

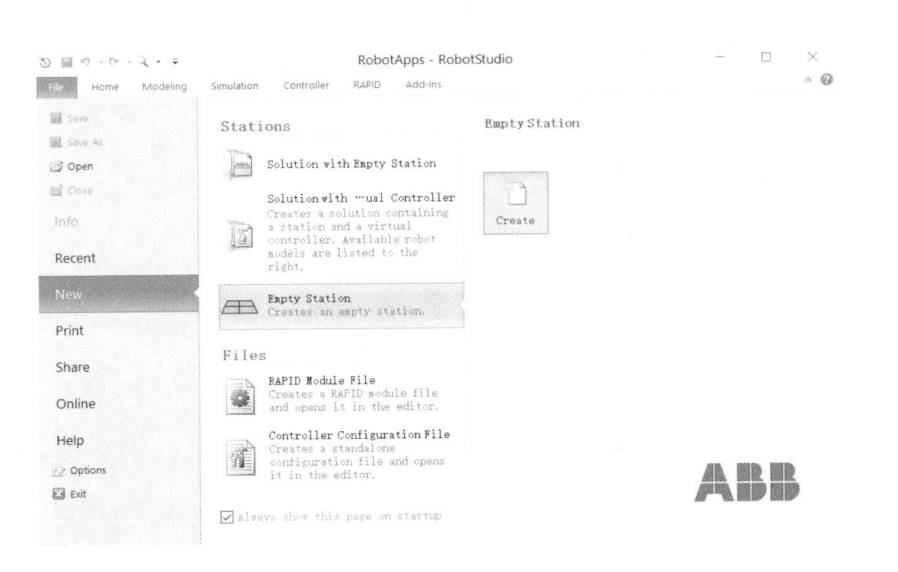

Fig. 5-18　New Workstation

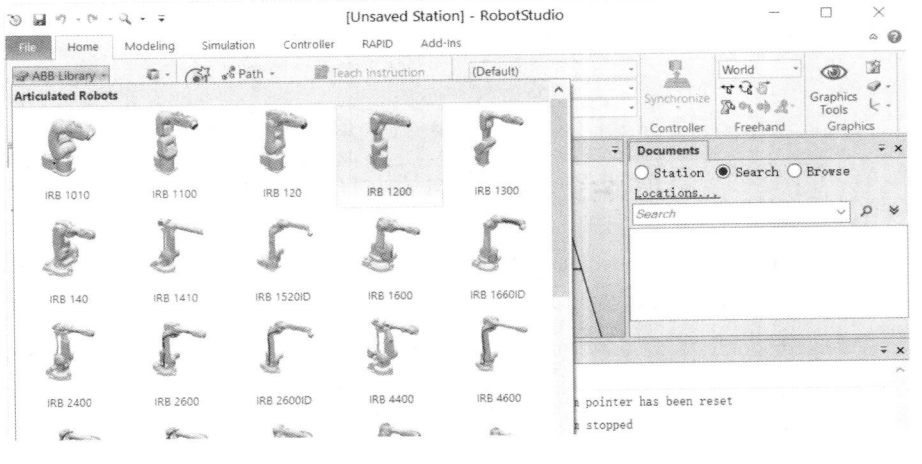

Fig. 5-19　Robot Selection

Industrial Robot Technology Application

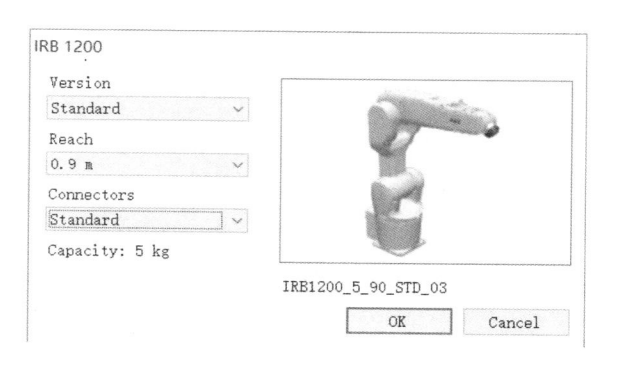

Fig. 5-20 Confirmation of Robot Version

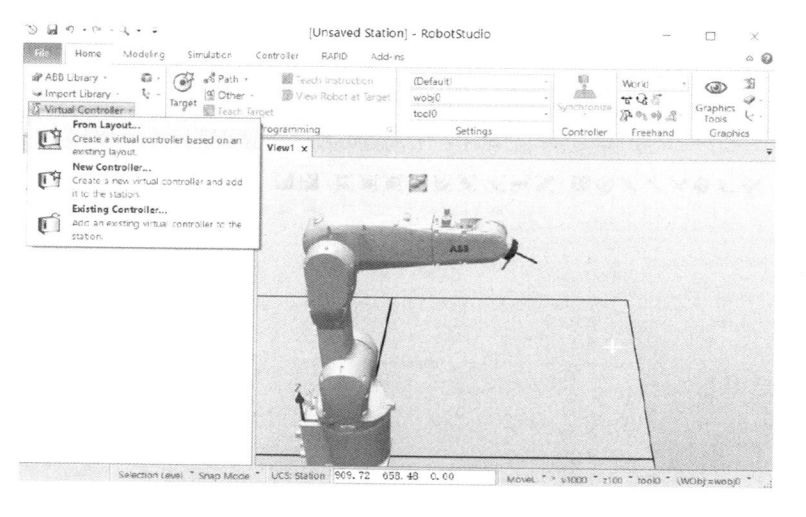

Fig. 5-21 Addition of Virtual Controller

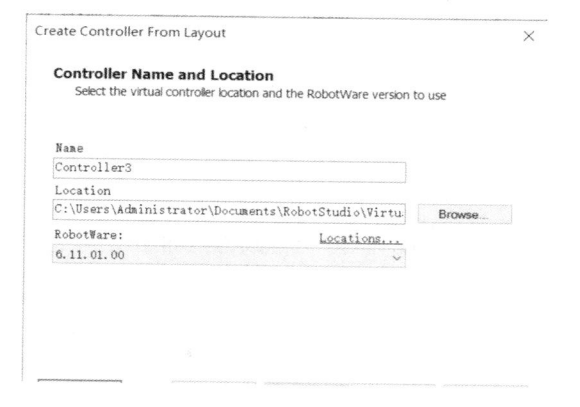

Fig. 5-22 RobotWare Selection Interface

5. RobotStudio Application

(6)Click the "Next" button to enter the setting interface of controller option, and click "Options–Default Language–Chinese" in turn to change the language of the virtual controller to Chinese, as shown in Fig. 5–23.

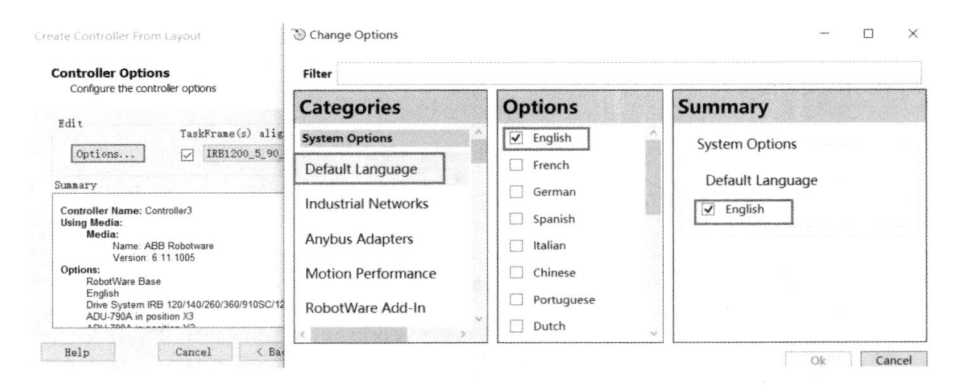

Fig. 5–23　Setting Interface of Controller Option

(7)Click "Basic–Import Model Library–Device–MyTool" in turn to add tools, as shown in Fig. 5–24.

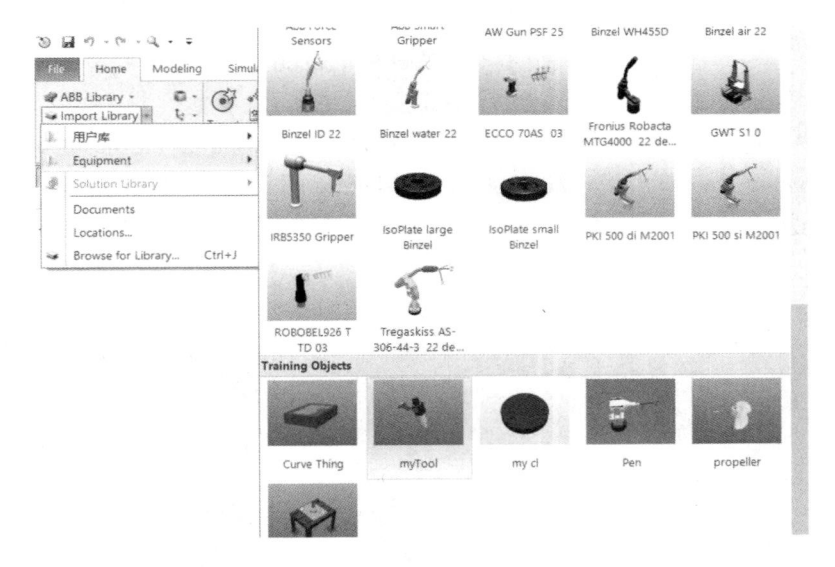

Fig. 5–24　Addition of Tools

81

Industrial Robot Technology Application

(8) After addition, the "MyTool" item will be added to the directory tree on the left. After clicking the item, the tool will be highlighted, and its default position is the origin position of the robot body coordinate system. To install the tool properly at the end of the robot flange, right-click the "MyTool" item bar, and select "Install to–IRB1200_5_90_STD_03 (T_ROB1)" in the pop-up edit menu, as shown in Fig. 5–25.

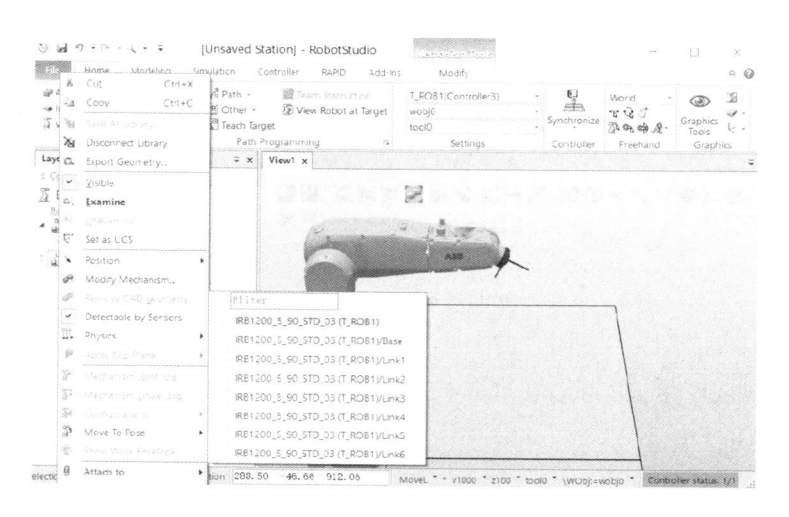

Fig. 5-25　Tool Relocation

(9) After confirmation, the "MyTool" is properly installed on the robot flange coordinates. The position change of the tool before and after relocation is shown in Fig. 5–26.

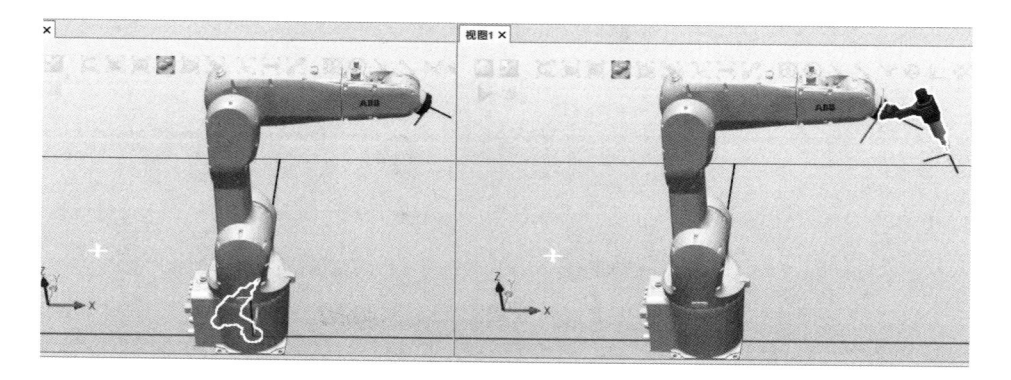

Fig. 5-26　Comparison of Tool Position Change

5. RobotStudio Application

(10) Click "Controller–Teach Pendant" in turn to open the virtual teach pendant, as shown in Fig. 5–27.

Fig. 5–27　Virtual Teach Pendant Button

(11) So far, the virtual robot workstation is created, and users can use the virtual teach pendant to control basic operation practices of the virtual robot in accordance with the basic operation method of industrial robot described in Chapter 4, as shown in Fig. 5–28.

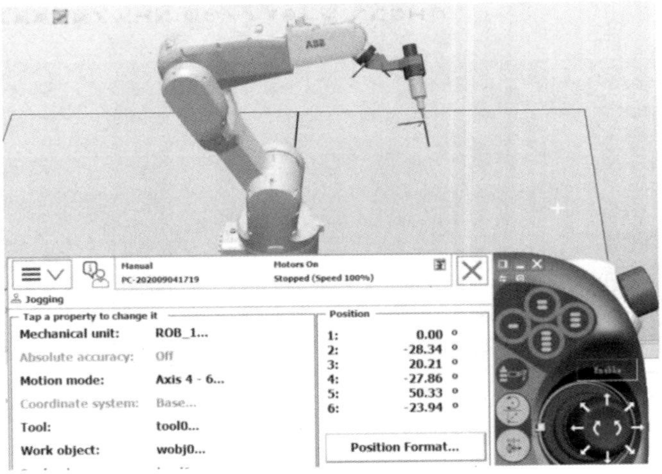

Fig. 5–28　Control of Robot Motion by Using Virtual Teach Pendant

6.

I/O Communication of ABB Robot

A standard ABB IRC5 controller consists of one control cabinet, which is indicated by C in Fig. 6–1, or is composed of two modules (A and B in the same figure): control module and drive module. In the latter case, the controller is called "Dual Cabinet Controller". The drive module comprises all electronic devices that supply power to the robot motor, while the control module comprises all electronic controls, such as the robot controller, I/O board and flash memory. I/O is the abbreviation of Input/Output, indicating the input/output port, through which the robot can interact with external devices. For example:

(1)Digital input: signal feedback of various switches, such as button switch, changeover switch and proximity switch; signal feedback of sensors, such as photoelectric sensor and optical fiber sensor; signal feedback of contactor and relay contact; switch signal feedback inside the touch screen.

(2)Digital output: control of various relay coils, such as contactor, relay and solenoid valve; control of various indication signals, such as signals of indicator

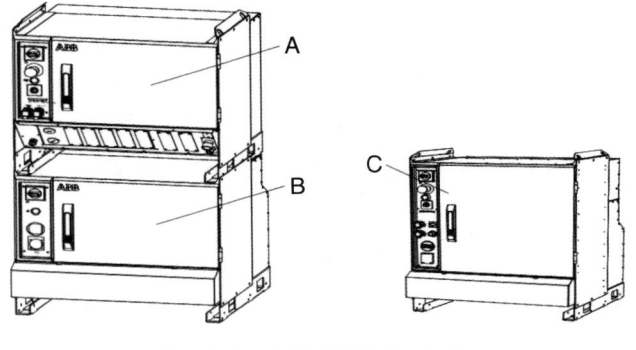

Fig. 6–1　ABB IRC5 Controller

Industrial Robot Technology Application

and buzzer.

Both input and output of the standard I/O board of ABB robot are of PNP type.

6.1 I/O communication types of ABB robot

ABB robots provide abundant I/O communication interfaces, which are given in Table 6-1. I/O communication interfaces include those for ABB standard communication, field bus communication with programmable logic controller(PLC)and data communication with personal computer(PC), so that communication with peripheral devices can be easily achieved.

Table. 6-1 I/O Communication Types of ABB Robot

ABB Standard	Field Bus	PC
Standard I/O board	Device Net	RS232
PLC	Profibus	OPC Server
	Profibus-DP	Socket Message
	Profinet	
	EtherNet IP	

The common signals processed by the standard I/O board of ABB robot include digital input(DI), digital output(DO), group input(GI), group output(GO), analog input(AI)and analog output(AO)signals.

ABB robots can be equipped with standard ABB PLCs. This avoids the trouble of making settings for communication with external PLCs and enables the completion of relevant operations with PLCs on the teach pendant.

I/O modules of the IRC5 control cabinet are shown in Fig. 6-2. These modules are internally connected to the I/O board (DSQC651 or DSQC652). See the robot product specification for their specific correspondence. In this chapter, the most commonly used ABB standard I/O board DSQC651 and Profibus -DP are taken as examples to explain in detail how to set relevant parameters.

6.2 Introduction to DSQC651

The commonly used ABB standard I/O boards are given in Table 6-2.

6. I/O Communication of ABB Robot

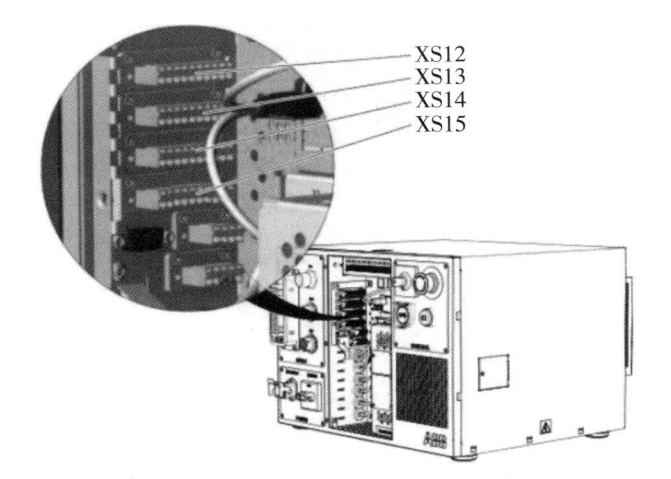

Fig. 6-2 I/O Modules of IRC5 Control Cabinet

Table 6-2 ABB Standard I/O Boards

Model	Description
DSQC 651	Distributed I/O module DI8\DO8\AO2
DSQC 652	Distributed I/O module DI16\DO16
DSQC 653	Distributed I/O module DI8\DO8 with relay
DSQC 355A	Distributed I/O module AI4\AO4
DSQC 377A	Conveyor chain tracking unit

The ABB standard I/O board DSQC651 that this chapter focuses on is shown in Fig. 6-3. This board mainly provides the processing of 8 DI signals, 8 DO signals and 2 AO signals. Its main components are:

(1)DO signal indicator.

(2)X1 digital output interface.

(3)X6 analog output interface.

(4)X5 DeviceNet interface.

(5)Module status indicator.

(6)X3 digital input interface.

(7)Digital input signal indicator.

87

Industrial Robot Technology Application

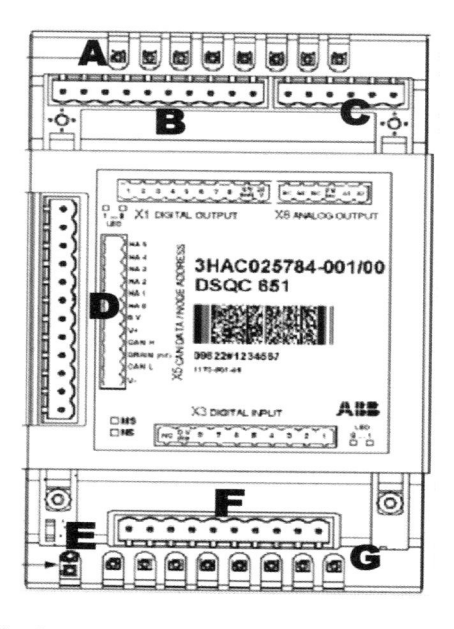

Fig. 6–3   ABB Standard I/O Board DSQC651

（1）The interface connection description of X1 module is given in Table 6–3.

Table 6–3   Terminal Connection Description of X1

X1 Terminal No.	Defined Use	Address Assignment
1	OUTPUT CH1	32
2	OUTPUT CH2	33
3	OUTPUT CH3	34
4	OUTPUT CH4	35
5	OUTPUT CH5	36
6	OUTPUT CH6	37
7	OUTPUT CH7	38
8	OUTPUT CH8	39
9	0V	
10	24V	

（2）The interface connection description of X3 module is given in Table 6–4.

（3）The interface connection description of X5 module is given in Table 6–5.

（4）The interface connection description of X6 module is given in Table 6–6. The voltage range of analog output is 0–10V.

88

6. I/O Communication of ABB Robot

Table 6–4 Terminal Connection Description of X3

X3 Terminal No.	Defined Use	Address Assignment
1	INPUT CH1	0
2	INPUT CH2	1
3	INPUT CH3	2
4	INPUT CH4	3
5	INPUT CH5	4
6	INPUT CH6	5
7	INPUT CH7	6
8	INPUT CH8	7
9	0V	
10	Not used	

Table 6–5 Terminal Connection Description of X5

X5 Terminal No.	Defined Use
1	0V BLACK
2	CAN signal line low BLUE
3	Shielded wire
4	CAN signal line high WHITE
5	24V RED
6	GND address selection, common terminal
7	Module ID bit 0(LSB)
8	Module ID bit 1(LSB)
9	Module ID bit 2(LSB)
10	Module ID bit 3(LSB)
11	Module ID bit 4(LSB)
12	Module ID bit 5(LSB)

Table 6–6 Terminal Connection Description of X6

X6 Terminal No.	Defined Use	Address Assignment
1	Not used	
2	Not used	
3	Not used	
4	0V	
5	AO1	0–15
6	AO2	16–31

89

Industrial Robot Technology Application

The ABB standard I/O board is connected to the DeviceNet network, so it is necessary to set the module address in the network. The 6–12 jumper of terminal X5 is used to determine the address of the module, and the available address range is 10–63. To obtain the address of 10, cut off the jumper connected to pin 8 and pin 10, as shown in Fig. 6–4. In this way, the address of 10(2+8=10)can be obtained.

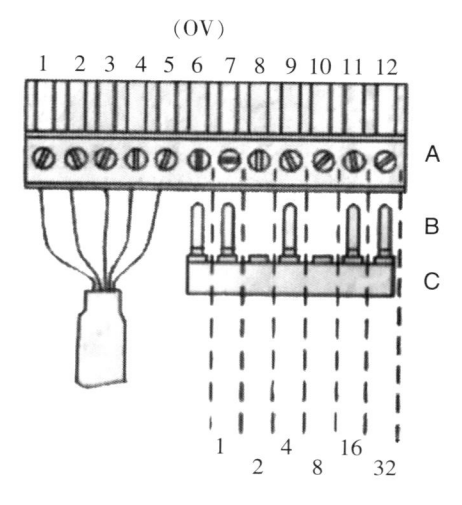

Fig. 6–4   Jumper of Terminal X5

6.3 DSQC651 configuration

ABB standard I/O board DSQC651 is the most commonly used module. Its configuration process is explained in detail below by taking the examples of creating DI, DO, GI, GO and AO signals.

All ABB standard I/O boards are devices connected to the DeviceNet field bus and communicate with the same bus through port X5. Refer to Table 6–7 for details of relevant parameters defining the bus connection of the DSQC651 board.

Table 6–7   Parameter Definitions

Parameter	Set Value	Description
Name	d651	Name of set I/O board in the system; the default name is d651 in the system
Type	d651	Type of set I/O board
Network	DeviceNet	Bus to which the I/O board connects
Address	10	Address of set I/O board on the bus

90

6.3.1 RobotStudio robot workstation configuration

ABB standard I/O board DSQC651 is not loaded by default, when the Robot-Studio software is used to establish a virtual robot workstation. When loading the RobotWare software, tap "Options" as shown in Fig. 6–5 for configuration.

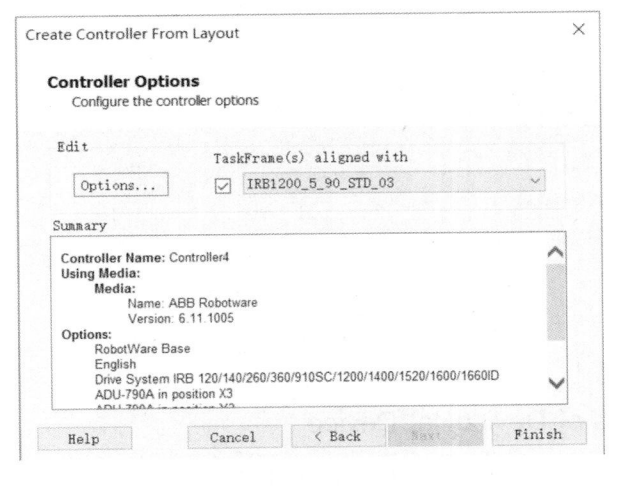

Fig. 6–5　Controller Option Button

(1)Tap "Industrial Networks" under category and select "709–1 DeviceNet Master/Slave" and "969–1 PROFIBUS Controller", as shown in Fig. 6–6.

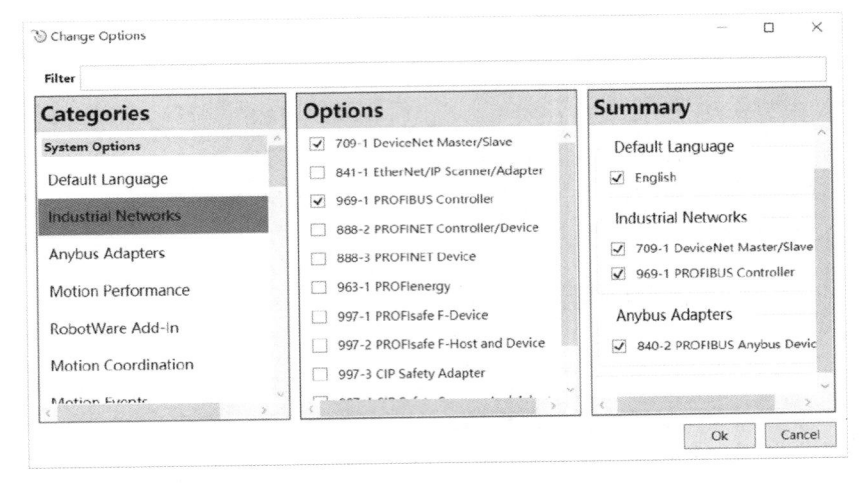

Fig. 6–6　Industrial Networks Option

(2)Tap "Anybus Adapters" under category and select "840-2 PROFIBUS Anybus Device", as shown in Fig. 6-7.

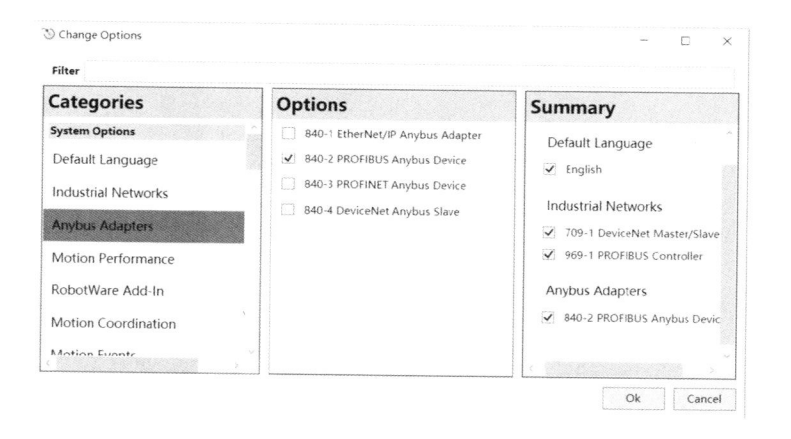

Fig. 6-7　Anybus Adapters Option

6.3.2 Settings of DeviceNet Device

The standard I/O board DSQC651 needs to be set in the DeviceNet Device module. The specific setting steps are as follows:

(1)Tap the ABB menu button in the upper left corner of the teach pendant screen, and select the "Control Panel" option and then the "Configuration" option, as shown in Fig. 6-8.(Note: The configuration module can be accessed in the manual mode only.)

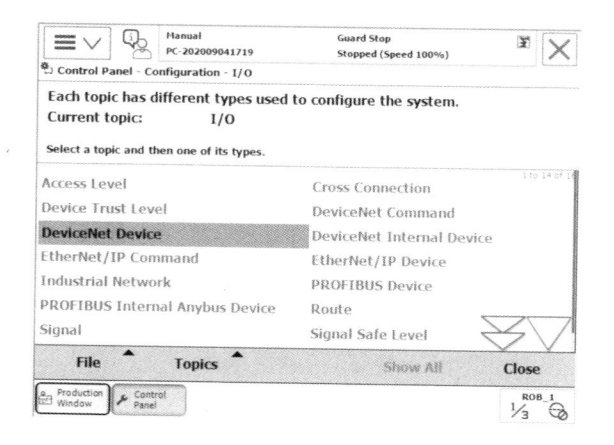

Fig. 6-8　DeviceNet Device Option

6. I/O Communication of ABB Robot

(2)Double tap the "DeviceNet Device" option to enter the setting interface, as shown in Fig. 6–9.

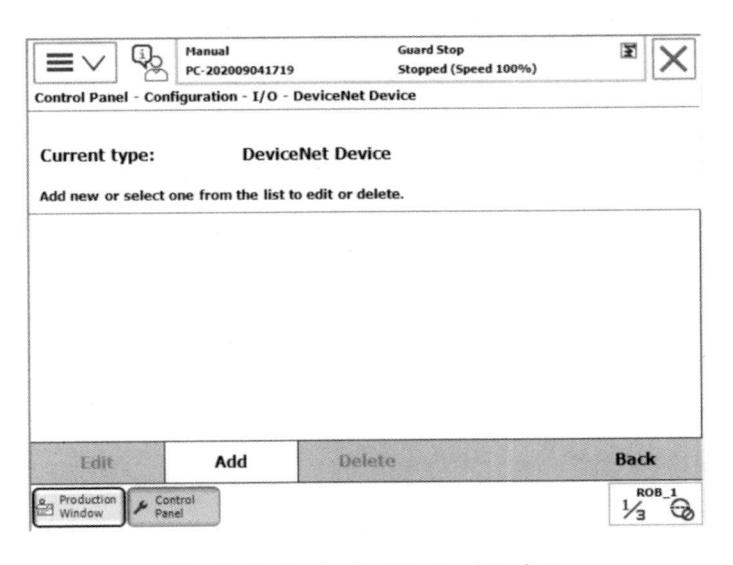

Fig. 6–9　DeviceNet Device Interface

(3)Tap the "Add" button and select the "DSQC 651 Combi I/O Device" template, as shown in Fig. 6–10.

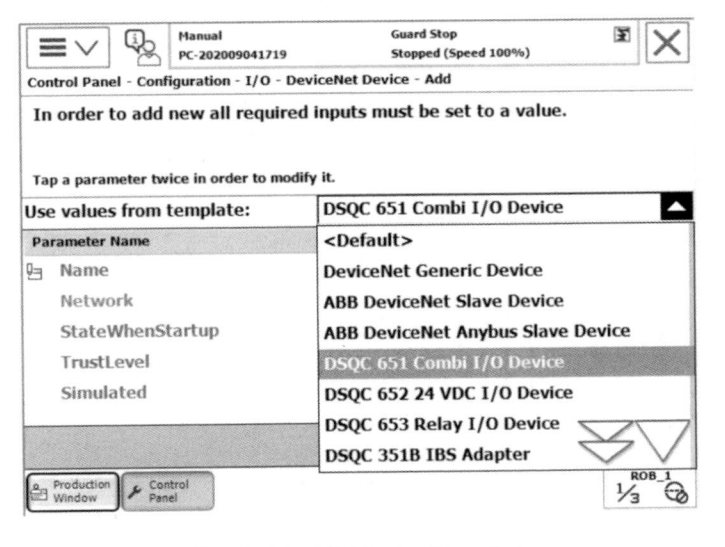

Fig. 6–10　I/O Device Template

93

Industrial Robot Technology Application

(4)After the template is selected, the default I/O device name will be "d651" in the system. In this case, double tap each option to modify the value of the corresponding parameter, as shown in Fig. 6-11, and set the "Address" to 10.

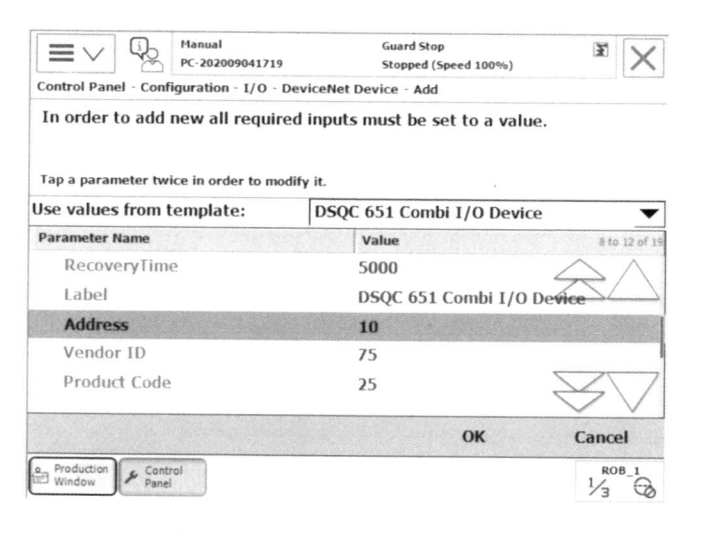

Fig. 6-11   Parameter Configuration of I/O Device

(5)After completing parameter settings, tap "OK" to pop up the restart dialog box as shown in Fig. 6-12. Tap "Yes" to restart the controller, so that the I/O board settings take effect.

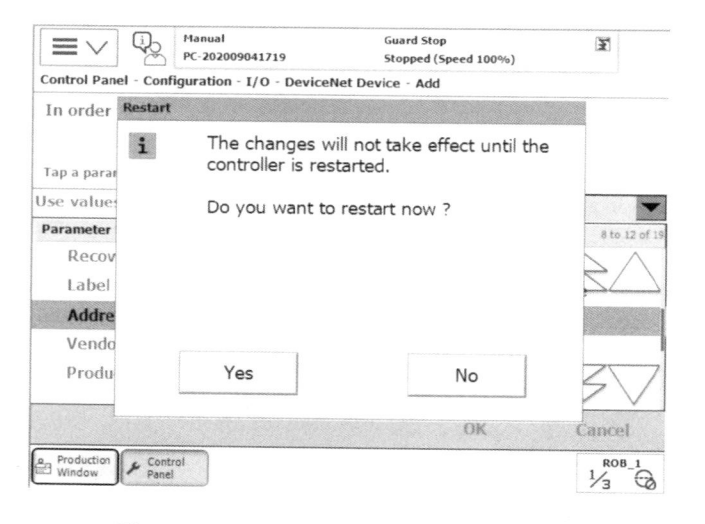

Fig. 6-12   Restart Confirmation Dialog Box

94

6. I/O Communication of ABB Robot

6.3.3 I/O signal configuration

(1)Tap the ABB menu button in the upper left corner of the teach pendant screen, select the "Control Panel" option and then the "Configuration" option, and double tap the "Signal" option, as shown in Fig. 6–13.

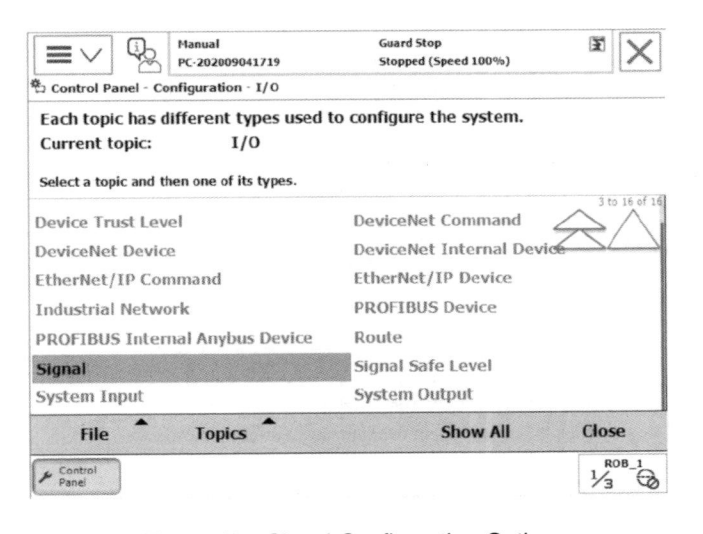

Fig. 6–13   Signal Configuration Option

(2)Enter the signal setting interface as shown in Fig. 6–14, and tap the "Add" button to define new signal variables.

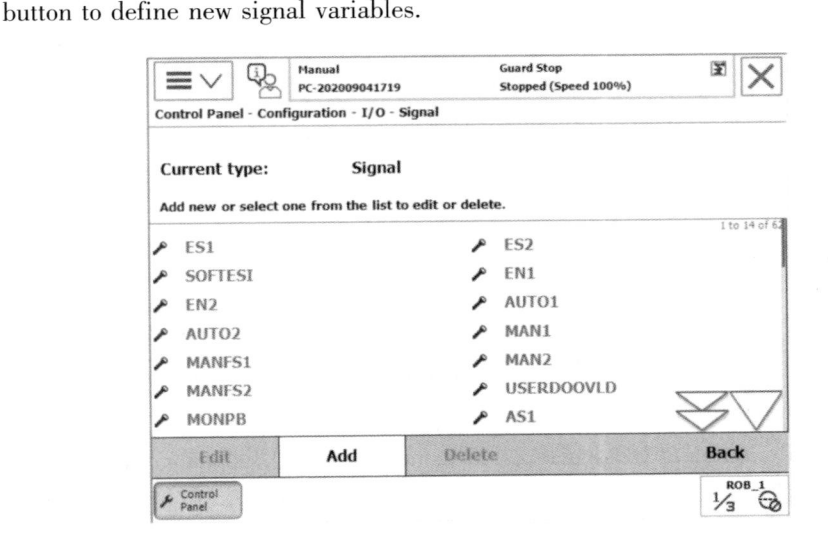

Fig. 6–14   Signal Setting Interface

Industrial Robot Technology Application

(3)Enter the signal parameter setting interface. The names and descriptions of parameters to be defined are given in Table 6–8. Tap the parameter name to modify it. The setting completion effect of digital input signal Di1 is shown in Fig. 6–15.

Table 6–8　DI Signal Parameters

Parameter	Set Value	Description
Name	Di1	Name of set DI signal
Type of Signal	Digital Input	Type of set signal
Assigned to Device	d651	I/O module where the set signal is
Device Mapping	0	Address occupied by the set signal

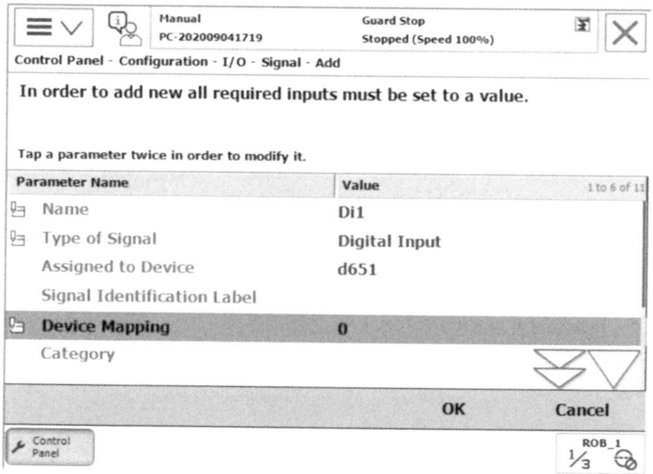

Fig. 6–15　Signal Parameter Setting Interface

(4)After completing the settings, tap "OK" to pop up the restart confirmation interface as shown in Fig. 6–16. Tap "Yes" to restart the controller, so that the newly set parameters take effect. Or, tap "No" to continue to set new signal variables, and then restart the controller after completing all settings.

(5)The definition of DO signals is the same as that of DI signals. The parameter names and descriptions of DO signals are given in Table 6–9, and the setting completion effect of Do1 signal is shown in Fig. 6–17.

96

6. I/O Communication of ABB Robot

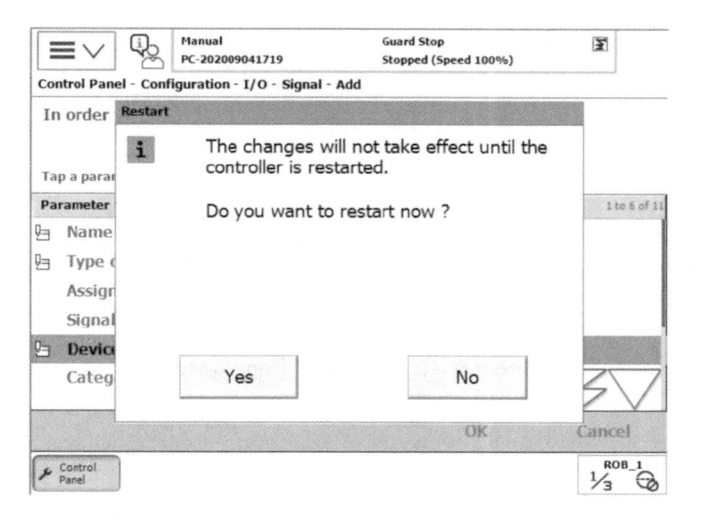

Fig. 6-16 Signal Setting Interface

Table 6-9 DO Signal Parameters

Parameter	Set Value	Description
Name	Do1	Name of set DO signal
Type of Signal	Digital Output	Type of set signal
Assigned to Device	d651	I/O module where the set signal is
Device Mapping	32	Address occupied by the set signal

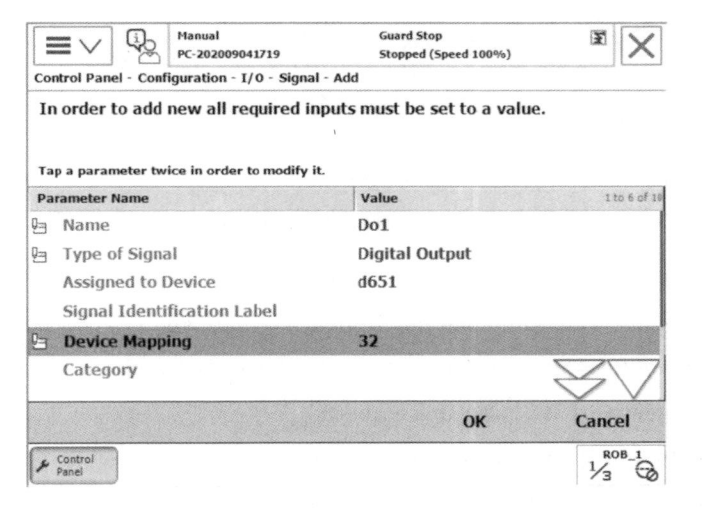

Fig. 6-17 Definition of DO Signal

97

Industrial Robot Technology Application

(6)Definition of GI signal Gi1: GI signal is a decimal number used to combine several DI signals for receiving BCD codes input by peripheral devices. The parameter names and descriptions of GI signals are given in Table 6-10. As for the "Device Mapping" address, "1-4" means that the GI signal Gi1 occupies a 4-bit DI port, and the formed 4-bit binary number can represent decimal numbers 0-15. By analogy, if an address occupies 5 bits, it can represent decimal numbers 0-31. The definition completion effect of GI signal Gi1 is shown in Fig. 6-18.

Table 6-10 GI Signal Parameters

Parameter	Set Value	Description
Name	Gi1	Name of set GI signal
Type of Signal	Group Input	Type of set signal
Assigned to Device	d651	I/O module where the set signal is
Device Mapping	1-4	Address occupied by the set signal

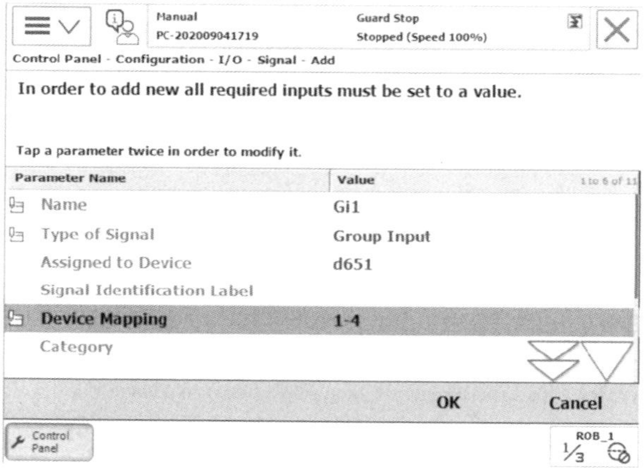

Fig. 6-18 Definition of GI Signal

(7)Definition of GO signal Go1: The definition of GO signals is similar to that of GI signals. The difference lies in that the data type is "Group Output" and the mapping address must be the DO port address. The definition completion effect of GO signal Go1 is shown in Fig. 6-19.

(1)Definition of AO signal Ao1: AO signal is commonly used to control the operating voltage of a device and further control the motor speed or device power. The parameter names and descriptions of the AO parameters are given in Table 6-11.

98

6. I/O Communication of ABB Robot

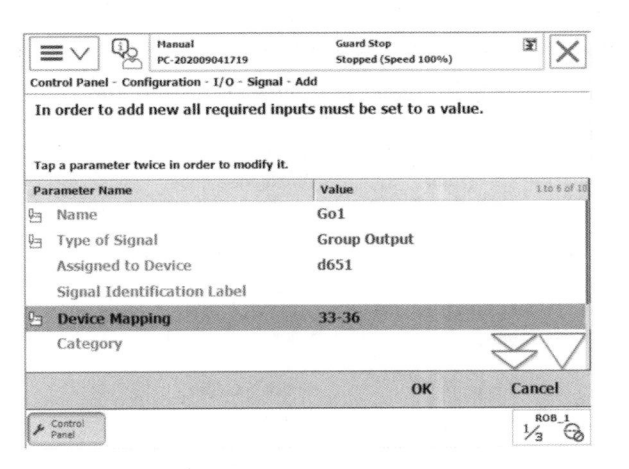

Fig. 6-19   Definition of GO Signal

Table 6-11   AO Signal Parameters

Parameter	Set Value	Description
Name	Ao1	Name of set AO signal
Type of Signal	Analog Output	Type of set signal
Assigned to Device	d651	I/O module where the set signal is
Device Mapping	0–15	Address occupied by the set signal
Default Value	12	Default value, not less than the minimum logic value
Analog Encoding Type	Unsigned	Default value, not less than the minimum logic value
Maximum Logical Value	24	Maximum logic value, e.g., the maximum output voltage of 24V
Maximum Physical Value	10	Maximum physical value, e.g., the maximum output voltage of the corresponding I/O board at the maximum output voltage
Maximum Physical Value Limit	10	Maximum physical limit, i.e., the maximum output voltage of I/O board port
Maximum Bit Value	65535	Maximum logic bit value, 16 bits
Minimum Logical Value	12	Minimum logic value, e.g., the minimum output voltage of 12V
Minimum Physical Value	0	Minimum physical value, e.g., the minimum output voltage of the corresponding I/O board at minimum output voltage of welding machine
Minimum Physical Value Limit	0	Minimum physical limit, i.e., the minimum output voltage of I/O board port
Minimum Bit Value	0	Minimum logic bit value

99

Industrial Robot Technology Application

The definition completion effect of AO signal Ao1 is shown in Fig. 6–20. After completing all signal definitions, tap "OK" to restart the controller, so that all settings take effect.

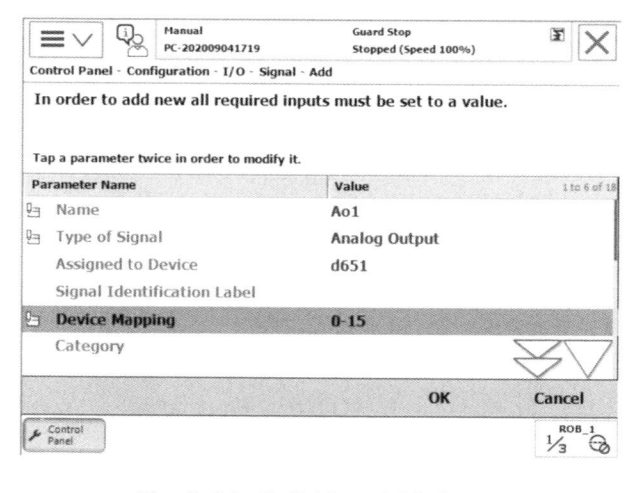

Fig. 6–20 Definition of AO Signal

6.3.4 I/O signal monitoring

All defined I/O signals can be monitored and simulated in the "I/O" module. The specific operation method is as follows:

(1)Tap the ABB menu button in the upper left corner of the teach pendant screen and select the "I/O" option, as shown in Fig. 6–21.

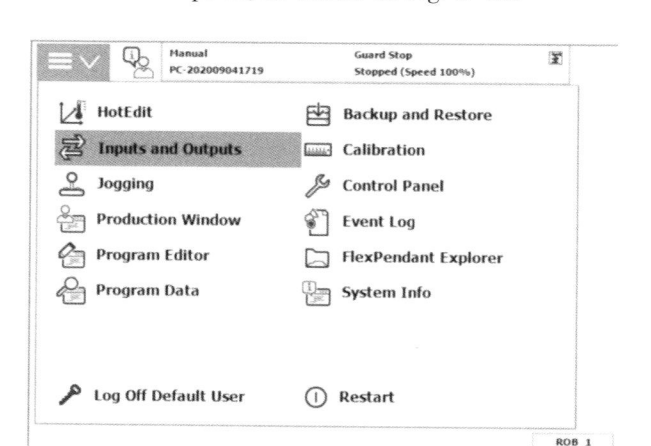

Fig. 6–21 I/O Option

6. I/O Communication of ABB Robot

(2) Enter the "I/O" setting interface as shown in Fig. 6–22, tap the "View" button in the lower right corner of the screen to pop up the view classification option menu, and select "I/O device".

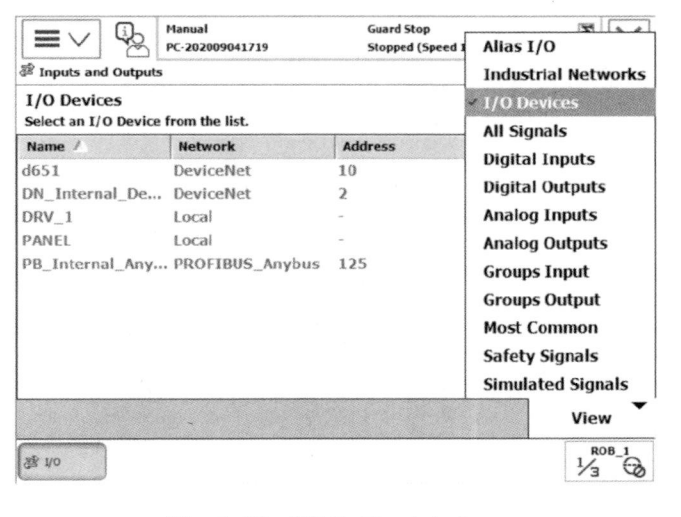

Fig. 6–22   I/O Setting Interface

(3) Switch the view display mode to display all "I/O devices", as shown in Fig. 6–23, select the "d651" device and tap the "Signal" button at the bottom.

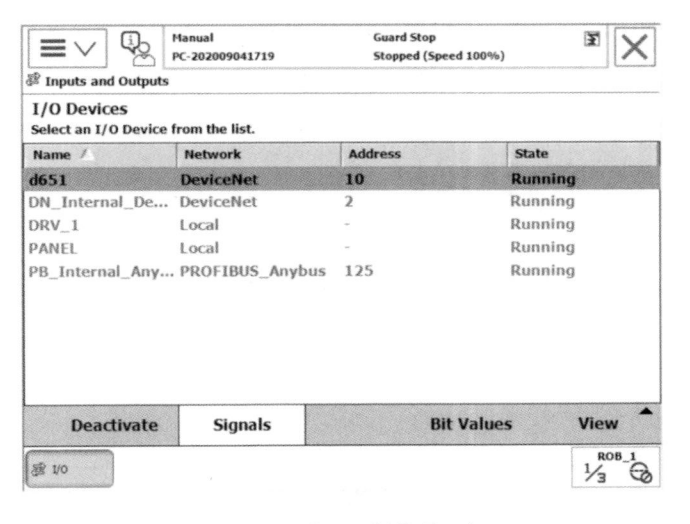

Fig. 6–23   View of I/O Devices

101

Industrial Robot Technology Application

(4)Enter the I/O signal list interface as shown in Fig. 6-24, in which all signals defined in Section 6.3.3 are displayed. In this interface, the signals can be monitored and the changes of corresponding signals can be seen if any external input signals have been modified.

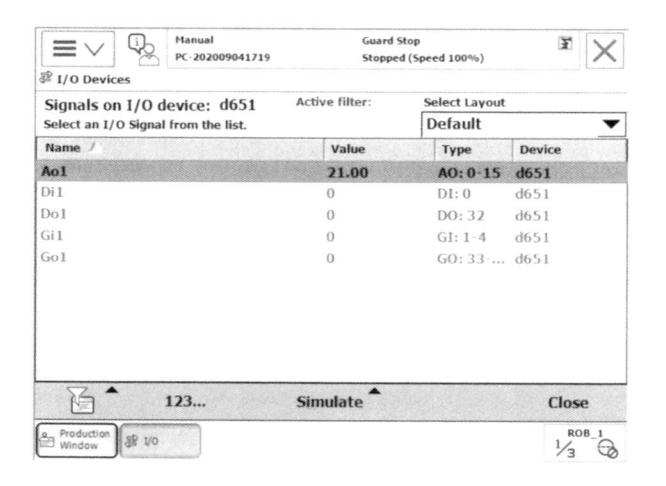

Fig. 6-24　I/O Signal List

(5)It is also possible to either forcibly modify output signals and observe the dynamics of external devices, or carry out simulation operations in this interface. To do this, select the output signal to be modified and tap "Simulation", and tap "0" or "1" to modify the value of the corresponding signal, as shown in Fig. 6-25.(Note: The simulation function can be used in the manual mode only.)

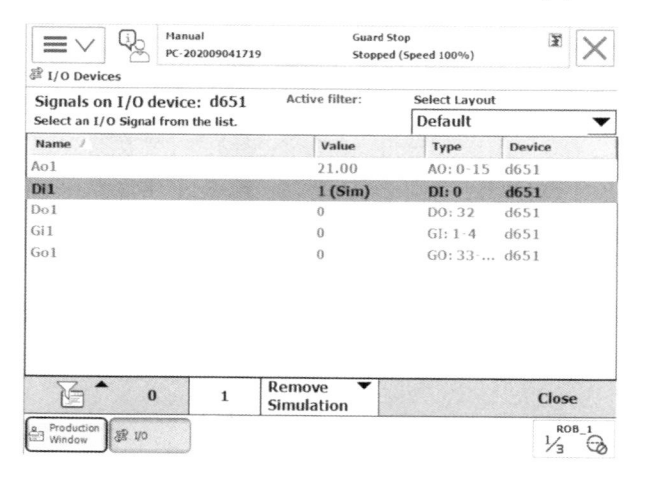

Fig. 6-25　Digital Signal Simulation

102

(6)The values of input signals may also be forcibly modified for debugging the program. The value setting of GI signal is shown in Fig. 6–26. Tap the "123..." button to call out the input keyboard, and the system will prompt the maximum and minimum values that are allowed to be set according to the bit number of the GI signal.

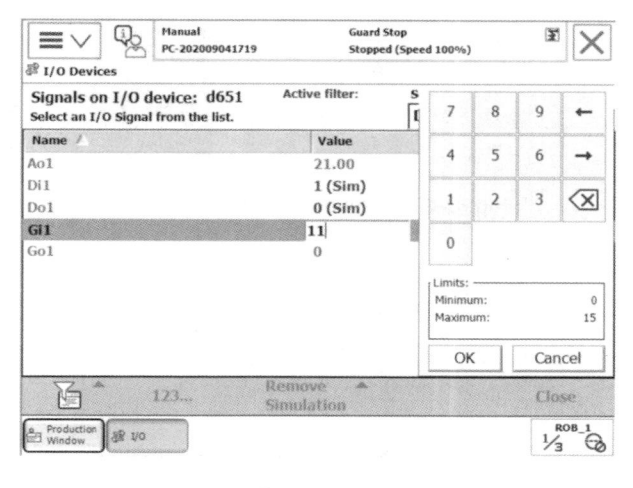

Fig. 6–26　Simulation of GI Signal

(7)The setting of an analog signal value is similar to that of GI and GO signals. The difference lies in that its maximum and minimum values depend on the settings made when the analog signal is defined.

6.4 Profibus communication

In addition to communicating with peripheral devices through its self–contained standard I/O board, an ABB robot can also use a DSQC667 module to communicate with the PLC through Profibus at a fast speed and with large data volume. Fig. 6–27 shows the schematic diagram of Profibus communication. The part in the red box shown in the figure is the DSQC667 module installed on the robot controller inside the electric cabinet. This module supports up to 512 DI signals and 512 DO signals. Other parts represent:

A: Master station of PLC

B: Slave station on the bus

C: Profibus adapter DSQC667 for robot

D: Robot control cabinet

Industrial Robot Technology Application

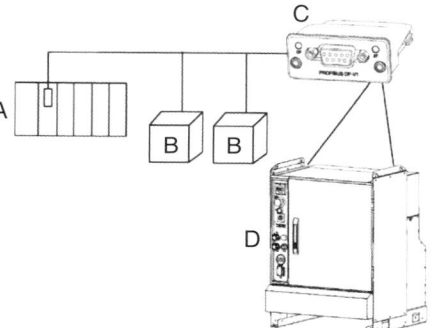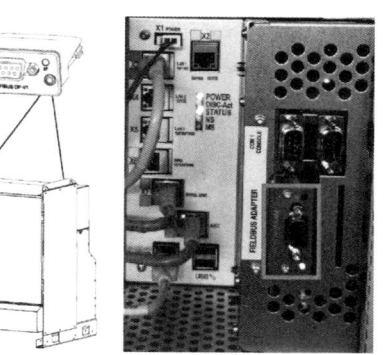

Fig. 6-27　Profibus Communication

The Profibus address set at the robot end needs to be consistent with the Profibus address set when the robot station is added to the PLC end. Refer to Table 6-12 for detailed parameter settings of the Profibus address set at the robot end of the slave station.

Table 6-12　Profibus Address Parameters

Parameter	Set Value	Description
Name	PROFIBUS_Anybus	Bus network(non-editable)
Identification Label	PROFIBUS Anybus Network	Identification label
Address	8	Bus address
Simulated	No	Simulated state

The parameter settings of Profibus input and output bytes at the robot end of the slave station are given in Table 6-13. The byte size is set to "4" to represent 32 bits, which indicates that the communication between the robot and PLC supports 32

Table 6-13　Profibus I/O Parameters

Parameter	Set Value	Description
Name	PB_Internal_Anybus	Name of board
Network	PROFIBUS_Anybus	Bus network
VendorName	ABB Robotics	Supplier name
ProductName	PROFIBUS Internal Anybus Device	Product name
Label		Label
Input Size(bytes)	4	Input size(bytes)
Output Size(bytes)	4	Output size(bytes)

6. I/O Communication of ABB Robot

DI signals and 32 DO signals. The maximum permissible set value of this parameter is 64, which indicates that up to 512 DI signals and 512 DO signals are supported.

The connection for Profibus communication is set as follows:

(1)Tap the ABB menu button in the upper left corner of the teach pendant screen, and select "Control Panel", "Configuration" and "Industrial Network" in turn, as shown in Fig. 6-28.

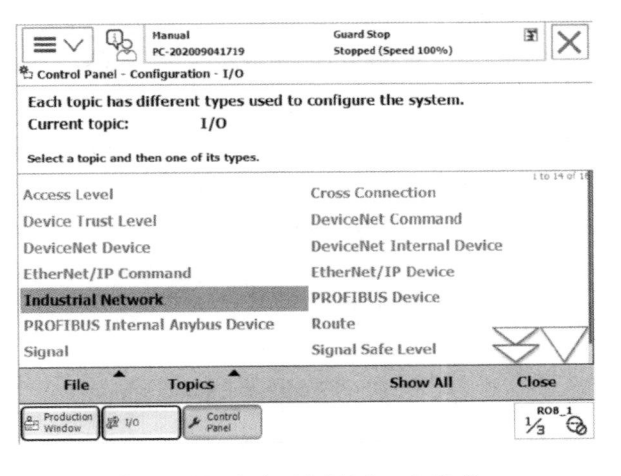

Fig. 6-28　Industrial Network Option

(2)Enter the "Industrial Network" setting interface, as shown in Fig. 6-29.

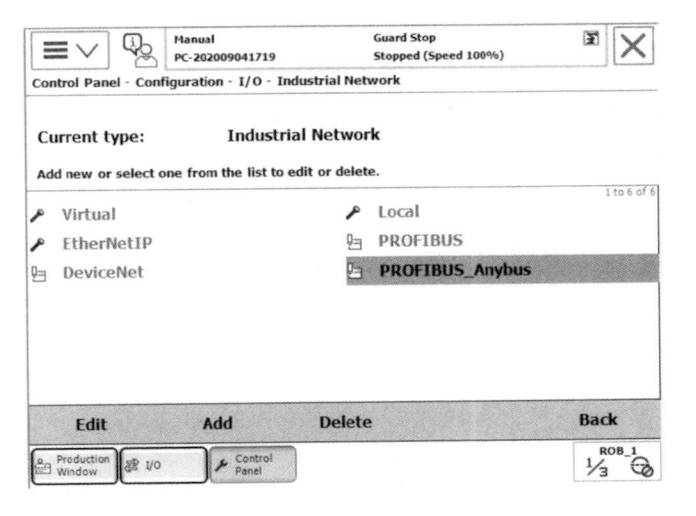

Fig. 6-29　Industrial Network Setting Interface

105

Industrial Robot Technology Application

(3)Double tap the "PROFIBUS_Anybus" option to enter the Profibus parameter setting interface, as shown in Fig. 6–30. Double tap the "Address" option to set the Profibus address at the robot end of the slave station to 8.

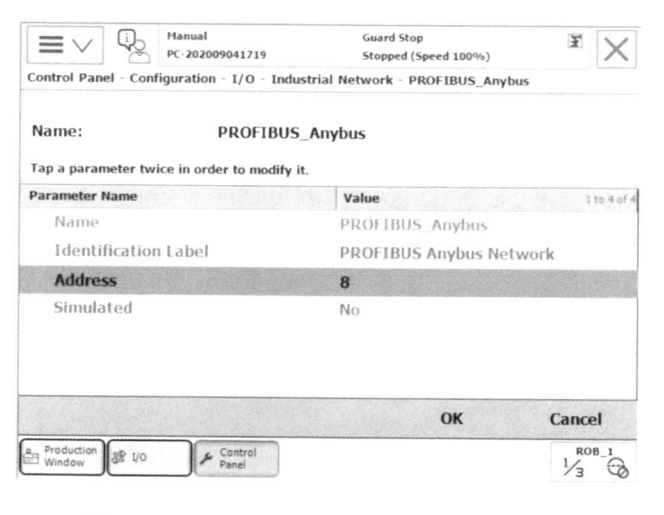

Fig. 6–30　Profibus Address Setting Interface

(4)Tap "OK" to complete the setting and pop up an inquiry dialog box, and select "No" in this dialog box to continue with I/O signal configuration. Tap the "Back" button to return to the "Configuration" interface, and double tap the "PROFIBUS Internal Anybus Device" option, as shown in Fig. 6–31.

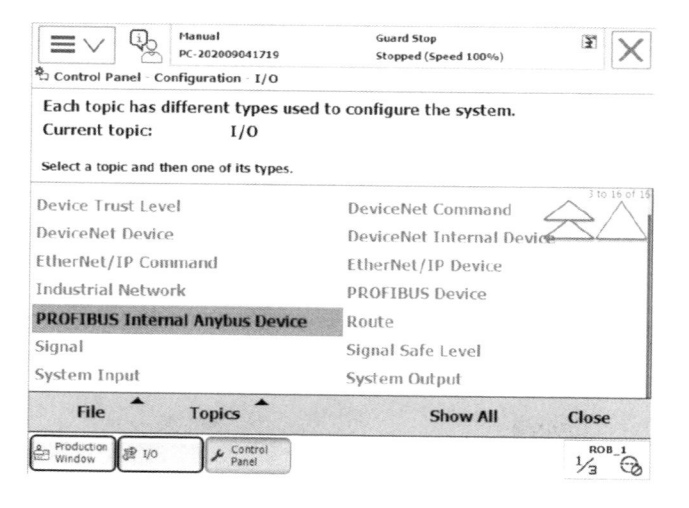

Fig. 6–31　Profibus Configuration Option

6. I/O Communication of ABB Robot

(5)Enter the Profibus parameter setting interface and double tap the "PB_Internal_Anybus" option, as shown in Fig. 6–32.

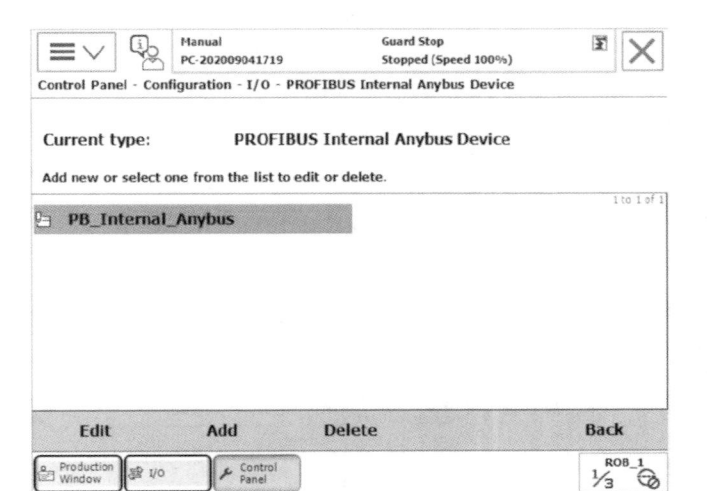

Fig. 6–32 Profibus Parameter Setting Interface

(6)Set both the "Input Size(bytes)" and "Output Size(bytes)" to "4", which corresponds to 32 DI signals and 32 DO signals, as shown in Fig. 6–33. Tap "OK" after setting, and tap "Yes" in the pop–up dialog box to restart the controller, so that all settings take effect.

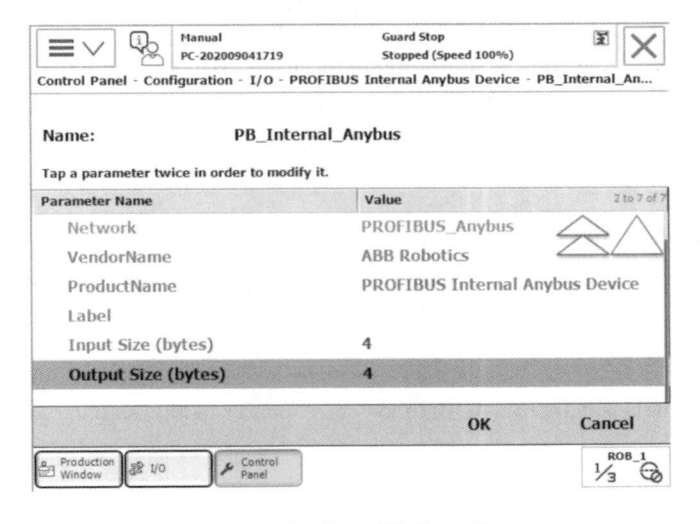

Fig. 6–33 Profibus I/O Byte Setting

Industrial Robot Technology Application

(7)The signal setting method based on Profibus is basically the same as that based on the ABB standard I/O board. The only difference to be noticed is that the I/O device should be selected as "PB_Internal_Anybus", as shown in Fig. 6-34.

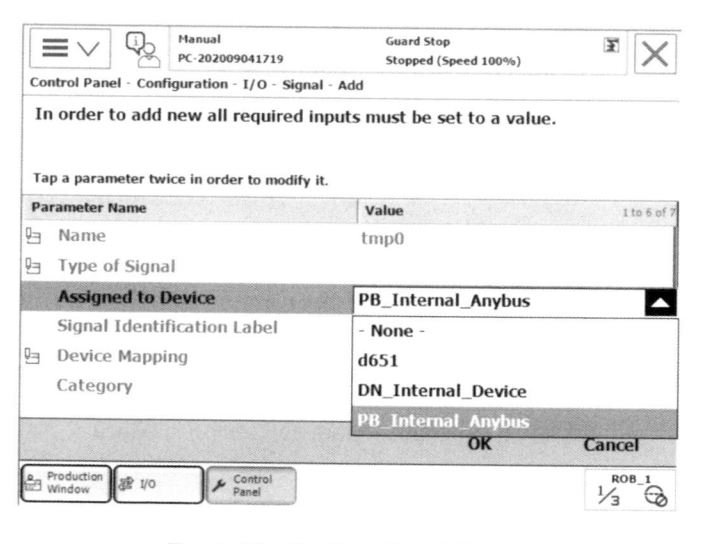

Fig. 6-34   Profibus Signal Setting

(8)After setting the Profibus slave station on the ABB robot, the following operations should be completed at the PLC end:

Install the DSQC667 configuration file of an ABB robot in the PLC configuration software(enter the "FlexPendant Explorer" and get the configuration file by following the path PRODUCTS/RobotWare_6** /utility/service/GSD/HMS_***. gsd).

Open the configuration software, add the newly added "Anybus–CC PROFIBUS DP–V1" to the workstation, and set the Profibus address(e.g., to 8).

Add an I/O module(e.g., an I/O module with 4 types for input size and output size each).

The signals set in the ABB robot corresponds to the signals set at the PLC end.

108

6.5 Programmable key settings of teach pendant

There are four programmable quickset keys on the teach pendant of the ABB robot, as shown in Fig. 6–35. During debugging, peripheral signal input can be simulated or forced signal output can be done by configuring these four quickset keys. Flexible use of such programmable keys can greatly improve the efficiency of debugging. The following are detailed operation steps for configuring DO signal Do1 for programmable key 1:

(1)Tap the ABB menu button in the upper left corner of the teach pendant screen, and select the "Control Panel" option and the "ProgKeys" option in turn, as shown in Fig. 6–36.

(2)Enter the programmable key configuration interface, select the "Key 1" tab, and select "Output" under type, as shown in Fig. 6–37.

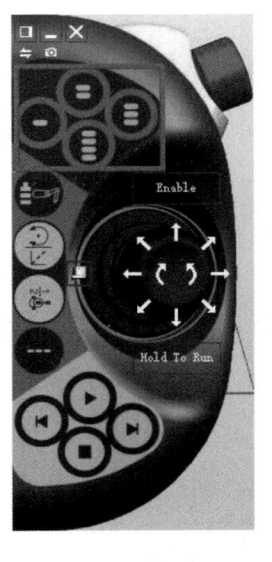

Fig. 6–35 Profibus Signal Setting

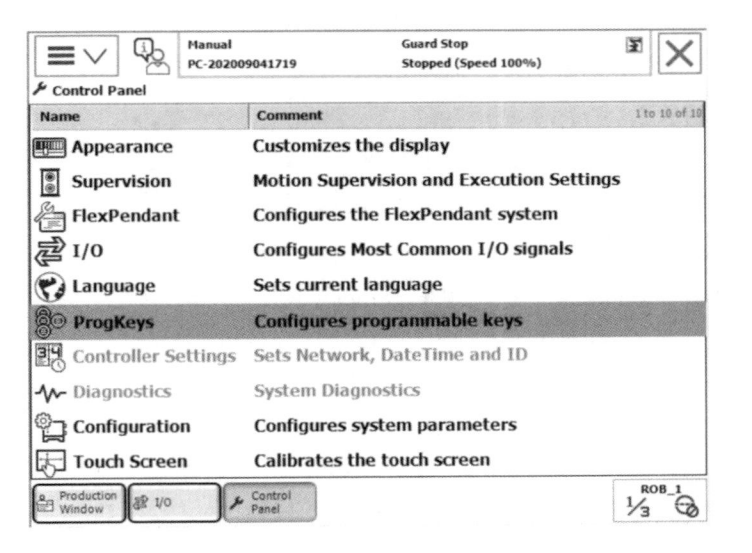

Fig. 6–36 ProgKeys Option

Industrial Robot Technology Application

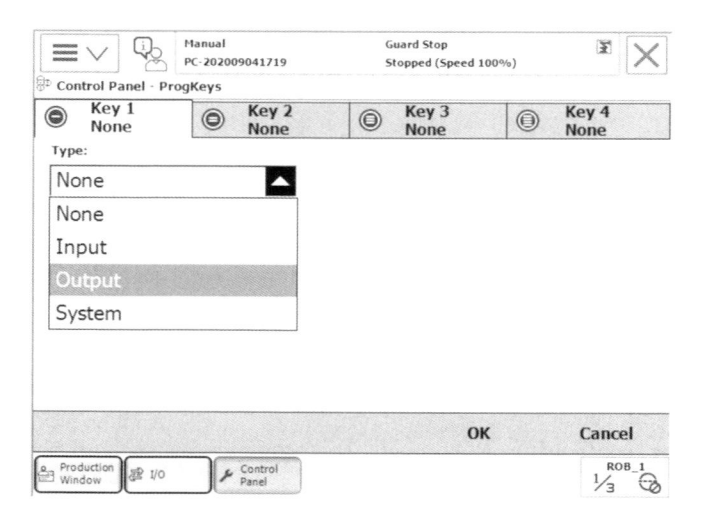

Fig. 6-37　Programmable Key Configuration Interface

(3)All DO signals will be displayed in the list box on the right. Select "Do1" and select the option corresponding to the pressing of the key, as shown in Fig. 6-38. The functions of relevant options differ as follows:

Toggle: When the key is pressed, the value of Do1 will be switched between 0 and 1;

Set to 1: When the key is pressed, the value of Do1 will be set to 1, which is

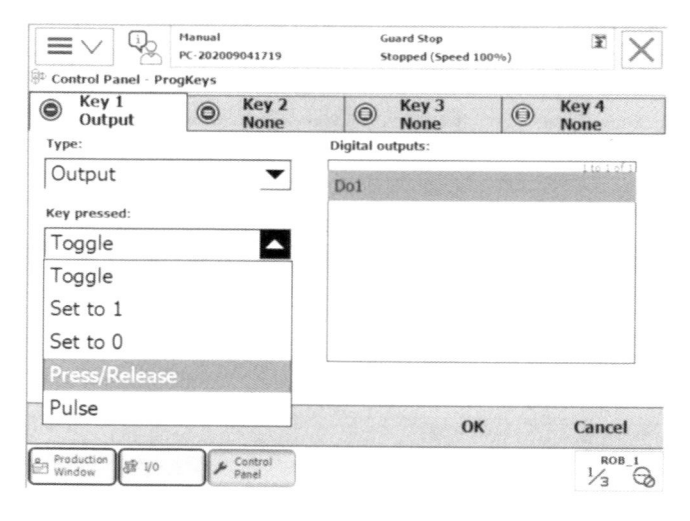

Fig. 6-38　Programmable Key Configuration Interface

110

6. I/O Communication of ABB Robot

equivalent to setting;

Set to 0: When the key is pressed, the value of Do1 will be set to 0, which is equivalent to resetting;

Press/Release: The value of Do1 will be set to 1 when the key is pressed and to 0 when the key is released;

Pulse: When the rising edge of the key is pressed, the value of Do1 will be set to 1.

(4) After completing the settings, enter the I/O signal monitoring interface, operate key 1 to check whether the corresponding signal changes, or directly run the program for testing.

7.

Programming Basics of ABB Robot

Programming basics of ABB robot mainly include robot system related terminology, ABB robot program structure and program data.

7.1 Robot system related terminology

A robot workstation mainly consists of the robot and its control system, auxiliary devices and other peripheral devices. The workstation includes all hardware and software necessary for operating the robot. The main concepts and terms related to the robot system are as follows:

7.1.1 Tool

A tool is an object that can be directly or indirectly installed on the rotating disk of the robot or can be assembled in a fixed position within the operating range of the robot. As shown in Fig. 7−1, when the flange plane of the robot is taken as the boundary, side A is the tool side and side B is the robot side. To complete various tasks, the tools installed at ends of industrial robots, such as spray gun, gripper and welding torch, are different in shape and size. After any replacement or adjustment is made the tool, the actual working point of the robot will change relative to the end of the robot. Therefore, all tools must be defined with tool center points (TCPs). In order to get accurate TCP positions, all tools used by the robot must be measured, and the measured data must be kept.

7.1.2 Tool center point(TCP)

To define a tool at the robot end, a common practice is to establish a tool coordinate system on the robot tool, and the system origin is called the Tool Center

Industrial Robot Technology Application

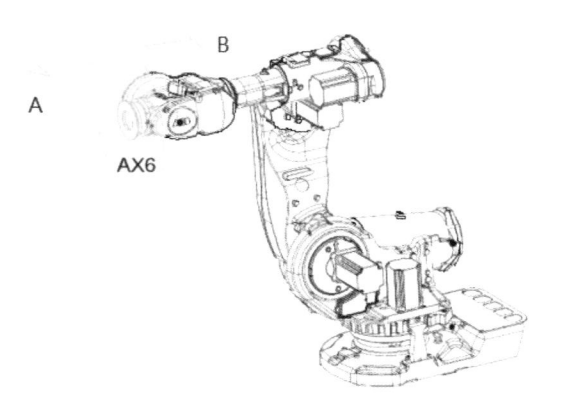

Fig. 7-1 Control of Robot Motion by Using Virtual Teach Pendant

Point (TCP). The robot is programmed in this coordinate system. After the tool is adjusted, the robot can come into service again only by recalibrating the position and pose of the working coordinate system.

Different TCPs can be defined for the same robot due to various tools used by it, but one robot can only have one valid TCP at one time. Two basic types of TCP are involved: Moving TCP and Stationary TCP.

Moving TCP is common and moves along with the movement of the robot arm. Examples include the welding torch of a welding robot and the fixture of a handling robot.

Stationary TCP is a point outside the robot body, and the work object carried by the robot makes trajectory motion around this point. For example, when a fixed spot welding torch is used, its TCP should be defined with reference to a stationary device rather than the moving manipulator.

Default TCP of robot system: Regardless of the brand of an industrial robot, a default tool coordinate system is defined for it in advance. Without exception, the XY plane of this coordinate system is bound to the flange plane of the robot's sixth axis, and the origin of the coordinates coincides with the center of the flange. Obviously, TCP is at the center of the flange. Different names are given to such tool coordinate systems for robots of different brands. As for an ABB robot, its tool coordinate system is called tool0.

114

7.1.3 Coordinate system

Coordinate system is a position and pose index system set on a robot or other spaces to determine the position and pose of the robot. Usually, the coordinate system defines a plane or space with axes starting from a fixed point called the origin, and the robot target and position are determined based on the measurements along the axes of the coordinate system. Multiple coordinate systems can be defined for an industrial robot, and each of such systems is suitable for certain type of jogging control or programming.

(1)World coordinate system: It is also known as geodetic coordinate system, which is a standard Cartesian coordinate system fixed in space. This coordinate system is fixed in a predetermined position, and the user coordinate system is set based on it. The world coordinate system has a zero point corresponding to a fixed position in the working unit or workstation. This helps with the processing of several robots or those robots moving along the outer axis. As shown in Fig. 7–2, A and C are robot base coordinate systems, and B is a world coordinate system. The world coordinate system is consistent with the base coordinate system by default.

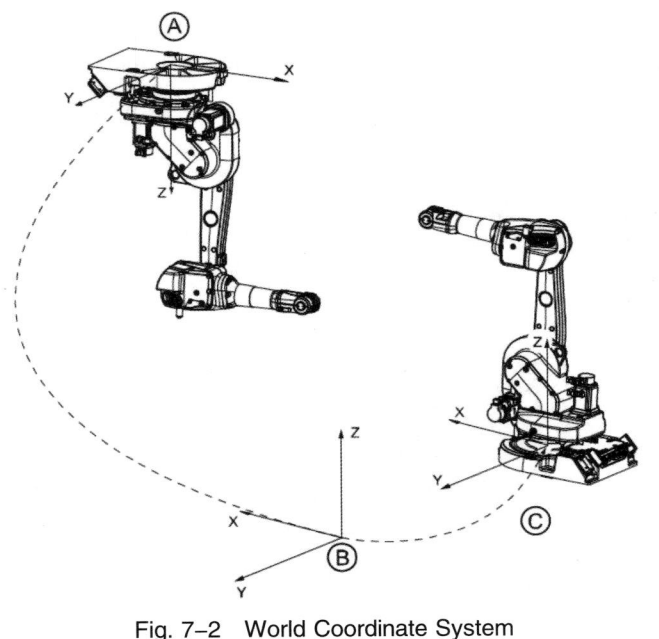

Fig. 7–2　World Coordinate System

Industrial Robot Technology Application

(2) Base coordinate system: It is also known as body coordinate system. As shown in Fig. 7-3, it is a Cartesian coordinate system established from the base point of the robot base. This coordinate system is used as the basis of other coordinate systems of the robot. The base coordinate system makes the movement of a fixed-mounted robot predictable, so it helps to move the robot from one position to another.

(3) Joint coordinate system: It is the coordinate system set in the robot joint and shows the absolute angle of each axis relative to its origin position. If the robot end is moved to the expected position, each joint can be driven by making operations in the joint coordinate system, thus guiding the robot end to the specified position.

(4) Tool coordinate system: It is used to determine the position and pose of a tool. This coordinate system consists of the tool center point (TCP) and the coordinate orientation, as shown in Fig. 7-4. The tool coordinate system must be set in advance and will be replaced by the default tool coordinate system if not defined. When executing a program, the robot moves its TCP to the programmed position. This means that to if the tool (and the tool coordinate system) is changed, the robot's movement will change accordingly, so that the new TCP gets to the target position. The default tool coordinate system of ABB robot is tool0, and the new tool coordinate system is defined as an offset value of tool0. The tool coordinate system is very useful if the operator does not want to change the orientation of the

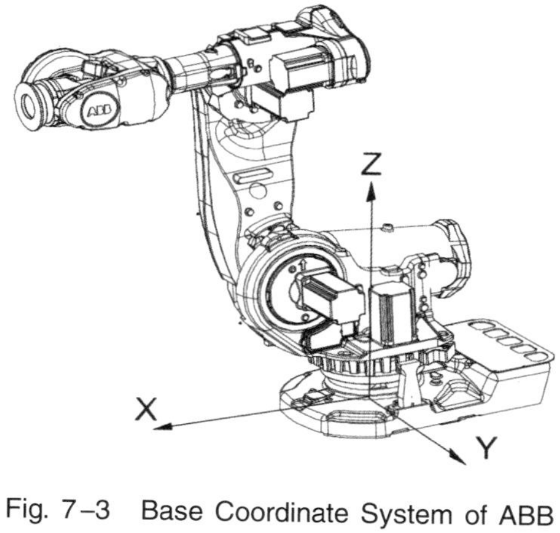

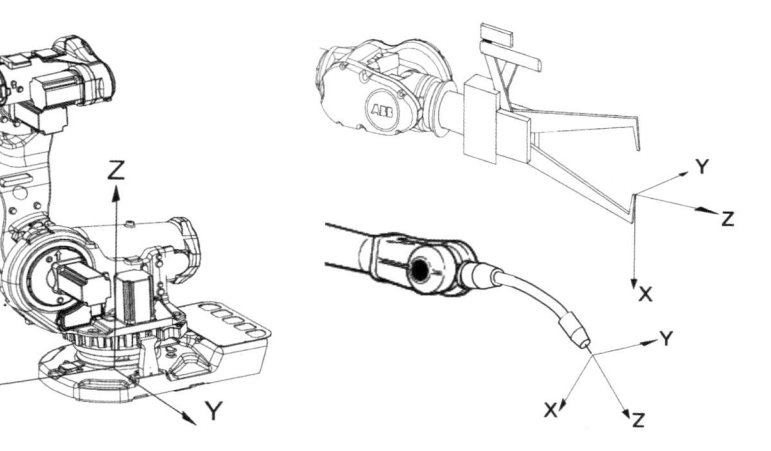

Fig. 7-3 Base Coordinate System of ABB Robot

Fig. 7-4 Tool Coordinate System

7. Programming Basics of ABB Robot

moving tool when performing jogging control of the robot (e.g., keeping translational motion when moving a welding torch).

(5)Work object coordinate system: It is used to determine the position and
pose of a work object and define the position of the work object relative to the
world coordinate system (or other coordinate systems). This coordinate system
consists of the origin of the work object and the coordinate orientation. The work
object coordinate system can be
determined by the three −point
method: the line connecting points
X1 and pX2 forms the X−axis, the
line passing through point Y1 and
perpendicular to the X−axis is the
Y−axis, and the direction of Z−axis
is determined by the right −hand
rule. As shown in Fig. 7−5, B and
C are work object coordinate systems. A robot can possess several
work object coordinate systems,
which either represent different
work objects or represent several

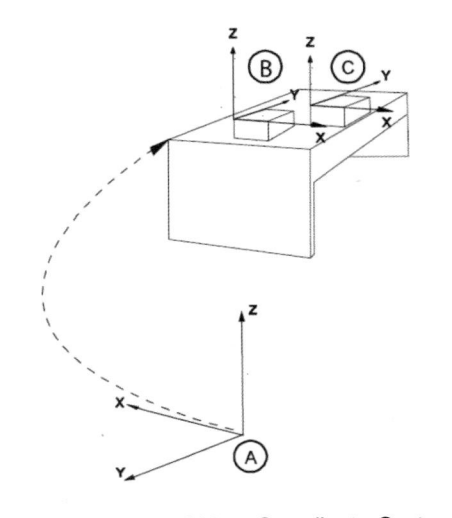

Fig. 7−5　Work Object Coordinate System

copies of the same work object at different positions. Programming a robot is to
create relevant objects and paths in the work object coordinate system. This brings
many advantages. One advantage is that when reorienting the work object in the
workstation, the only thing to do is changing the position of the work object coordinate system, and all paths will update accordingly.

(6)User coordinate system: It is a Cartesian coordinate system customized by
the user for each workspace. This coordinate system is used for the teaching and
execution of position registers, the execution of position compensation instructions,
etc. As shown in Fig. 7−6, A is the user coordinate system, B is the world coordinate system, C is the base coordinate system, D is the user coordinate system after
moving, and E is the work object coordinate system moving along with the user
coordinate system. If not defined, the user coordinate system will be replaced by
the world coordinate system.

117

Industrial Robot Technology Application

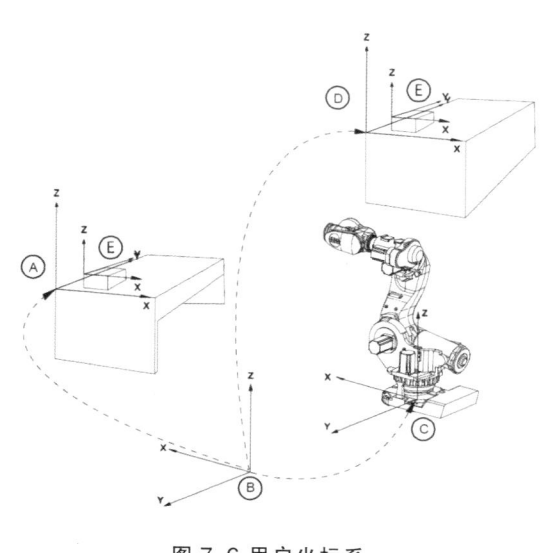

图 7-6 用户坐标系

7.2 Program structure of ABB robot

Inside an ABB robot, the program run by the robot is called RAPID, which comprises a series of instructions for controlling the robot. The robot can be controlled by executing these instructions. The RAPID program is divided into tasks, which are subdivided into several modules. These modules are the carriers of the robot program and data, and they are classified into system modules and task modules. Fig. 7-7 shows the schematic diagram for the program structure of ABB robot.

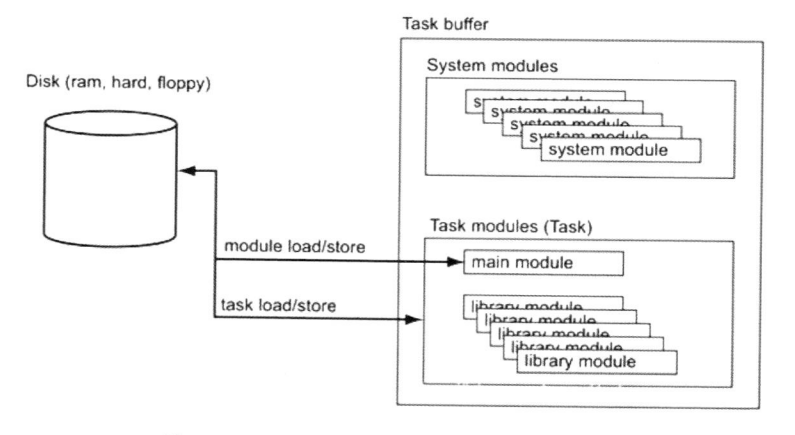

Fig. 7-7　Program Structure of ABB Robot

118

7. Programming Basics of ABB Robot

(1)Task: Usually, each task contains a RAPID program and a system module and aims at implementing a specific function (e.g., spot welding or manipulator movement). A RAPID application contains a task. Multiple tasks may be included if the Multitasking option is installed.

(2)Program: Usually, each program contains several program modules with RAPID codes indicating different functions. Executable main routines must be defined for all programs.

(3)Module(program module): Each program module contains function–specific program data and routine. By dividing a program divided into different modules, the structure of the program can be improved, and the processing of the program becomes easier. Each module represents a specific robot motion or a similar motion. If a program is deleted from the controller program memory, all program modules that are usually written by users will also be deleted.

(4)Routine: A routine contains some instruction sets. It defines the tasks actually executed by the robot system and also contains the data needed by these instructions.

(5)Main(main program/main routine): A special routine is sometimes called "main", which is defined as the start point for the execution of a program. Each program must contain a main routine named "main"; otherwise, the program will not be executable.

(6)Instruction: Instruction is a request for executing a specific event, e.g., "Run the controller TCP to the specific position MoveJ" or "Set a specific digital output Set".

The following concepts should be known for program editing and testing:

(1)Program pointer(PP): Program pointer refers to the instruction to start a program by pressing the "Start", "Step Forward" or "Step Backward" button on FlexPendant. The program will continue to be executed starting from the PP instruction. However, if the cursor moves to another instruction when the program stops, the program pointer may move to the cursor position (or the cursor may move to the program pointer position), and the program execution can also be restarted from this position. The program pointer is displayed as a yellow arrow to the left of the program code in the "Program Editor" and the "Production Window".

119

Industrial Robot Technology Application

(2)Motion pointer(MP): Motion pointer refers to the instruction currently being executed by the robot. Usually, the motion pointer lags one or more instructions behind the program pointer, because executing and computing the robot path are much faster than executing and computing the robot motion in the system. The motion pointer is displayed as a small robot to the left of the program code in the "Program Editor" and the "Production Window".

(3)Cursor: It can represent a complete instruction or a variable. It is highlighted in blue at the program code in the "Program Editor".

(4)Program editor: In case of switching between the "Program Editor" and other views and returning again, the same code part will be displayed on the "Program Editor" as long as the program pointer has not been moved. If the program pointer has been moved, a code will be displayed at the program pointer position on the "Program Editor".

7.3 Program data of ABB robot

It is necessary to make sure that a base coordinate system and a world coordinate system have been set during the installation of the robot system and define three types of program data(i.e., the tool coordinate system, the work object coordinate system and the payload)as required before starting programming. Program data refers to the data declared in the program and quoted by the instruction(s)in the same module or other modules. As shown in Fig. 7-8, MoveJ invokes 4 types of program data:

p01: target movement position of the robot(robtarget);

v1000: moving speed data of the robot(speed data);

z50: moving turning data of the robot(zonedata);

tool1: tool data TCP of the robot(tooldata).

There are more than 100 data types of the ABB robot controller. Common data types include basic data, I/O data and motion-related data, as given in Table 7-1.

7. Programming Basics of ABB Robot

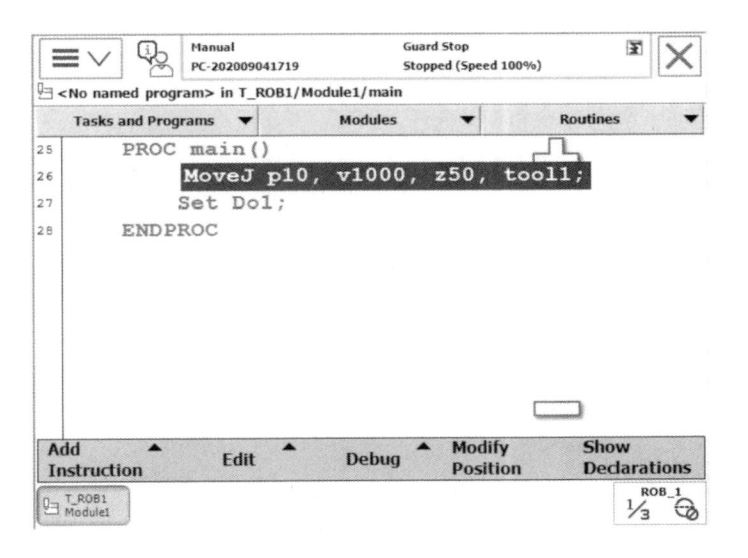

Fig. 7–8 Program Instruction Example

Table 7–1 Common Data Types of ABB Robot Controller

Classifi-cation	Description	Data Type	Instructions for Use
Basic data	bool	Logic value	True or False assigned in the logic state. There are two logic values: True or 1 if the situation is true; False or 0 if the situation is not true
	byte	Byte value	A unit of measurement used to measure the storage capacity, with the value range of (0–255)
	num	Numerical value	A variable, storable integer or decimal, with the integer value range of (−8388607~8388608)
	dnum	Dual numerical value	A storable integer or decimal, with the integer value range of (−4503599627370495~+4503599627370496)
	string	Character string	A string of characters consisting of numbers, letters, and underscores and used to represent the data type of the text in the programming language
	stringdig	String containing digits only	A string able to process positive integers not greater than 4294967295

121

Industrial Robot Technology Application

Table 7-1(续)

Classification	Description	Data Type	Instructions for Use
I/O data	dionum	Numerical value	With the value of 0 or 1 for processing digital I/O signals: 0 for low level 0~0.7V and 1 for high level 3.4~5.0V
	signaldi/do	Digital input/output	Binary value for input and output: 1 for connected switch and 0 for disconnected switch
	signalgi/go	Digital input/output signal group	Combination of multiple digital inputs or outputs
	signalai	Analog input	For example, if a temperature value is collected by a temperature sampler, it will be converted into a binary number that can be recognized by PLC through a transmitter
	signalao	Analog output	Data-transmitter-actuator
Motion data	robtarget	Position data	For defining positions of the manipulator and additional axes
	jointtarget	Joint data	For defining each individual axis position for the manipulator and the outer axis
	robjoint	Joint data	For defining the position of each joint of the manipulator
	speeddate	Speed data	For defining the movement rate of the manipulator and axes, including four parameters in total
	zonedata	Zone data	Generally referred to as the turning radius and used to define how the robot axis approaches the programmed position before moving towards the next target movement position
	tooldata	Tool data	For defining the characteristics of the tool, including the position and orientation of the tool center point (TCP) and the load on the tool
	wobjdata	Work object data	For defining the position and state of a work object
	loaddata	Load data	For defining the load on the manipulator mounting interface

7. Programming Basics of ABB Robot

7.3.1 Program data creation method

In general, there are two ways to create program data. One is to create program data directly in the program data screen of the teach pendant, and the other is to allow the automatic generation of corresponding program data in the system during the creation of program instructions. Usually, teach programming is done by using a FlexPendant, which is a teach pendant and the most suitable for modifying various data in the program. Take the data type of bool as an example. The method to create program data in the program data screen of the teach pendant is as follows:

(1)Tap the ABB menu button in the upper left corner of the teach pendant screen and select the "Program Data" option, as shown in Fig. 7–9.

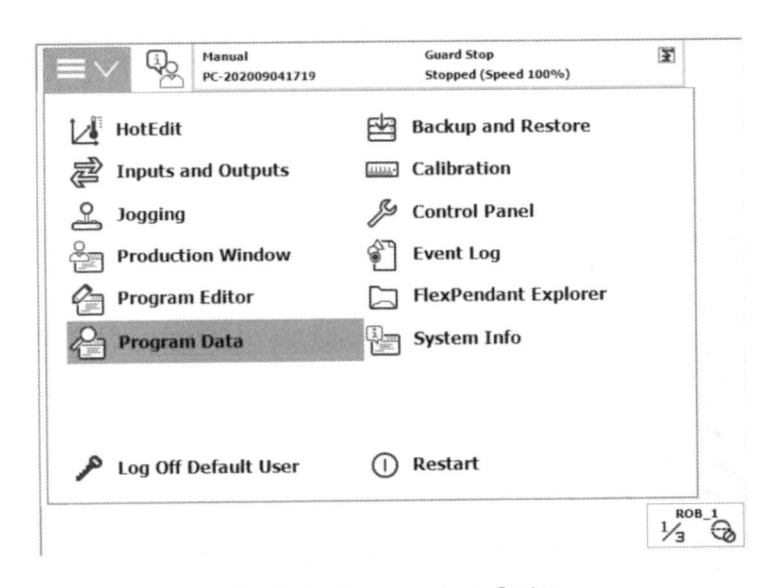

Fig. 7–9   Program Data Option

(2)Enter the data type display interface as shown in Fig. 7–10, select the data type "bool", and double tap or tap the "Display data" button at the bottom.(Note: If the data type "bool" cannot be found in the current interface, the reason is that no data corresponding to the defined data type "tool" exists in the current task domain. In this case, tap the "View" button in the lower right corner to switch to the "All data types" option.)

123

Industrial Robot Technology Application

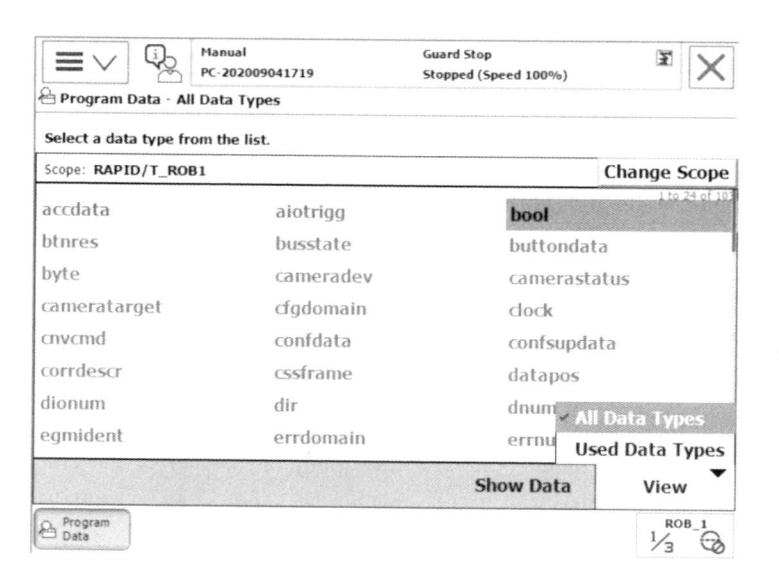

Fig. 7-10　Data Type Display Interface

　　(3)Enter the bool data list interface as shown in Fig. 7-11, and tap "New" button to define new data.

　　(4)Set the variable name, application scope, task and module in turn, then

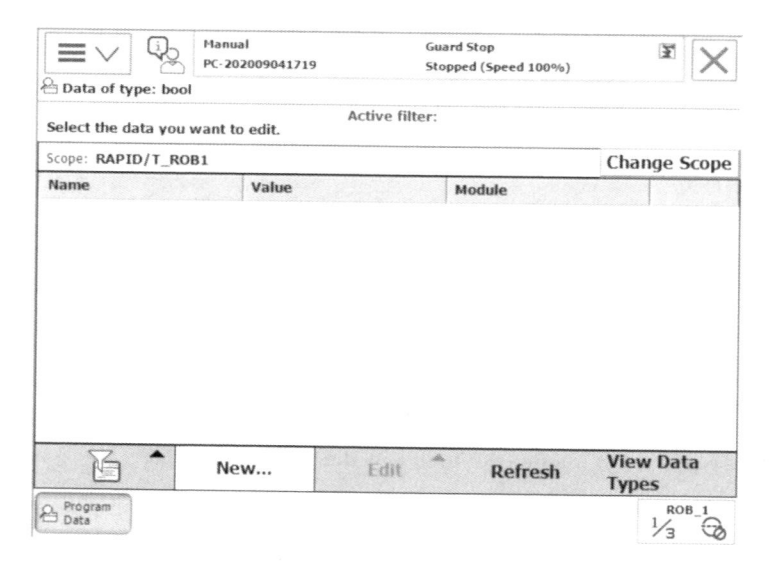

Fig. 7-11　Create New "bool" Data

7. Programming Basics of ABB Robot

tap "OK" to complete the definition of new data, as shown in Fig. 7–12. There are three options for "Scope": "Global" may be used for any module; "Local" is used for a specific module; "Task" is used for a specific task of the robot, or a default task of a single robot. For "Task" and "Module" options, the application scope of the data can be set for easy data management.

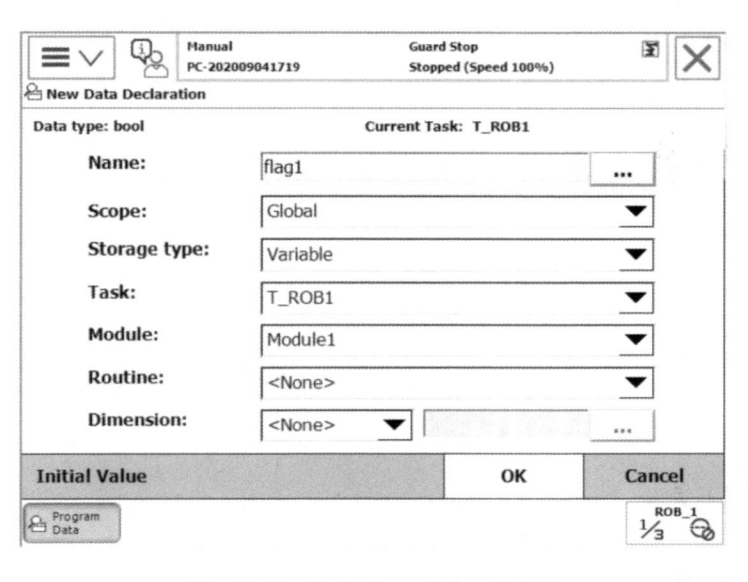

Fig. 7–12   Definition of "bool" Data

(5)There are three options for "Storage Type": VAR, PERS and CONST. The main differences among these options are as follows:

VAR: The current value will be maintained when the program is being executed or stops. However, if the program pointer is reset or the robot controller is restarted, the value will restore to the initial value assigned when the variable is declared.(Example: VAR num Distance: = 0;)

PERS: PERS data will remain the last value assigned, regardless of the change of program pointer or the restart of robot controller.(Example: PERS num Height: = 0;)

CONST: CONST is characterized in that its value is assigned during definition and cannot be modified in the program, but can be manually modified only. (Example: CONST num Pi: = 3.14159;)

125

Industrial Robot Technology Application

(6)By default, the initial values of all new bool variables are false, which may be modified by tapping the "Initial Value" button in the lower left corner, as shown in Fig. 7-13.

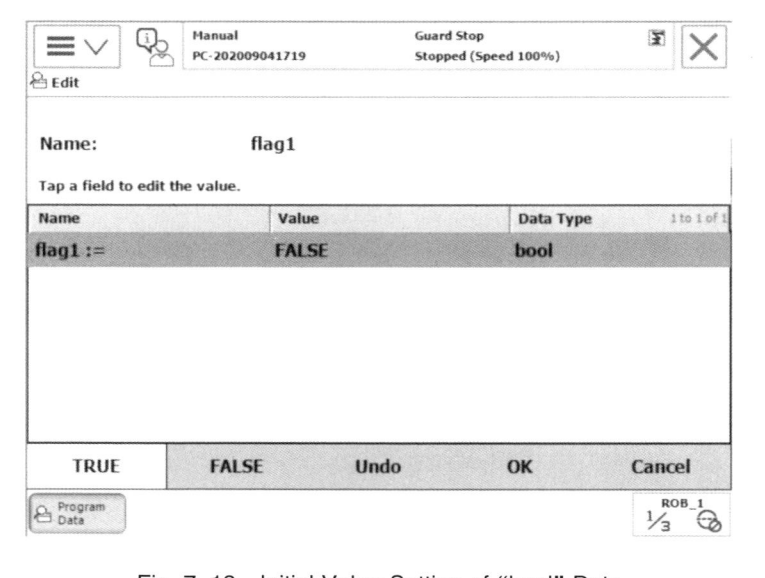

Fig. 7-13　Initial Value Setting of "bool" Data

7.3.2 Setting of tool data(TOOLDATA)

The concepts of robot tool coordinate system and TCP have been described in Section 7.1 above. Tooldata is the concept describing TCP, pose, mass, center of gravity and other parameters of the tool installed on the sixth axis of the robot. For a default ABB robot tool(tool0), its tool center point(TCP)is at the center of the robot mounting flange and its X, Y and Z directions are shown in Fig. 7-14. The Z-axis is perpendicular to the flange mounting surface, the X-axis is opposite to the Z -axis of the base coordinate system, and the coordinate origin, i.e. the TCP, is at the center of the flange.

Tooldata affects the robot control algorithms (e.g., calculated acceleration speed, acceleration monitoring, torque monitoring, collision monitoring, energy monitoring, etc.), so the robot tool data should be set correctly. A new tool has initial default values for mass, frame, orientation, etc. These values must be defined before the tool is used. The positions of all newly defined tool coordinate

126

7. Programming Basics of ABB Robot

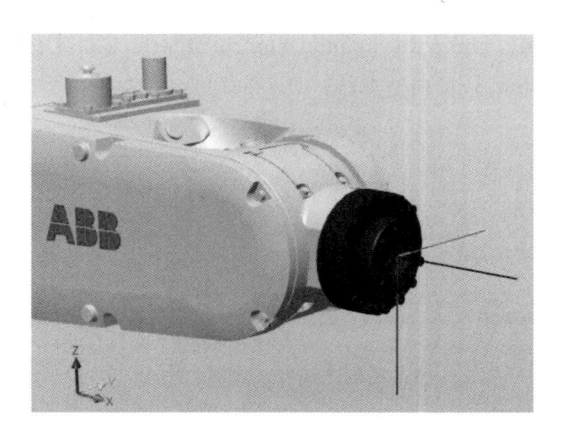

Fig. 7-14　Tool Coordinate System of "tool0"

systems can be understood as offset and rotation on the basis of tool0. The setting steps are detailed as follows:

(1)Tap the ABB menu button in the upper left corner of the teach pendant screen, select the "Program Data" option, then select the "tooldata" data type option and tap "Display Data", or directly double tap the "tooldata" data type option, as shown in Fig. 7-15;

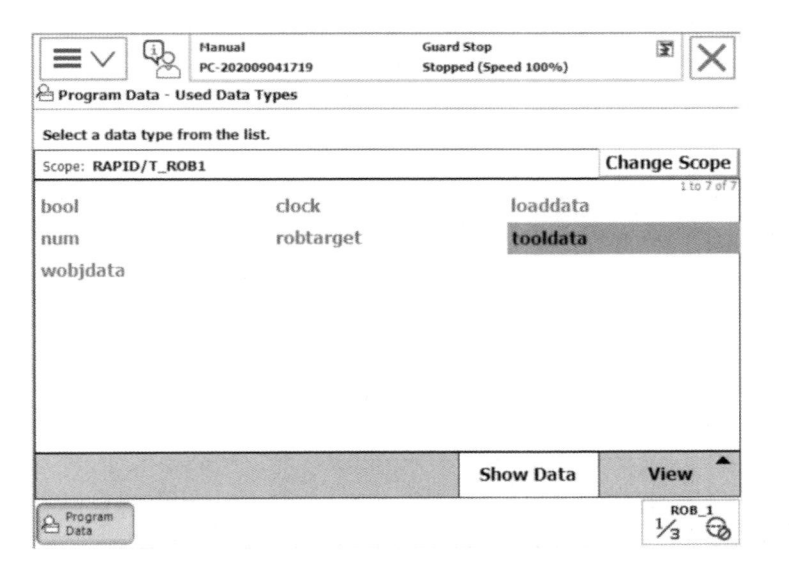

Fig. 7-15　"tooldata" Data Type Option

127

Industrial Robot Technology Application

(2)Alternatively, enter the manual control interface and tap the "Tool Coordinate" option, as shown in Fig. 7–16.

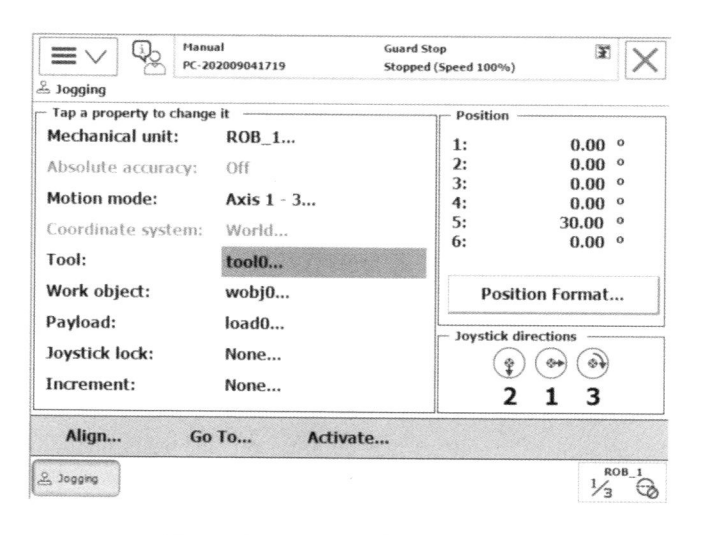

Fig. 7–16   Tool Coordinate Option

(3)Enter the tooldata display interface as shown in Fig. 7–17, and tap the "New" button to create data, or tap the "Edit" menu to delete tool coordinates, change the declaration, change values, copy or define tool coordinates.

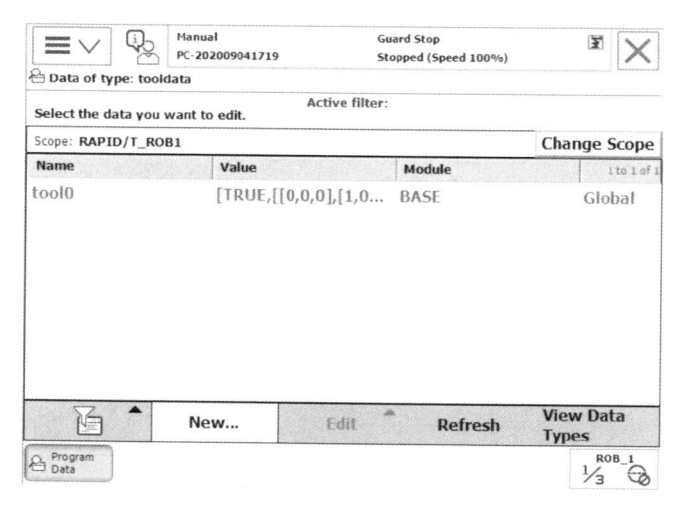

Fig. 7–17   "tooldata" Display Interface

128

7. Programming Basics of ABB Robot

(4)Enter the tooldata definition interface as shown in Fig. 7-18. The system will automatically name the tool as "tool + number" in sequence, and for easy use, the tool should be renamed as per the specific tool (e.g., welding torch, fixture). The scope of tool should always remain "Global" for application to all modules in the program. The tool data must always be a persistent variable. Finally, tap "OK" to complete the new creation operation.

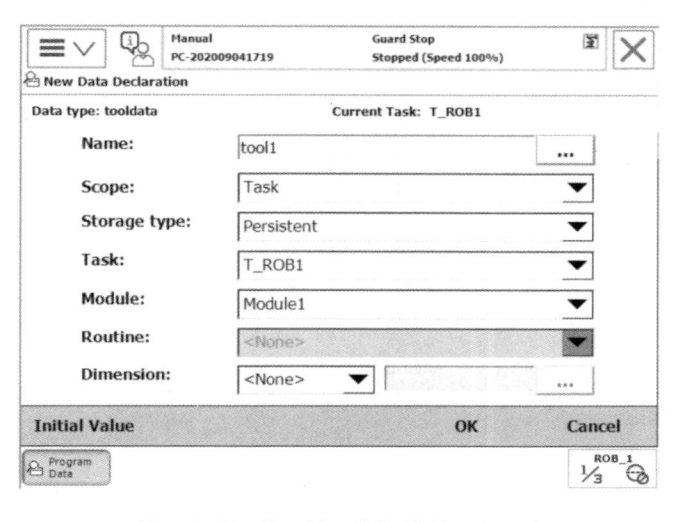

Fig. 7-18 "tooldata" Definition Interface

(5)All parameters of the new coordinate system are consistent with those of tool0 by default. If the data (i.e., parameters)of the tool coordinate system are known, tap the "Initial Value" button to enter the tooldata parameter setting interface to set all parameters, as shown in Fig. 7-19.

Table 7-2 shows the detailed parameters of "tooldata" data and specifies the meaning and value of each parameter.(Note: In graphics, the most commonly used spatial pose representation methods are Quaternions and Euler angles. Euler angles are three angles, which are intuitive and easy to understand, but with a locking problem. Quaternions are free of locking problem for the universal joint, with small storage space and high computation efficiency, but their representation is not intuitive enough. The reciprocal conversion of Euler angles and quaternions can be done with the aid of the special software or website: https://quaternions. online/)

129

Industrial Robot Technology Application

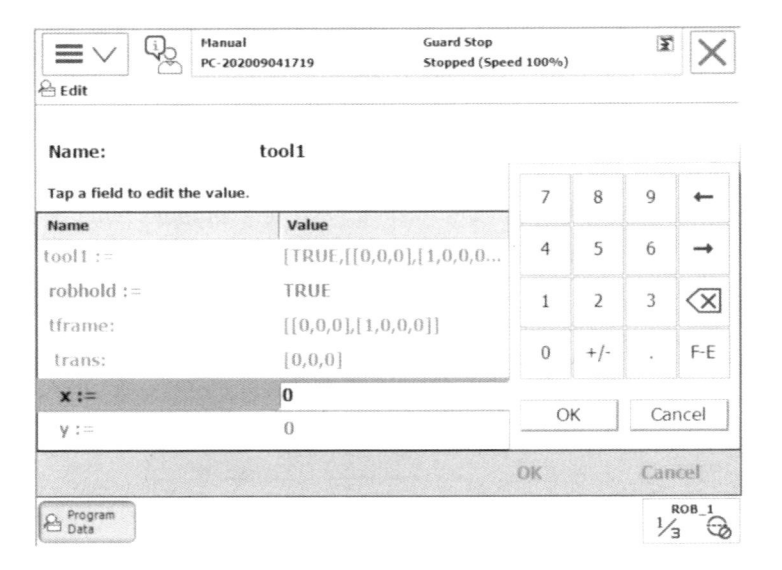

Fig. 7–19　"tooldata" Parameter Setting Interface

Table 7–2　"tooldata" Data Parameters

Parameter	Parameter
robhold	Full name: robot hold
	Data type: bool
	For defining whether the manipulator holds the tool:
	• TRUE: The manipulator is holding the tool.
	• FALSE: The manipulator does not hold the tool, i.e. the tool is not fixed.
tframe	Full name: tool frame
	Data type: pose
	For defining the position and pose of the tool coordinate system:
	• trans[x y z]: position of TCP relative to tool0, in mm.
	• rot [q1 q2 q3 q4]: quaternions representing the poses of the tool coordinate system relative to tool0.
tload	Full name: tool load
	Data type: loaddata
	For defining the load of the tool held by the manipulator, i.e.:
	• mass: mass of the tool, in kg.
	• cog[x y z]: center of gravity (x, y, z) of the tool load relative to tool0, in mm.
	• aom [q1 q2 q3 q4]: orientation of the principal inertial axis of tool torque relative to tool0.
	• [ix iy iz]: moment of inertia around the moment inertia axis, in $kg \cdot m^2$. If all inertial components are defined as 0, the tool should be deemed as a point mass.

7. Programming Basics of ABB Robot

(6)Return to the tooldata display interface, select "tool1", and tap the "Definition" option in the "Edit" menu to set the TCP of the tool coordinate system using the teach calibration method, as shown in Fig. 7–20.

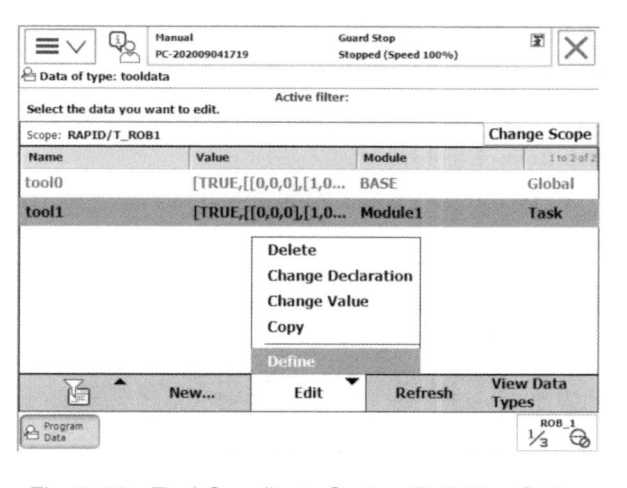

Fig. 7–20   Tool Coordinate System Definition Option

(7)Enter the tool coordinate definition interface as shown in Fig. 7–21, in which three different methods are available. TCP –based Cartesian coordinates should be defined for all these methods, and different methods correspond to dif-

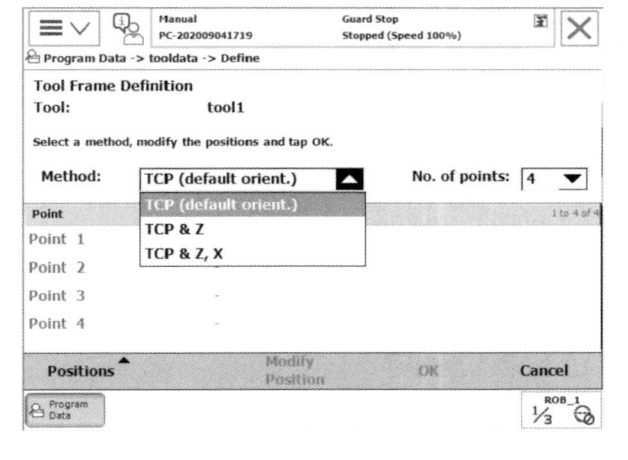

Fig. 7–21   Tool Coordinate Definition Interface

131

Industrial Robot Technology Application

ferent tool pose definition methods. The TCP collides with a fixed point by N different poses to get multiple sets of solutions, thus the corresponding positions of the current TCP and the default tool coordinate system tool0 can be calculated. The three methods are described below:

TCP(with default orientation): to be used when the orientation of the tool coordinate system is consistent with that of tool0.

TCP & Z: to be used when the Z-axis direction of the tool coordinate system is inconsistent with that of tool0.

TCP & Z, X: to be used when Z-axis and X-axis directions need to be changed for the orientation of the tool coordinate system.

(8)Usually, 4 approach points are sufficient to define the TCP. However, if more approach points are selected to get a more accurate result, the same care should be taken in defining each approach point. If the four-point method is selected to define the TCP, the differences among the four poses should be maximized to improve the accuracy of fitting, as shown in Fig. 7-22.

(9)The specific operation method is:

Move the robot to an appropriate position A to get the first approach point. Manually control the teach pendant to move the

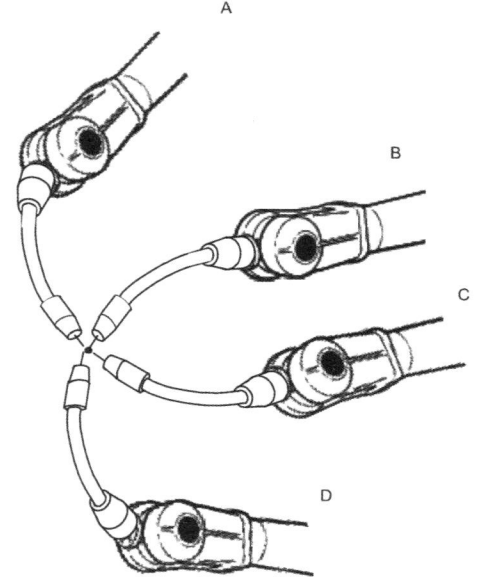

Fig. 7-22   4-Point Method for Defining TCP

robot in small increments, in order to bring the tool apex as close to the reference point as possible. Tap "Modify Position" as shown in Fig. 7-23 to complete the setting of the first approach point.

Repeat the above steps to define other approach points and get positions B, C and D. Note that the robot should be moved away from fixed world coordinate

7. Programming Basics of ABB Robot

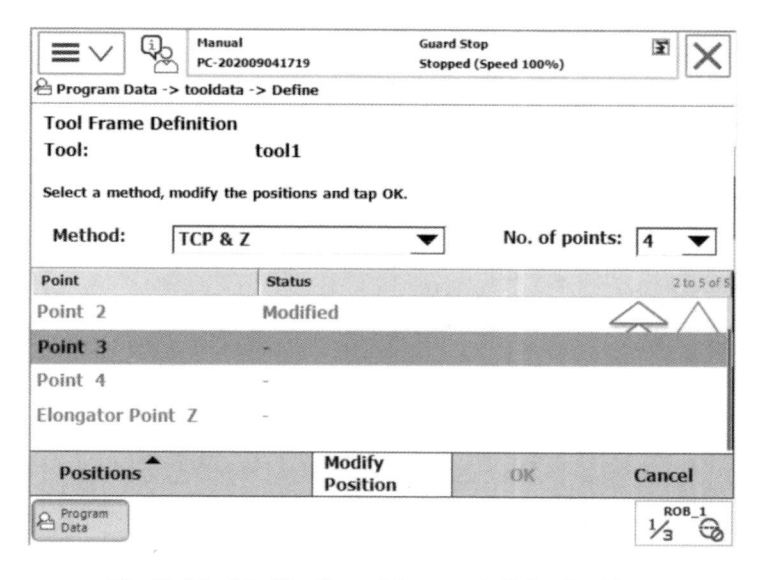

Fig. 7–23 Modification of Approach Point Position

points to obtain the best results. No good results can be obtained by modifying the tool orientation alone.

The orientation must also be defined if the "TCP & Z" or "TCP & Z, X" method is used. Defining an extension point is to move the robot without changing the tool orientation to make a reference world coordinate point be a point on the positive axis of the desired rotating tool coordinate system.

If it is necessary to reset all approach points for some reason, tap the "Position" button and select the "Reset All" option for recalibration.

After defining all points, tap "OK" to complete the definition of the tool co-ordinate system.

(10)Upon calibration completion of tool1, the system will pop up the error information as shown in Fig. 7–24, and the operator needs to confirm the error. Generally, the smaller the error, the better the result. However, it is difficult to get a high accuracy by manual operation, so the error should be evaluated based on the actual application and the operator's experience.

(11)Finally, set the mass and the center of gravity data of the tool(relative to tool0)based on the actual situation to complete the setting of tooldata.

133

Industrial Robot Technology Application

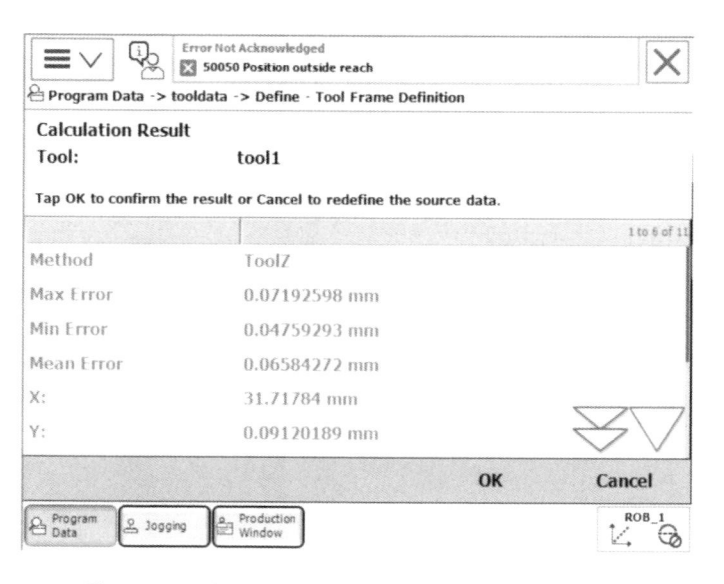

Fig. 7-24　Setting Error of Tool Coordinate System

7.3.3 Setting of work object coordinate data(WOBJDATA)

The concept of the work object coordinate system has been described in Section 7.1 above. If the work object coordinate system is used for programming, when reorienting the work object, the only thing to do is changing the definition of the work object coordinate system, and all paths will update accordingly. As shown in Fig. 7-25, if another work object C on the workbench needs a path the same as that of work object A, it is only necessary to set the work object coordinate D, copy the program for the work object coordinate B and change the work

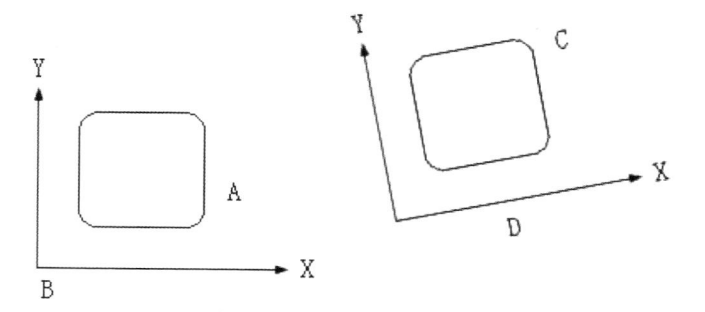

Fig. 7-25　Relationship between Program Trajectory and Work Object Coordinate

7. Programming Basics of ABB Robot

object coordinate of the new program from B to D. In this way, a new path can be obtained without repeating the programming.

The work object coordinate data(WOBJDATA)can be defined as follows:

(1)Tap the ABB menu button in the upper left corner of the teach pendant screen, select the "Program Data" option, then select the "wobjdata" data type option and tap "Display Data", or directly double tap the "wobjdata" data type option, as shown in Fig. 7-26;

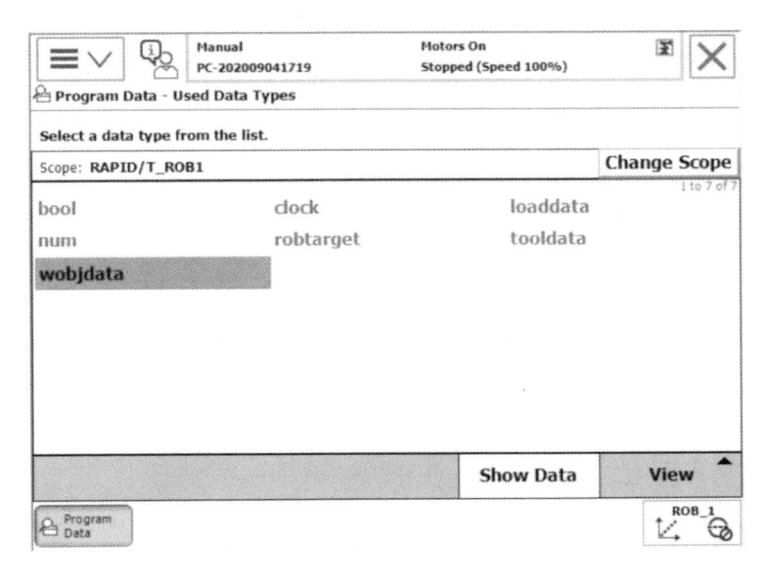

Fig. 7-26 "wobjdata" Data Type Option

(2)Alternatively, tap the "Work Object Coordinate" option in the "Manual Operation" interface to enter the work object coordinate system display interface as shown in Fig. 7-27. The default work object coordinate system(wobj0)is the same as the world coordinate system. The user needs to tap the "New" button to define the work object coordinate system.

(3)Enter the wobjdata definition interface as shown in Fig. 7-28. The system will automatically name the work object as "wobj + number" in sequence. To be more clear, the work object can be renamed(once a work object coordinate system is created, it should not be renamed at will; otherwise, all programs using the work object coordinate system have to be modified). In all modules, the work ob-

135

Industrial Robot Technology Application

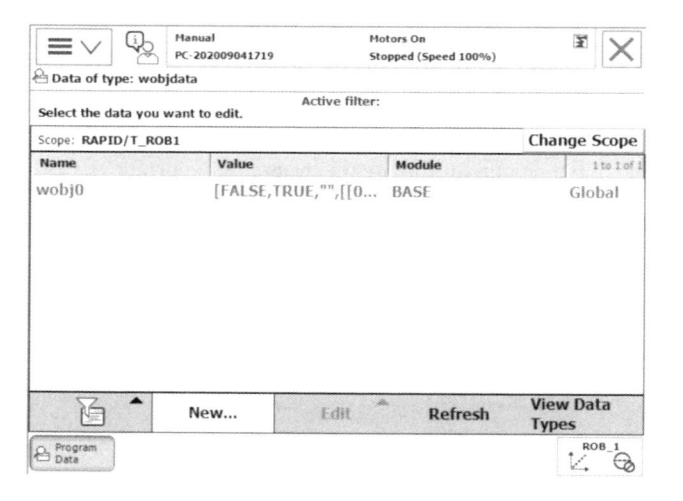

Fig. 7-27 "wobjdata" Data Display Interface

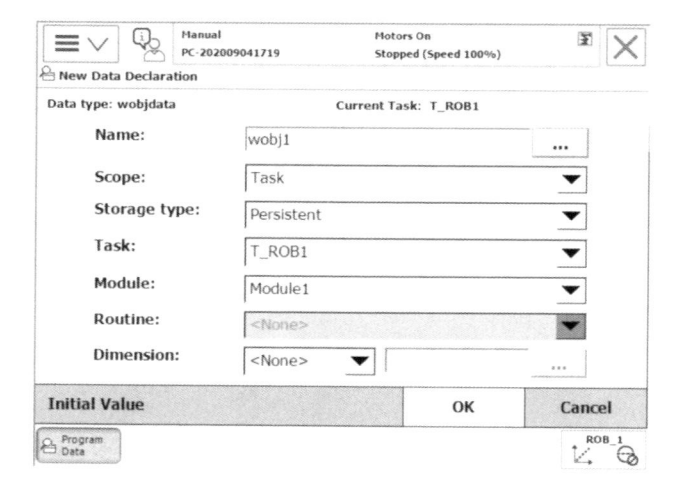

Fig. 7-28 "wobjdata" Definition Interface

ject should be a global variable in the program and the work object data must always be a persistent variable. Finally, tap "OK" to complete the new creation operation.

(4)All parameters of the new coordinate system are consistent with those of wobj0 by default. If the data (i.e., parameters)of the tool coordinate system are known, tap the "Initial Value" button to enter the wobjdata parameter setting interface to set all parameters, as shown in Fig. 7-29.

136

7. Programming Basics of ABB Robot

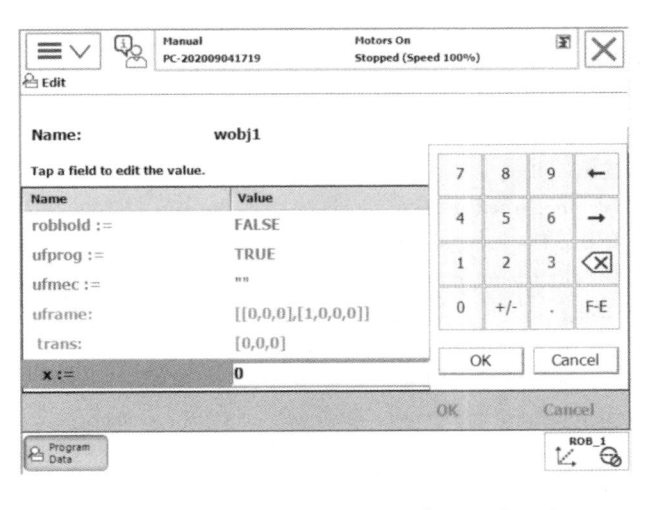

Fig. 7-29 "wobjdata" Parameter Setting Interface

Table 7-3 shows the detailed parameters of "wobjdata" data and specifies the meaning and value of each parameter.

(5)Alternatively, return to the wobjdata display interface, select "wobj1", and tap the "Definition" option in the "Edit" menu to set the work object coordinate system using the teach calibration method, as shown in Fig. 7-30.

(6)Enter the work object coordinate definition interface as shown in Fig. 7-31, select "3-point" in the drop-down menu of "User Method", so that the system will display such options as user points X1, X2 and Y1. .

(7)In the work object plane, only three points need to be defined to determine the origin and the positive directions of X-axis and Y-axis, so that a work object coordinate system can be set according to the right-hand rule, as shown in Fig. 7-32. Where, point X1 is used for determining the origin, points X1 and X2 for determining the positive direction of X-axis, and point Y1 for determining the positive direction of Y-axis of the work object coordinate system.

(8)Manually move the robot's tool reference point close to positions X1, X2 and Y1 on the work object in turn, and tap "Modify Position" to record the corresponding coordinates respectively. After that, tap "OK" to complete the creation of the work object data(wobjdata).

137

Industrial Robot Technology Application

表 7-3 wobjdata 数据参数表

Parameter	Description
robhold	Full name: robot hold Data type: bool For defining whether the manipulator holds the work object: • TRUE: The manipulator is holding the work object, i.e., a fixing tool is used. • FALSE: The robot does not hold the work object, i.e., the manipulator holds the tool.
ufprog	Full name: user frame programmed Data type: bool For defining whether to use a fixed user coordinate system: • TRUE: fixed user coordinate system. • FALSE: movable user coordinate system, i.e., using a coordinated outer axis. It is also used for the semi−coordination or synchronous coordination mode of the MultiMove system.
ufmec	Full name: user frame mechanical unit Data type: string For defining the mechanical unit that moves in coordination with the manipulator. It is used to specify the name of a mechanical unit (such as orbit_a) defined in system parameters only when (ufprog is FALSE) is assigned in a movable user coordinate system.
uframe	Full name: user frame Data type: pose For defining the user coordinate system, i.e., the current working face or the fixture position: • trans[x y z]: position of origin of the user coordinate system, in mm. • rot [q1 q2 q3 q4]: quaternions representing the rotation of the user coordinate system. Note: If the manipulator is holding the tool, the user coordinate system will be defined in the world coordinate system (if a fixing tool is used, the user coordinate system will be defined in the wrist coordinate system). For a movable user coordinate system (ufprog is FALSE), the user coordinate system is continuously defined by the system.
oframe	Full name: object frame Data type: pose For defining the target coordinate system (i.e., the position of the current work object) in the user coordinate system: • trans[x y z]: position of origin of the target coordinate system, in mm. • rot[q1 q2 q3 q4]: quaternions representing the rotation of the target coordinate system.

7. Programming Basics of ABB Robot

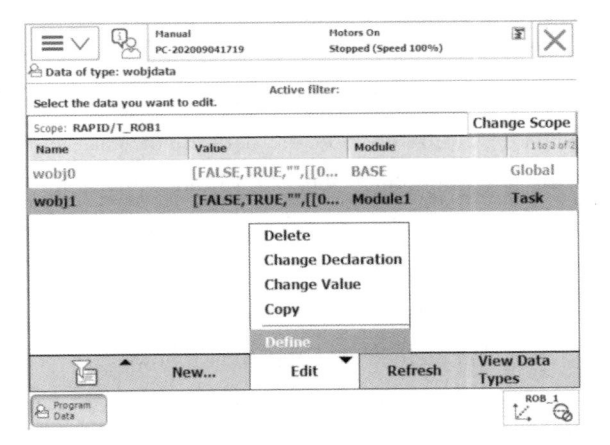

Fig. 7-30 "wobjdata" Data Definition Option

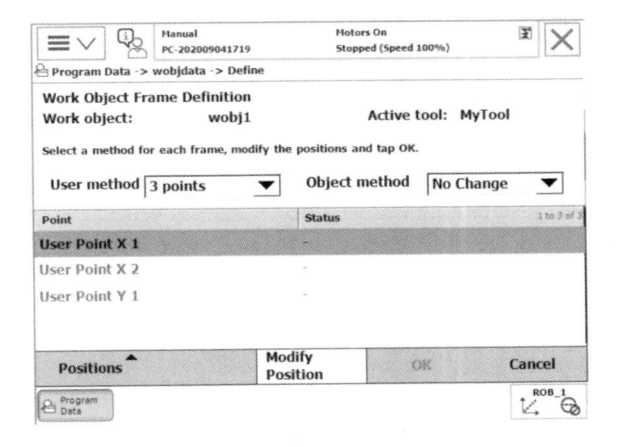

Fig. 7-31 "wobjdata" Parameter Setting Interface

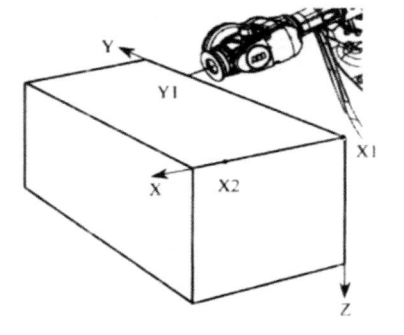

Fig. 7-32 Setting Work Object Coordinate System by Three-point Method

139

Industrial Robot Technology Application

7.3.4 Setting of payload(loaddata)

For an industrial robot used for handling, the data about mass and center of gravity (i.e., tooldata)of the fixture and the data about mass and center of gravity (i.e., loaddata)of the handling object must be set correctly, and the loaddata is set based on tool0. Setting the loaddata to define the payload of the robot or the grip load (which is set by the instruction GripLoad or MechUnitLoad), i.e., the load gripped by the robot fixture. The loaddata is also a part of the tooldata to describe the tool load. The operation method is detailed as follows:

(1)Tap the ABB menu button in the upper left corner of the teach pendant screen, select the "Program Data" option, then select the "loaddata" data type option and tap "Display Data", or directly double tap the "loaddata" data type option, as shown in Fig. 7-33.

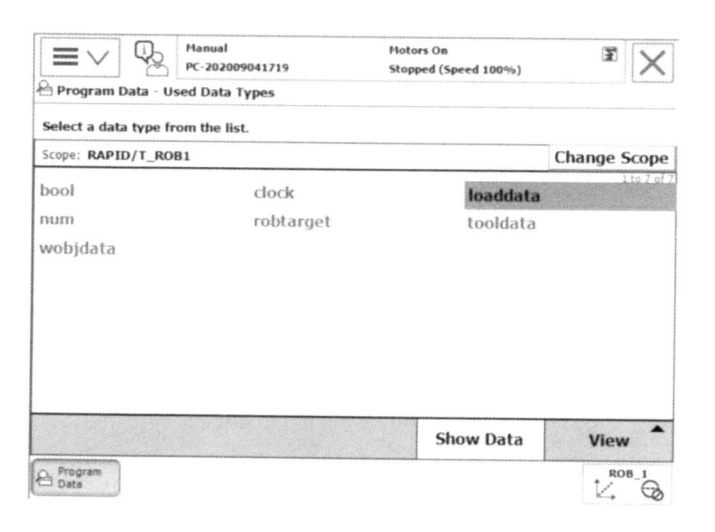

Fig. 7-33 "loaddata" Data Type Option

(2)Alternatively, tap the "Payload" option in the "Manual Control" interface to enter the payload display interface as shown in Fig. 7-34, and tap the "New" button.

(3)Enter the loaddata definition interface as shown in Fig. 7-35. The system will automatically name the load as "load + number" in sequence. To be more clear, the load can be renamed (once the loaddata is created, it should not be renamed

140

7. Programming Basics of ABB Robot

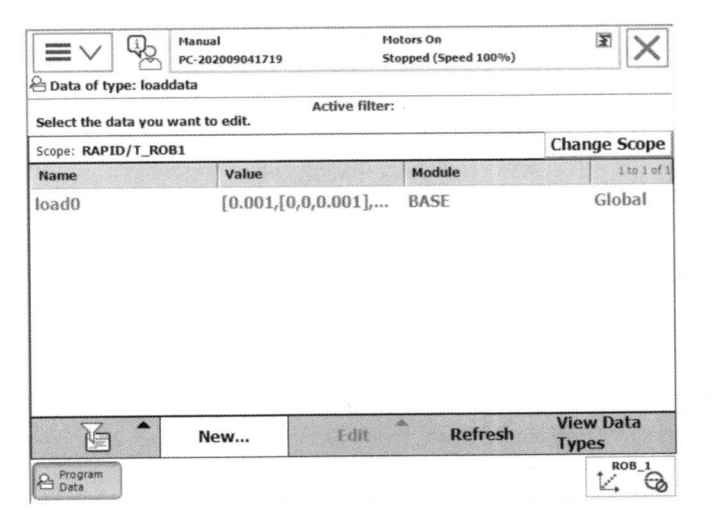

Fig. 7–34 "loaddata" Data Display Interface

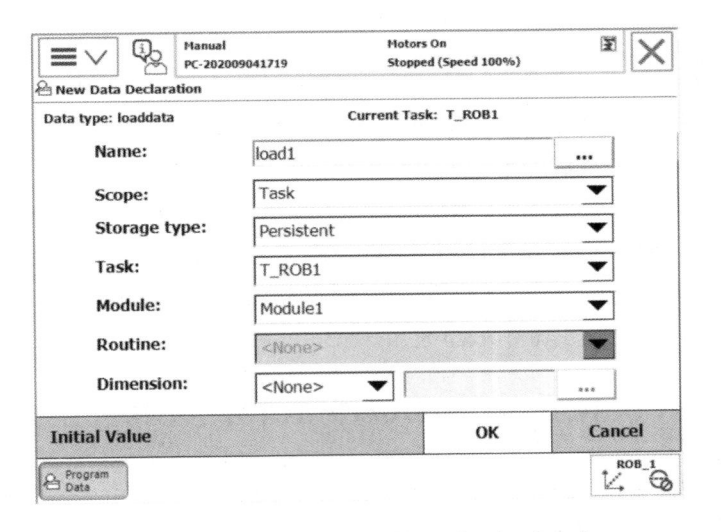

Fig. 7–35 "loaddata" Parameter Setting Interface

at will; otherwise, all programs using the payload have to be modified). In all modules, the payload should be a global variable in the program and the payload variable must always be a persistent variable.

(4)Tap the "Initial Value" button to enter the loaddata parameter setting interface as shown in Fig. 7–36. In the figure, load1 is defined as a load with the mass of 5 kg and the center of gravity at coordinate [50,0,40], and the payload is

141

Industrial Robot Technology Application

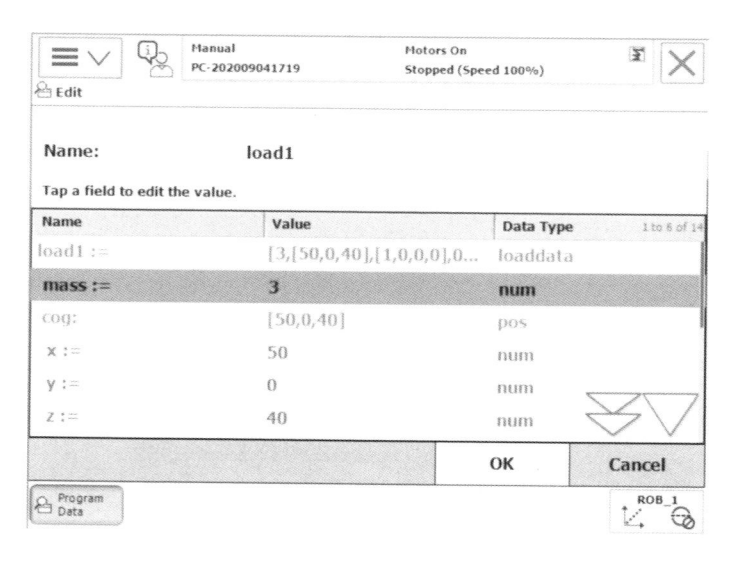

Fig. 7–36 "loaddata" Parameter Setting Interface

a point mass.

Table 7–4 shows the detailed parameters of "loaddata" data and specifies the meaning and value of each parameter.

(5)During programming, the payload needs to be adjusted in real time according to load changes, as shown in Fig. 7–37. Payload (load1)will be applied when the fixture grips the load, and the payload will be switched to default payload(load0)when the fixture releases the load.

7.4 Commonly used instructions for ABB robot

In ABB robot programs, all robot actions are described and controlled by RAPID language or RAPID instructions. Different operations correspond to different instructions, which further require varied types of instruction syntax and usage. Numerous instructions are applied to ABB robots. Some commonly used instructions are mainly introduced herein.

7.4.1 Basic motion instructions

The basic motion instructions of ABB industrial robot are divided into 5 parts: motion mode, target position, moving speed, turning radius and TCP. The turning radius of the robot is shown in Fig. 7–38 below, and the instructions cor-

7. Programming Basics of ABB Robot

Table 7-4 "loaddata" Data Parameters

Component	Description
mass	Data type: num Mass of load, in kg.
cog	Full name: center of gravity Data type: pos For defining the position of the center of gravity of the load, in mm. If the manipulator is holding the tool, the center of gravity of the payload will be relative to the tool coordinate system; if a fixing tool is used, the center of gravity of the payload will be relative to the movable work object coordinate system on the manipulator.
aom	Full name: axes of moment Data type: orient For defining the orientation of the axis of moment, where the axis refers to the principal axis for the inertia moment of the payload in the "cog" position. If the manipulator is holding the tool, the orientation will be relative to the tool coordinate system; if a fixing tool is used, the orientation will be relative to the movable work object coordinate system.
ix	Full name: inertia x Data type: num For defining the moment of inertia of the load around the X-axis, in $kg \cdot m^2$. Correct definition of the moment of inertia helps to reasonably use the path planner and the axis controller. This parameter is particularly important during the handling of large metal plates, etc. Any moment of inertia equal to 0, e.g., ix, iy or iz, refers to a point mass.
iy	Full name: inertia y Data type: num For defining the moment of inertia of the load around the Y-axis
iz	Full name: inertia z Data type: num For defining the moment of inertia of the load around the Z-axis

responding to the path shown in the figure are as follows:

MoveL p10,v1000,z50,tool0; ! The robot TCP reaches 50 mm away from the target point p10 and then bypasses this point.

MoveL p20,v1000,fine,tool0; ! The movement speed reduces to zero when the robot TCP reaches the target point p20.

Industrial Robot Technology Application

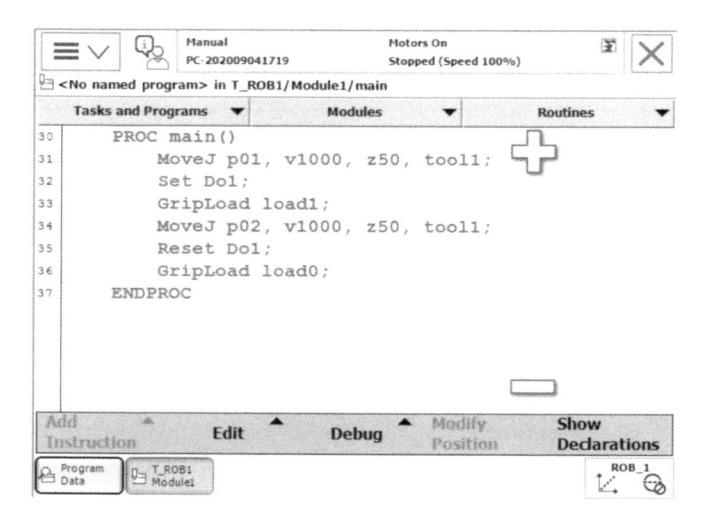

Fig. 7-37   Application Example of Payload

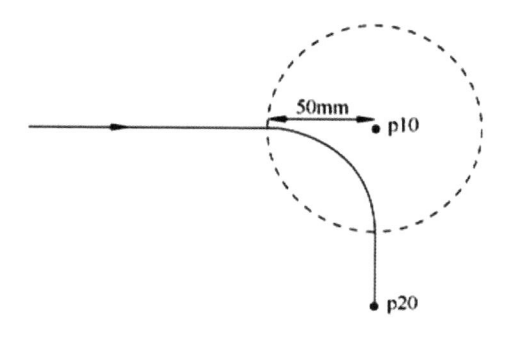

Fig. 7-38   Turning Radius of Robot

In the program instruction, "fine" means that the movement speed reduces to zero when the robot TCP reaches the target point, where the robot stops for a while and then continues to move. The zone data "z50" means that the robot TCP does not reach the target point, but smoothly bypasses this point at the set length (of 50 mm)away from it. Generally, the last point of a path is set as the parameter "fine" in the turning zone.

Among basic movements of the robot, most involve linear or circular motion tracks, which are also combined to form relatively complex motion tracks. Commonly used instructions include the following:

(1)MoveJ(joint motion instruction): used to direct the manipulator to quickly move from one point to another. According to this instruction, both the manipulator and the outer axis move to the target position along a nonlinear path, as shown in Fig. 7-39. All axes reach the target position at the same time, thus the accuracy of the motion track is not high.

144

7. Programming Basics of ABB Robot

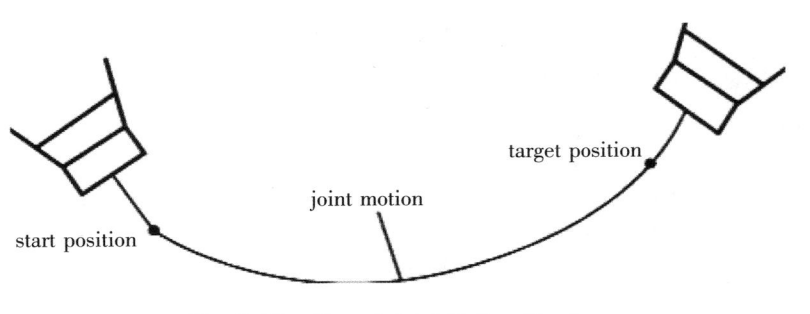

Fig. 7–39 MoveJ Joint Motion Track

The instruction syntax is given as follows. Table 7–5 specifies the meaning and value of each parameter contained in the MoveJ instruction.

MoveJ [\Conc] ToPoint [\ID] Speed [\V] | [\T] Zone [\Z] [\Inpos] Tool [\WObj] [\TLoad]

Table 7–5 MoveJ Instruction Parameters

Parameter	Description
[\Conc]	Full name: Concurrent Data type: switch Subsequent instructions are executed while the manipulator is moving. Generally, this parameter is not used. However, it may be used to avoid unnecessary stops caused by an overloaded CPU if any fly–by point is used. This parameter applies when programmed points are very close to each other at a high speed. This parameter also applies when communication with other external and synchronizing devices between an external device and the motion track of the manipulator is not required. The number of continuous motion instructions can be limited to 5 by using the parameter \Conc. The use of \Conc is not allowed in program segments including StorePath–RestoPath motion instructions and parameters. If this parameter is omitted and ToPoint is not the stop point, subsequent instructions will be executed for a period before the manipulator reaches the programmed zone. This parameter cannot be used for coordinated synchronous motions in the MultiMove system.
ToPoint	Data type: robtarget Target point of the robot and the outer axis. It is defined as a named position or directly stored in the instruction (marked with * in the instruction).

145

Industrial Robot Technology Application

Table 7-5(续)

Parameter	Description
[\ID]	Full name: Synchronization id Data type: identno In the case of coordinated synchronous motion, which is not allowed in any other circumstance, this parameter must be used in the MultiMove system. The assigned id must be the same for all cooperated tasks. The id ensures that all motions are not confused at run time.
Speed	Data type: speeddata Speed data suitable for motion. The speed data specifies the TCP, the tool orientation adjustment and the speed of outer axis.
[\V]	Full name: Velocity Data type: num This parameter specifies the velocity (in mm/s) of TCP in the instruction. It is used to replace the relevant speed designated in the speed data.
[\T]	Full name: Time Data type: num This parameter specifies the total movement time (in s) of the manipulator. It is used to replace the relevant speed data.
Zone	Data type: zonedata It defines the zone data for relevant movement, which describes the range of the generated corner path.
[\Z]	Full name: Zone Data type: num This parameter specifies the position accuracy of the manipulator TCP in the instruction. The length of a corner path is measured in mm and is used to replace the relevant zone designated in the zone data.
[\Inpos]	Full name: In position Data type: stoppointdata This parameter specifies the convergence criterion for the manipulator TCP position at the stop point, and the stop point data is used to replace the designated zone in the Zone parameter.
Tool	Data type: tooldata This parameter indicates the tool being used during movement of the manipulator, and the TCP refers to the point that moves to the specified target point.
[\WObj]	Full name: Work Object Data type: wobjdata This parameter refers to the work object coordinate system related to robot position in the instruction. If this parameter is omitted, the position will be related to the world coordinate system. Moreover, this parameter must be designated if a fixed TCP or coordinated outer axis is used.

7. Programming Basics of ABB Robot

Table 7-5(续)

Parameter	Description
[\TLoad]	Full name: Total load
	Data type: loaddata
	\TLoad of the master axis describes the total load used during movement. If the \TLoad argument is used, no loaddata in the current tooldata will be considered.
	If the \TLoad argument is set to load0, it will not be considered and will be replaced with loaddata in the current tooldata.

Application example 2 of MoveJ instruction:

MoveJ p5, v2000, fine \Inpos := inpos50, grip3;

The TCP of the tool (grip3)moves along a nonlinear path to the stop point (p5). The manipulator deems that the tool is in position when the conditions of 50% position and 50% speed regarding the stop point fine are met. The manipulator waits for 2s at most for the satisfaction of these conditions.

(2)MoveL(TCP linear motion instruction): used to make the TCP move linearly to the given target point while keeping the motion track straight, as shown in Fig. 7-40. This instruction can also be used to adjust the tool orientation when the TCP remains fixed.

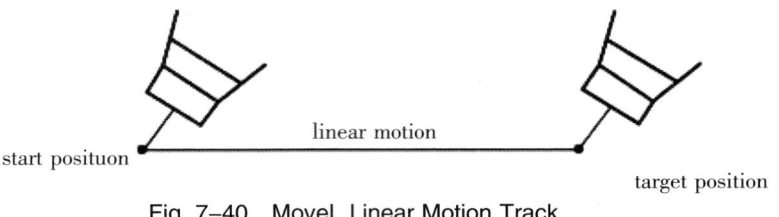

start posituon   linear motion   target position

Fig. 7-40   MoveL Linear Motion Track

The instruction syntax is given as follows. The meaning and value of each parameter in this instruction are similar to those in the MoveJ instruction.

MoveL [\Conc] ToPoint [\ID] Speed [\V] I [\T] Zone [\Z] [\Inpos] Tool [\WObj] [\Corr] [\TLoad]

Application example 1 of MoveL instruction:

MoveL *, v2000 \V:=2200, z40 \Z:=45, grip3;

The TCP of the tool(grip3)moves linearly to the position contained in the in-

147

Industrial Robot Technology Application

struction. The tool starts moving when the relevant data are set to v2000 and z40. The TCP rate and zone radius are 2,200 mm/s and 45 mm respectively.

Application example 2 of MoveL instruction:

MoveL p5, v2000, fine \Inpos := inpos50, grip3;

The TCP of the tool(grip3)moves linearly to the stop point(p5). The manipulator deems that the tool is in position when the conditions of 50% position and 50% speed regarding the stop point fine are met. The manipulator waits for 2s at most for the satisfaction of these conditions.

Application example 3 of MoveL instruction:

MoveL start, v2000, z40, grip3 \WObj:=fixture;

The TCP of the tool (grip3)moves linearly to the specified position in the work object coordinate system of the fixture.

(3)MoveC(TCP circular motion instruction): used to make the TCP move along the circumference to a given target point. The orientation of the cycle usually keeps relatively unchanged during movement. It is well known that three points define an arc, and the actual motion track according to the MoveC instruction is closely related to the start point. The codes shown in Fig. 7–41 are as follows:

MoveL p40, v200, z50, tool1;

MoveC p50, p60, v200, fine, tool1;

The above codes mean that the robot

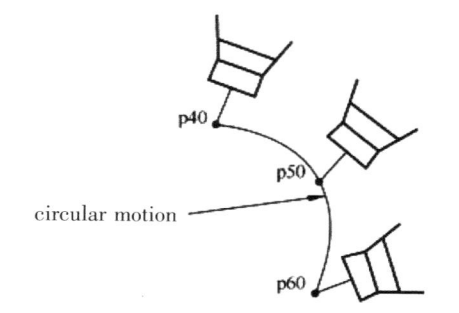

Fig. 7–41   MoveC Circular Motion Track

TCP moves linearly from the previous position to the arc start point(p40), and then makes circular motion from this point to the end point(p60). Where, P50 is the middle point of the arc, the running speed is 200 mm/s, the radius of the turning zone is fine, and the tool used is tool1. Accordingly, to determine a circular motion track, there must be the combination of a motion instruction strictly specifying the start point of the arc and a MoveC instruction.

The instruction syntax is given as follows. Compared with MoveJ and MoveL instructions, this instruction contains an additional CirPoint, which indicates a position on the arc between the relevant start and end points. To get the best accura-

148

7. Programming Basics of ABB Robot

cy, this point should be in the middle of the relevant start and end points. If this point is too close to the start or end point, the robot may give an alarm. The meaning and value of other parameters in this instruction are similar to those in the MoveJ instruction.

MoveC [\Conc] CirPoint ToPoint [\ID] Speed [\V] | [\T] Zone [\Z] [\Inpos] Tool [\WObj] [\Corr] [\TLoad]

Application example of MoveC instruction:

MoveL p1, v500, fine, tool1;

MoveC p2, p3, v500, z20, tool1;

MoveC p4, p1, v500, fine, tool1;

The above examples show how to implement a complete cycle with two MoveC instructions (note that the start point of the arc is given in the MoveL instruction). The motion track is shown in Fig. 7–42.

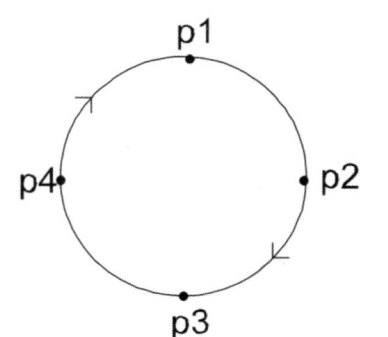

Fig. 7–42   Full Circle Motion Track

(4)MoveAbsJ(instruction for motion to absolute axis angle and position): used to make the manipulator and the outer axis move to the specified absolute position among axis positions. When a MoveAbsJ motion instruction is used, the position of the manipulator will not be affected by a given tool and work object and by the effective programmed displacement. The manipulator uses this data to calculate the load, TCP speed and corner path.

The instruction syntax is given as follows, where ToJointPos is the data indicating the target joint point. If the parameter \NoEOffs is set, the movement related to MoveAbsJ will not be affected by the effective offset of the outer axis.

MoveAbsJ [\Conc] ToJointPos [\ID] [\NoEOffs] Speed [\V] | [\T] Zone [\Z] [\Inpos] Tool [\WObj] [\TLoad]

Application example of MoveAbsJ instruction:

MoveAbsJ p50, v1000, z50, tool2;

With the speed data(v1000)and the zone data(z50), the manipulator and the tool(tool2)are able to move along the non–linear path to the absolute axis position (p50).

149

Industrial Robot Technology Application

7.4.2 Logic control instructions in the program

(1)Compact IF: If conditions are met, an instruction will be executed. The instruction syntax is given as follows. The control logic is: judge the whether the Condition is TRUE or FALSE; if TRUE, execute the subsequent instructions; if FALSE, do not execute the subsequent instructions and move the program pointer to the next statement(the selection instruction has no branch options).

IF <conditional expression>(<instruction> | <SMT>)´;´

Application example of Compact IF instruction:

IF counter > 10 Set do1;

If the counter is greater than 10, set the do1 signal.

(2)IF: condition selection instruction. This instruction indicates that if a condition is met, then ...; otherwise, ... The instruction syntax is given as follows:

IF Condition THEN ...

{ELSEIF Condition THEN ...}

[ELSE ...]

ENDIF

Application example of IF instruction:

IF counter > 100 THEN

counter := 100;

ELSEIF counter < 0 THEN

counter := 0;

ELSE

counter := counter + 1;

ENDIF

If the counter is within 0–100, increase the value of the counter by 1; however, if the value of the counter goes is out of 0–100, assign a corresponding limit to the counter.

(3)FOR: loop instruction. A FOR instruction should be used when one or more instructions have to be repeated for many times. The instruction syntax is given as follows. It is necessary to set the start value, end value and increment of the loop counter(the default value is 1 if omitted).

FOR Loop counter FROM Start value TO End value [STEP Step value]

DO ... ENDFOR

150

7. Programming Basics of ABB Robot

Application example of FOR instruction:

FOR i FROM 1 TO 10 DO

routine1;

ENDFOR

Repeat routine 1 without return value 10 times.

(4)WHILE: loop instruction. A WHILE instruction should be used if some instructions have to be repeated, as long as a given conditional expression is evaluated as TRUE. The instruction syntax is given as follows:

WHILE Condition DO ... ENDWHILE

Application example of WHILE instruction:

WHILE reg1 < reg2 DO

...

reg1 := reg1 + 1;

ENDWHILE

Where, reg1 is a conditional variable. The instruction in the WHILE block is repeated if reg1 < reg2.

(5)TEST: condition selection instruction. A TEST instruction should be used when different instructions are to be executed according to the expression or the data value. IF..ELSE instruction may also be used if there are not too many alternatives. Its application example is as follows:

TEST reg1

CASE 1,2,3 :

routine1;

CASE 4 :

routine2;

DEFAULT :

TPWrite "Illegal choice";

Stop;

ENDTEST

Different instructions will be executed according to the value of reg1: if the value is 1, 2 or 3, execute routine 1; if the value is 4, execute routine 2; in other cases, print an error message and stop execution.

(6)GOTO: jump instruction, which is used to transfer program execution to

151

Industrial Robot Technology Application

another thread(label)in the same program. Its application example is as follows:

reg1 := 1;

next:

...

reg1 := reg1 + 1;

IF reg1<=5 GOTO next;

Program execution is transferred to next four times(reg1= 2, 3, 4, 5).

(7)Label: used to name a program in another. It is usually used as a jump label in combination with the GOTO instruction. The label, e.g., "next:" in the program given above, can then be used to transfer program execution in the same program. The syntax of Label instruction is < identifier>´:´ and should not be the same as all other tags or all data names in the same program.

7.4.3 Stop program execution

(1)Stop: used to stop program execution. All currently performed movements will be completed before the Stop instruction is ready. The instruction syntax is given as follows:

Stop [\NoRegain] | [\AllMoveTasks]

[\NoRegain]: With the data type of switch, this parameter specifies the start point of the next program, whether the affected mechanical unit should return to the stop position or not. If parameter \NoRegain is set, the manipulator and the outer axis will not return to the stop position(provided that they have been far from the stop position). If this parameter is omitted and the manipulator or outer axis slowly moves away from the stop position, a question about the manipulator will be displayed on the FlexPendant. Next, the user answers whether the manipulator should return to the stop position.

[\AllMoveTasks]: with the data type of switch, this parameter specifies all common tasks being executed and the programs that should be stopped in actual tasks. If this parameter is omitted, only the program in the task in which the instruction is executed will be stopped.

Application example of Stop instruction:

MoveL p1, v500, fine, tool1;

TPWrite "Jog the robot to the position for pallet corner 1";

Stop \NoRegain;

152

p1_read := CRobT(\Tool:=tool1 \WObj:=wobj0);

MoveL p2, v500, z50, tool1;

Program execution is stopped by the manipulator at p1. As the operator jogs, the manipulator moves to p1_read. If the manipulator does not return to p1 when the next program starts, the position p1_read can be stored in the program.

(2)EXIT: used to terminate program execution. An EXIT instruction should be used when a fatal error occurs or permanent stop of the program execution is required. After execution of the EXIT instruction, the program pointer will disappear. To continue with program execution, the program pointer must be set again.

Application example of EXIT instruction:

ErrWrite "Fatal error","Illegal state";

EXIT;

Program execution stops and cannot be restarted from this position in the program.

(3)Break: used to immediately interrupt program execution. A Break instruction can immediately stop the program execution without waiting for the manipulator and the outer axis to reach the target point specified in the program at that time. After that, program execution can be restarted from the next instruction. If a Break instruction is present in some program events, execution of the program will be interrupted and no program events will be stopped.

7.4.4 Assignment instruction

":=": assignment instruction, which is used to assign a new value to a specific data. This value may be a constant value or an arithmetic expression. The instruction syntax is Data: = Value. The following are some application examples:

reg1 := 5;

counter := counter + 1;

7.4.5 Wait instructions

(1)WaitTime: used to direct waiting for a given time before continuous execution of the program. This instruction can also direct waiting until the manipulator and the outer axis become stationary. The minimum time (in s)for waiting before program execution is 0s. The maximum time for waiting is unlimited and the resolution is 0.001s. The instruction syntax is given as follows:

153

Industrial Robot Technology Application

WaitTime [\InPos] Time

Application example of WaitTime instruction:

WaitTime \InPos,0;

This instruction makes the program execution wait until the manipulator and the outer axis become stationary.

(2)WaitDI: used to direct waiting until a digital signal input value is set. The instruction syntax is given as follows:

WaitDI Signal Value [\MaxTime] [\TimeFlag]

Table 7-6 specifies the meaning and value of each parameter contained in the WaitDI instruction. During execution of this instruction, if the signal value is correct, the program will continue with subsequent instructions only. If the signal value is incorrect, the manipulator will wait, and program execution will continue only after the signal value is corrected. If the wait time exceeds the maximum time, program execution will continue at the specified TimeFlag; otherwise, an error will be caused. If a TimeFlag is specified, the condition will be set to TRUE when the maxi-

Table 7-6　WaitDI Instruction Parameters

Parameter	Description
Signal	Data type: signaldi
	Name of signal.
Value	Data type: dionum
	Expected value of signal.
[\MaxTime]	Full name: Maximum Time
	Data type: num
	Maximum allowed wait time, in s. If the time is out before the condition is met, an incorrect processor will be called out, with the error code of ERR_WAIT_MAXTIME. If no incorrect processor exists, program execution will stop.
[\TimeFlag]	Full name: Timeout Flag
	Data type: bool
	If the maximum allowed time is out before the condition is met, the output parameter containing this value will be TRUE. If this parameter is included in the instruction, it should not be deemed as an error about maximum time out. This parameter will be ignored if no MaxTime parameter is included in the instruction.

mum time is exceeded, and set to FALSE when the same is not exceeded. If program execution is stopped and then restarted, this instruction will evaluate the current value of the signal. Any change during program execution stop will be denied.

Application example of WaitDI instruction:

WaitDI di4, 1;

Program execution continues only after the input di4 has been set.

(3)WaitDO: used to direct waiting until a digital signal output value is set. The syntax and usage of this instruction are similar to those of the WaitDI instruction.

(4)WaitGI: used to direct waiting until a group of digital input signal is set to the specified value. The instruction syntax is given as follows:

WaitGI Signal [\NOTEQ] | [\LT] | [\GT] Value | Dvalue [\MaxTime] [\ValueAt-Timeout] | [\DvalueAtTimeout]

Table 7-7 specifies the meaning and value of each parameter contained in the WaitGI instruction.

Application example of WaitAI instruction:

WaitAI ai1, \GT, 1;

Program execution continues only after gi1 is less than 1.

(5)WaitGO: used to direct waiting until a group of digital output signals is set to the specified value. The syntax and usage of this instruction are similar to those of the WaitGI instruction.

(6)WaitAI: used to direct waiting until an analog signal input value is set. The instruction syntax is given as follows. The syntax and usage of this instruction are similar to those of the WaitGI instruction.

WaitAI Signal [\LT] | [\GT] Value [\MaxTime] [\ValueAtTimeout]

Application example of WaitAI instruction:

WaitAI ai1, \GT, 5;

Program execution continues only after the analog signal input ai1 has a value greater than 5.

(7)WaitAI: used to direct waiting until an analog signal output value is set. The syntax and usage of this instruction are similar to those of the WaitAI instruction.

7.4.6 I/O signal value setting instructions

(1)Reset: used to reset the value of a digital output signal to 0. For example:

Reset do15;

155

Industrial Robot Technology Application

Table 7-7 WaitGI Instruction Parameters

Parameter	Description
Signal	Data type: signalgi
	Name of digital group input signal.
[\NOTEQ]	Full name: NOT EQual
	Data type: switch
	If this parameter is used, the WaitGI instruction will wait until the digital group signal value is divided by the number contained in the parameter "Value".
[\LT]	Full name: Less Than
	Data type: switch
	If this parameter is used, the WaitGI instruction will wait until the digital group signal value is less than the number contained in the parameter "Value".
[\GT]	Full name: Greater Than
	Data type: switch
	If this parameter is used, the WaitGI instruction will wait until the digital group signal value is greater than the number contained in the parameter "Value".
Value	Data type: num
	Expected value of signal. It must be an integer within the operating range of the digital group input signal used, and its allowed value depends on the number of signals in the group.
Dvalue	Data type: dnum
	Expected value of signal, with the same application as Value but having a different data type.
[\MaxTime]	Full name: Maximum Time
Data type: num	
	Maximum allowed wait time, in s. Its application is the same as that of the WaitDI instruction.
[\ValueAt	
Timeout]	Data type: num
	In case of instruction timeout, the current signal value will be stored in this variable. This variable will be set only when the system variable ERRNO is set to ERR_WAIT_MAXTIME.
[\DvalueAtTimeout]	Data type: dnum
	In case of instruction timeout, the current signal value will be stored in this variable. This variable will be set only when the system variable ERRNO is set to ERR_WAIT_MAXTIME.

7. Programming Basics of ABB Robot

(2)Set: used to set the value of a digital output signal to 1. For example: Se do15;

(3)SetDO: used to change the value of a digital output signal. The instruction syntax is given as follows:

SetDO [\SDelay]|[\Sync] Signal Value

Table 7−8 specifies the meaning and value of each parameter contained in the SetDO instruction.

Table 7−8   SetDO Instruction Parameters

Parameter	Description
[\SDelay]	Full Name: Signal Delay
	Data type: num
	Change in delay time (in s; up to 2,000s). Program execution continues directly by executing the next instruction. The signal is changed after the given time delay and subsequent program execution is not affected.
[\Sync]	Full name: Synchronization
	Data type: switch
	If this parameter is used, program execution will wait until the signal is physically set to the specified value.
Signal	Data type: signaldo
	Name of signal to be changed.
Value	Data type: dionum
	Expected value of signal, being 0 or 1. .
Dvalue	Data type: dnum
	Expected value of signal, with the same application as Value but having a different data type.
[\MaxTime]	Full name: Maximum Time
	Data type: num
	Maximum allowed wait time, in s. Its application is the same as that of the WaitDI instruction.
[\ValueAt Timeout]	Data type: num
	In case of instruction timeout, the current signal value will be stored in this variable. This variable will be set only when the system variable ERRNO is set to ERR_WAIT_MAXTIME.
[\DvalueAt Timeout]	Data type: dnum
	In case of instruction timeout, the current signal value will be stored in this variable. This variable will be set only when the system variable ERRNO is set to ERR_WAIT_MAXTIME.

157

Industrial Robot Technology Application

Application example 2 of SetDO instruction:

SetDO \Sync ,do1, 0;

The signal do1 is set to 0. Program execution will wait until the signal is physically set to the specified value.

(4)SetGO: used to change the value of a group of digital output signals. The instruction syntax is given as follows:

SetGO [\SDelay] Signal Value | Dvalue

(5)SetAO: used to change the value of an analog output signal. The instruction syntax is given as follows:

SetAO Signal Value

Before the programmed value is sent to the physical channel, the programmed value will be calculated according to the upper and lower limits defined by the analog signal. Fig. 7–43 shows how to measure analog signal values.

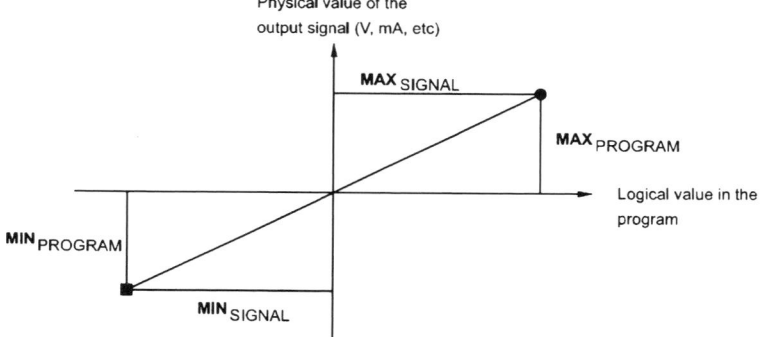

Fig. 7–43   Correspondence between Programmed Value and Physical Value

7.4.7 Program call instruction

ProcCall: used to call a new program without return value. Procedure call is used to transfer program execution to another program without return value. When the program without return value is fully executed, the program will continue to execute instructions after the procedure call. A set of parameters can also be sent to the new program without return value. There are two types of routines:

Routine without parameters, which is the most commonly used;

158

7. Programming Basics of ABB Robot

PROC global_routine( );

...

ENDPROC;

Routine with parameters.

PROC routine1(num in_par,INOUT num inout_par, VAR num var_par,PERS num pers_par);

...

ENDPROC;

Corresponding actual parameters must be provided when a routine with parameters is called. Four access modes are available for routine parameters.

(1)INPUT: Generally, this parameter is used as routine input only; if the parameter type is omitted, the access mode will be INPUT by default. Changing variables in a routine does not affect the corresponding input parameters, which can be of CONST, VAR or PERS type.

(2)INOUT: If routine parameters are set to this access mode, these parameters must be VAR and PERS data that can be modified by the program, and relevant input parameters can be modified during the call. However, the initial value will not be changed in VAR access mode, but may be changed in PERS access mode.

(3)VAR: If routine parameters are set to this access mode, these parameters must be VAR data that can be modified by the program. Actual parameters can be modified, but the initial values of VAR actual parameters will not be changed.

(4)PERS: If routine parameters are set to this access mode, these parameters must be PERS data that can be modified by the program. Actual parameters can be modified, and if so, the initial values of PERS actual parameters will be changed at the same time.

Optional parameters can be defined for a routine, and a routine parameter can be set as an optional parameter. Each optional parameter is provided with a "\" identifier and may be default during the call of a routine. For example:

PROC routine(num required_par \num optional_par)

"Switch" can be used to select a specific optional parameter, for example:

PROC routine4(\switch on|switch off)

End of routine: Running of routine can be terminated by giving a "Return"

159

Industrial Robot Technology Application

instruction or by providing a routine end identifier (ENDPROC, BACKWARD, ERROR, UNDO).

Call of routine: Procedure is called on the teach pendant by giving a ProCall instruction. When a routine with parameters is called, mandatory parameters must be specified and the sequence of these parameters must be correct; optional parameters can be default. The following is an application example. When the program without return value is ready, the program will execute the instruction "Set do1" after the return to procedure call.

Routine1 10;

Set do1;

...

PROC Routine1(num Count1)

...

ENDPROC

7.4.8 Functions

(1)Offs: used to realize the offset of a manipulator position in the work object coordinate system. The syntax is given as follows:

Offs(Point XOffset YOffset ZOffset)

The following instruction directs the manipulator to move to a position 10 mm away from position p2(in the z direction):

MoveL Offs(p2, 0, 0, 10), v1000, z50, tool1;

(2)RelTool: used to increase the displacement(and/or rotation)expressed in an effective tool coordinate system to the manipulator position. The syntax is given as follows. Where, Dx, Dy and Dz are translational amounts of the tool coordinate system, and [\Rx], [\Ry] and [\Rz] are the angles of rotation around the x, y, z axes of the tool coordinate system, respectively. If two or three rotations are specified at the same time, the rotation will be made first around the x−axis, then around the new y−axis, and finally around the new z−axis.

RelTool(Point Dx Dy Dz [\Rx] [\Ry] [\Rz])

The following instruction directs the tool to rotate around its z axis by 25°:

MoveL RelTool(p1, 0, 0, 0 \Rz:= 25), v100, fine, tool1;

8.

ABB Robot Programming

Industrial robot programming refers to program design for the robot to complete a certain operation. Efficient and reasonable programs are an important guarantee for the robot to complete complex, accurate action and automatic operation. There are mainly two following ways for the programming:

(1)Online programming: It is also called teach programming, in which the operator controls the end of the manipulator tool to reach the designated pose and position by the teach pendant, records the robot pose data and writes the robot motion instructions to complete the path planning of the robot. It is the programming mode for most industrial robot applications at present and characterized by the following:

- Simple operation, easy to use, and fast response to simple applications;
- Inefficient programming;
- The actual robot system needs to be operated on site. The programming operation will cause the robot to stop production and reduce equipment utilization. Furthermore, the on-site environment may be dangerous;
- Difficult to realize complicated robot motion track;
- The programming quality depends on the experience of the programmer.

(2)Offline programming: It refers to that the computer graphics programming system based on CAD data constructs a simulated work scene by 3D modeling of the work unit, performs robot control and path planning by algorithm, conducts 3D graphics animation simulation on the programming results to test the programming reliability, and finally generates the code actually produced by the robot. Offline programming is characterized by the following compared with teach programming:

- It does not occupy the production time of the robot and improves the e

quipment utilization;

  • Improved programming environment to keep the programmer away from dangerous working environment;

  • Easy to combine with CAD/CAM system, high programming efficiency;

  • Guaranteed and consistent programming quality;

  • It can realize complicated robot motion track and facilitate program modification;

  • Complicated technology, high requirements for programmers.

It can be seen from the comparison that online programming and offline programming have their own advantages and disadvantages, but they are not contradictory but complementary. The user shall evaluate the work efficiency, programming quality and economy comprehensively in accordance with actual application fields and scenarios, and select a reasonable programming mode.

8.1 Teach programming

RAPID program is composed of program modules and system modules. Generally, the robot program is constructed only by creating new program modules, while system modules are mainly used for system control.

8.1.1 Creation of routine

Its steps are as follows:

(1)Tap the ABB menu button in the upper left corner of the teach pendant screen and select the "Program Editor" option. When entering for the first time, the main program editing interface shown in Fig. 8-1 will be displayed, and the console will give a prompt of "Unnamed program".

(2)Tap the "Task and program" tab to enter the interface as shown in Fig. 8-2. A task corresponds to a program name. The file menu in the lower left corner can be used to perform operations such as creation, loading, saving as, renaming, and deletion for the program. As shown in the figure, the program is named as Transfer.

(3)Select the task "T_ROB1" and tap "Display module" option in the lower right corner to enter the task module interface as shown in Fig. 8-3. Creating multiple modules in the Program is conducive to classification management of the

8. ABB Robot Programming

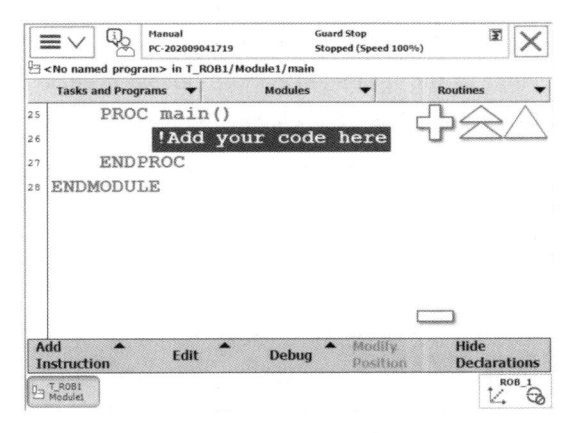

Fig. 8-1　Main Program Editing Interface

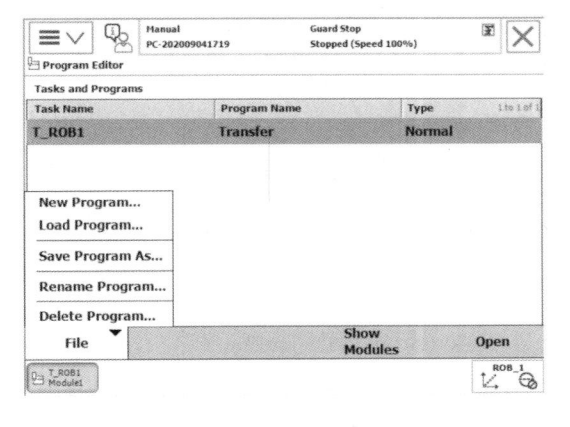

Fig. 8-2　Task and Program Interface

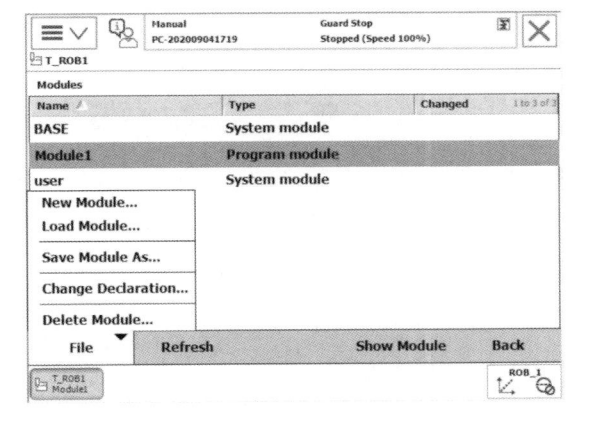

Fig. 8-3　Task Module Interface

163

Industrial Robot Technology Application

program. The system has three modules by default: BASE, MainModule and user modules. The file menu in the lower left corner can be used to perform operations such as creation, loading, saving as, changing declarations, and deletion for modules.

(4)Select the BASE module and tap the "Display module" button to view the definitions of tool0, wobj0 and load0 by the BASE module (Note: Do not modify the system module), as shown in Fig. 8-4.

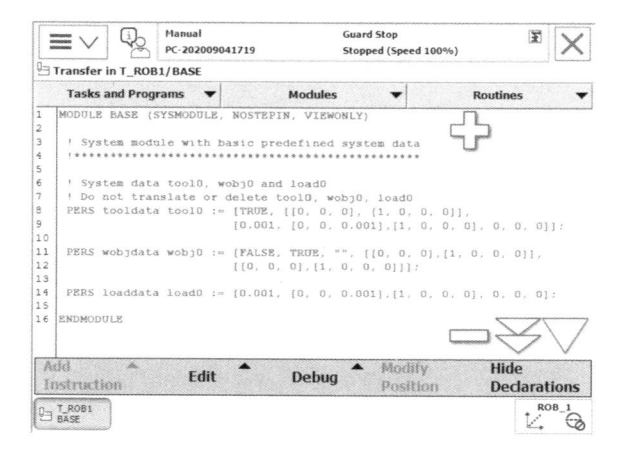

Fig. 8-4   Definition of Base Module

(5)Select the MainModule module and tap the "Display module" button to see that the main program is under the module by default. Tap the "File" menu and then the "New routine" option, as shown in Fig. 8-5.

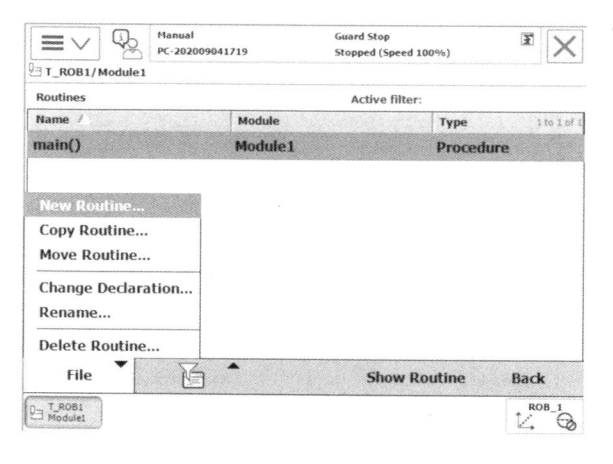

Fig. 8-5   MainModule Interface

164

8. ABB Robot Programming

(6)Enter the routine declaration interface as shown in Fig. 8–6, modify the name, set the parameters and select the module as required, and then tap "OK".

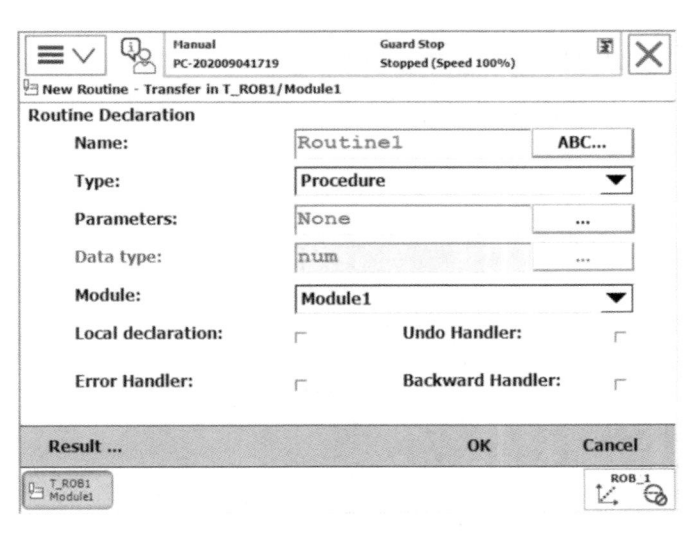

Fig. 8–6　New Routine Interface

(7)Return to the MainModule display interface as shown in Fig. 8–7, select Routine1()and tap "Display routine" button.

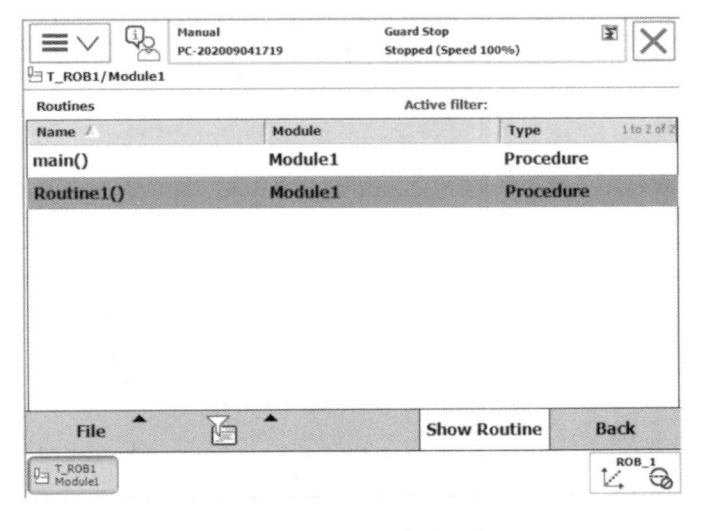

Fig. 8–7　MainModule Interface

165

Industrial Robot Technology Application

(8)Enter the routine editing interface, as shown in Fig. 8–8.

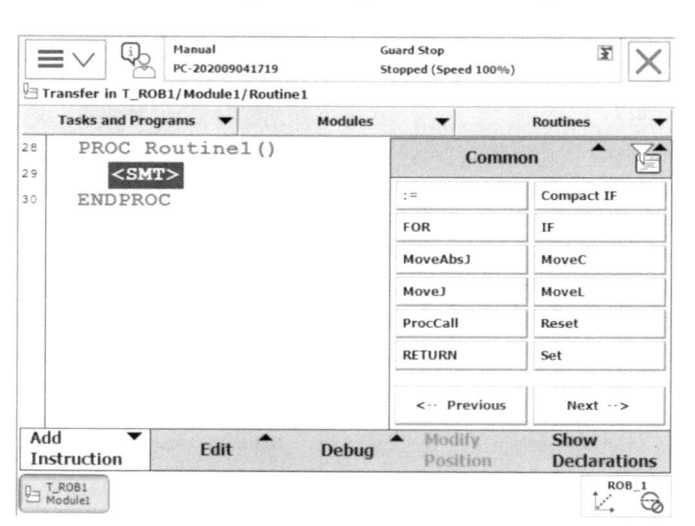

Fig. 8–8   Routine Editing Interface

8.1.2 Routine writing

The teach programming of industrial robots can be conducted in the following three ways: teaching points before programming, programming before teaching points, and teaching points while programming. In general, the mode of programming before teaching has high requirements on operators, who shall be clear in thinking and skilled in programming. However, it is also the most time –saving mode.

If the robot needs to move a work object from point A to point B, at least three points need to be defined: starting point, point A and point B. It is necessary to define DO signals for fixture clamping. After defining the program data, programming will be more convenient. The specific steps are as follows:

(1)First define jointtarget data jpos10, as shown in Fig. 8–9.

(2)Generally, the first point of the robot program is best defined as the joint type, and MoveAbsJ motion mode is adopted, which is conducive to pose adjustment of industrial robots. The robot pose corresponding to the current state of jpos10 is shown in Fig. 8–10.

166

8. ABB Robot Programming

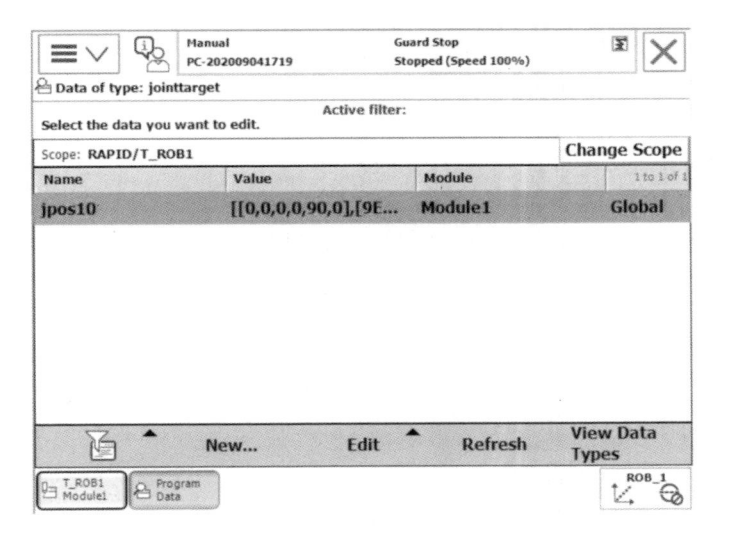

Fig. 8-9　Jointtarget Data

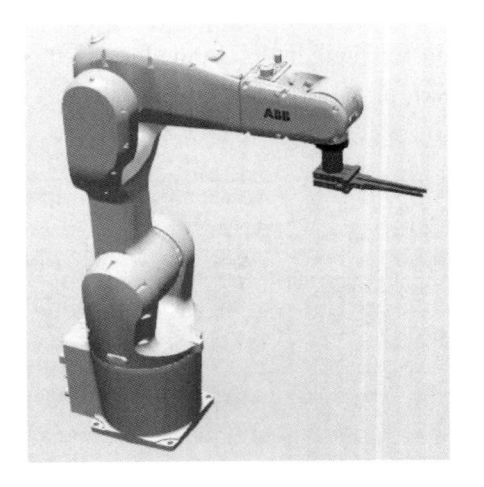

Fig. 8-10　Robot Pose Corresponding to robjoint1

(3)Define the robtarget type data p10, p20, corresponding to the starting point A and end point B during handling, as shown in Fig. 8-11. The newly defined data is [X, Y, Z, A, B, C] corresponding to the current position and pose of the robot by default. Use p10 and p20 for programming first, and then teach its coordinates.

167

Industrial Robot Technology Application

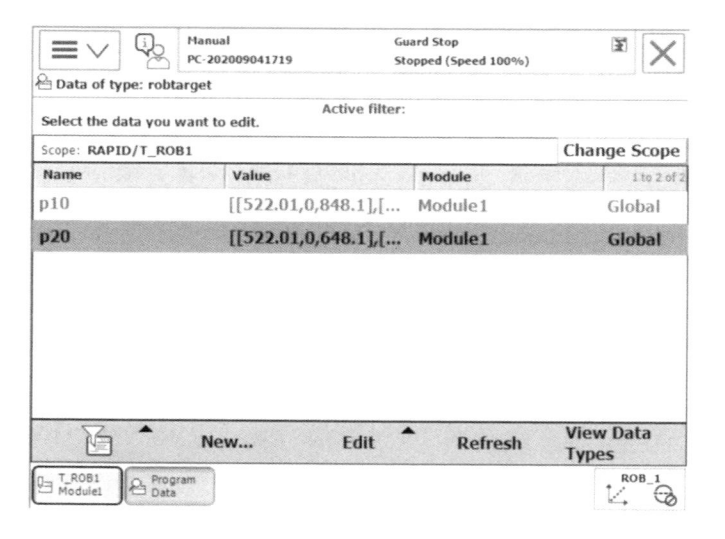

Fig. 8-11　Definition of Robtarget Data

(4)Tap the "Add instruction" button in the lower left corner to add the Move-AbsJ instruction, as shown in Fig. 8–12.

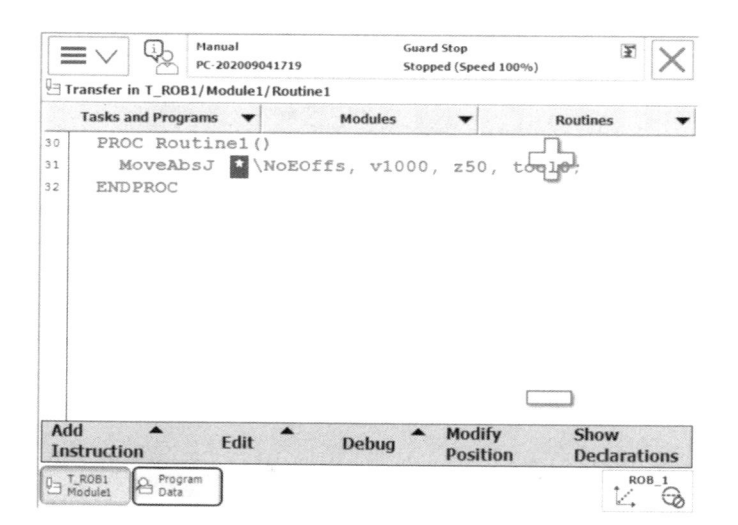

Fig. 8-12　Adding MoveAbsJ Instruction

8. ABB Robot Programming

(5)Select the "*" field and double-tap to enter the instruction parameter editing interface, as shown in Fig. 8–13. The system default joint data is the joint angle value corresponding to the current robot pose.

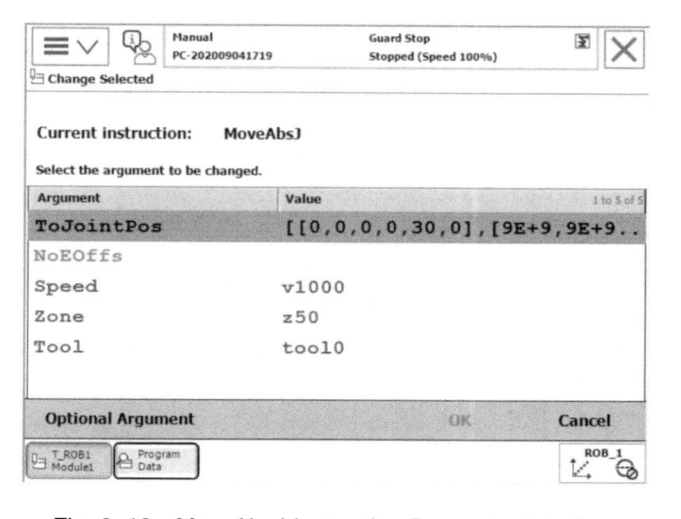

Fig. 8–13　MoveAbsJ Instruction Parameter Interface

(6)Tap the "ToJointPos" option. The joint position data selection interface is shown in Fig. 8–14. Select the previously created jpos10 to complete parameter replacement.

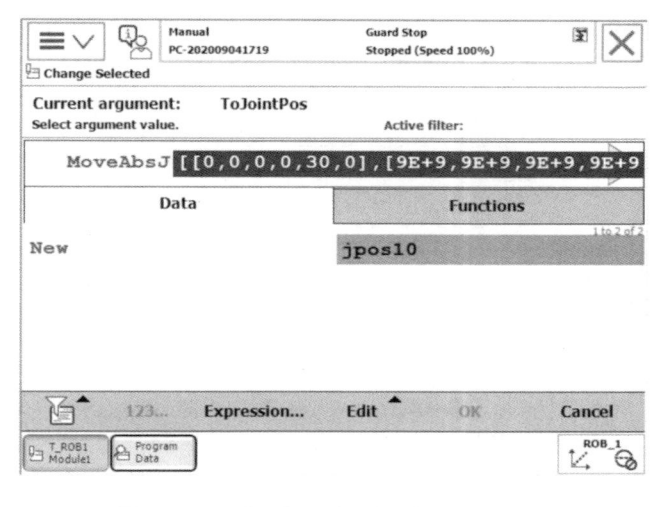

Fig. 8–14　Setting of Joint Position Data

169

Industrial Robot Technology Application

(7)After selecting jpos10 data, the effect is shown in Fig. 8–15. You can also modify parameters such as speed and tool in the current interface, and tap "OK" to complete parameter setting.

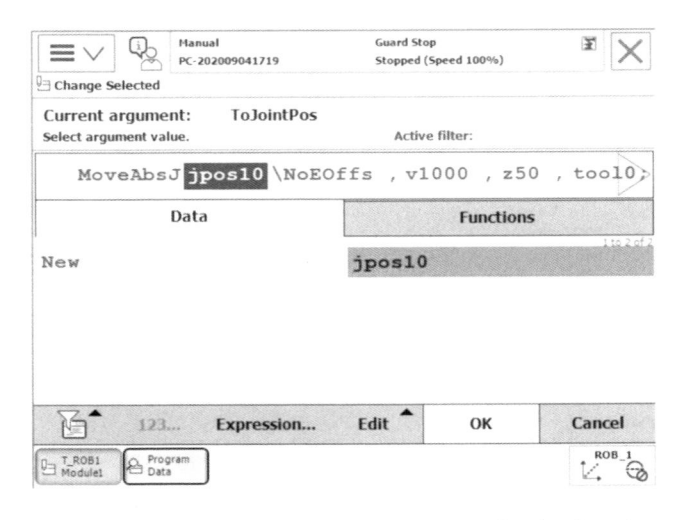

Fig. 8–15   MoveAbsJ Instruction Parameter Setting Interface

(8)Continue to add the MoveL instruction, double–tap to enter the parameter editing interface, select data p10, tap the "Function" tab and then the "Offs" option to add the position offset, as shown in Fig. 8–16.

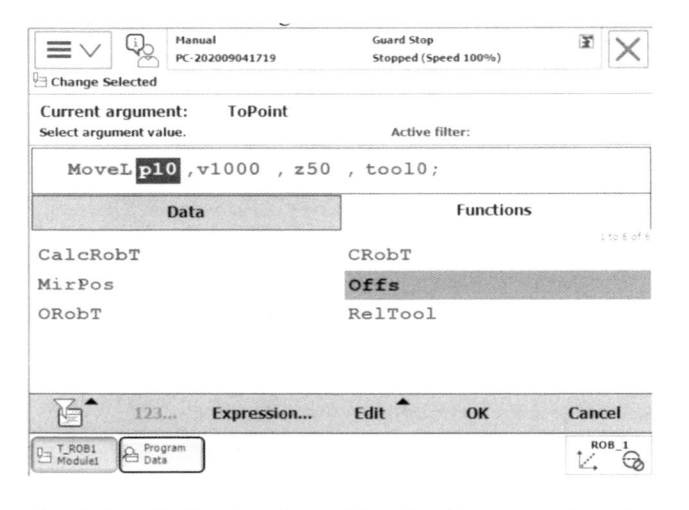

Fig. 8–16   Adding Interface of Position Parameter Function

8. ABB Robot Programming

（9）Enter the Offs parameter setting interface as shown in Fig. 8-17. Three offsets <EXP> can be represented by variables, or the value can be directly input by selecting the "Selected content only" in "Edit" menu.

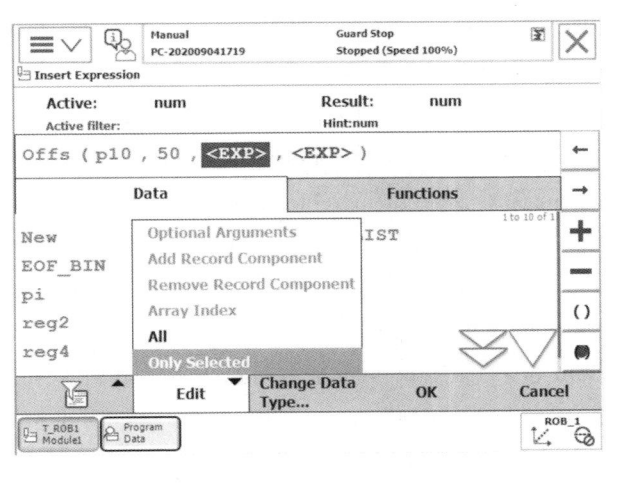

Fig. 8-17　Offs Parameter Setting Interface

（10）Continue to input instructions in accordance with the above methods, and complete the writing of Routine1（）as follows:

PROC Routine1（）

MoveAbsJ jpos10\NoEOffs, v1000, z50, tool1; /程序 home 点

MoveL Offs(p10,-60,0,0), v500, fine, tool1; /接近工件位置

MoveL p10, v30, fine, tool1; /低速运动至夹紧位置

Set Do1; /气动手爪夹紧

WaitTime 2; /等待 2s 确保夹紧动作完成

MoveL Offs(p10,0,0,100), v100, z50, tool1;/低速向上夹起工件

MoveL Offs(p20,0,0,100), v300, z50, tool1;/运动至放置点上方

MoveL p20, v20, fine, tool1; /低速运动至工件放置点

Reset Do1; /气动手爪松开

WaitTime 2; /等待 2s 确保夹紧动作完成

MoveL Offs(p20,-60,0,0), v100, z50, tool1; /机械手水平退出

MoveL Offs(p20,-60,0,100), v300, z50, tool1;/机械手向上抬起

MoveAbsJ jpos10\NoEOffs, v1000, z50, tool1;/返回 home 点

ENDPROC

171

Industrial Robot Technology Application

8.1.3 Teaching of program points

The advantage of programming before teaching is that the program thinking is clear, and can avoid teaching of many useless points. For example, multiple point targets are achieved by using point offset Offs in Routine1 ( ), while there are only two points that really need to be taught. After defining the program data, programming will be more convenient. The specific steps are as follows:

(1) Manually control the robot to approach the target position with a correct pose, gradually adjust the gear in incremental mode from far to near, and control the tool to the appropriate working position. Select the position data "p10" in the routine, and tap the "Modify position" button in the lower right corner, as shown in Fig. 8−18. The system will assign the current position data of the robot to p10, which is the origin of teach programming. After teaching at P10, tap the "PP to cursor" option in the "Debugging" menu, and tap the single−step operation button on the teach pendant to debug the program while teaching.

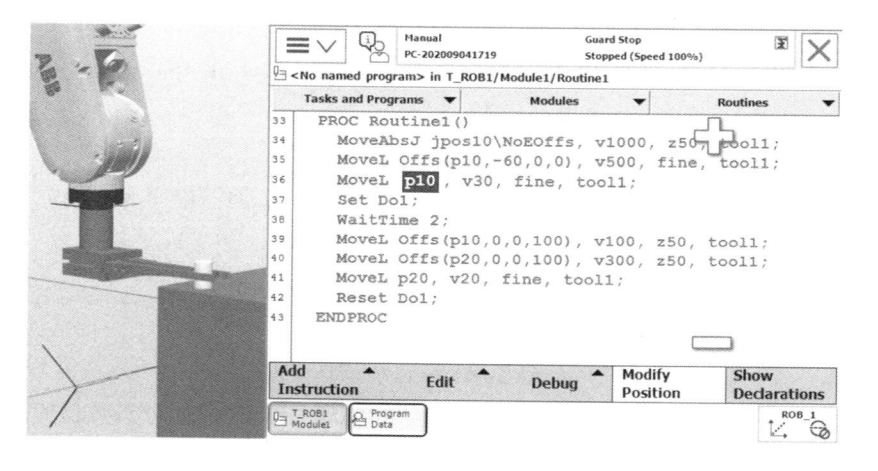

Fig. 8−18   Teaching at p10

(2) Manually control the manipulator in the same way to clamp the work object to the target point, and select the position data p20 for point teaching, as shown in Fig. 8−19.

(3) After teaching, the program shall be debugged to check whether the point teaching and programming are correct. Tap the "Debugging" menu to move the PP

8. ABB Robot Programming

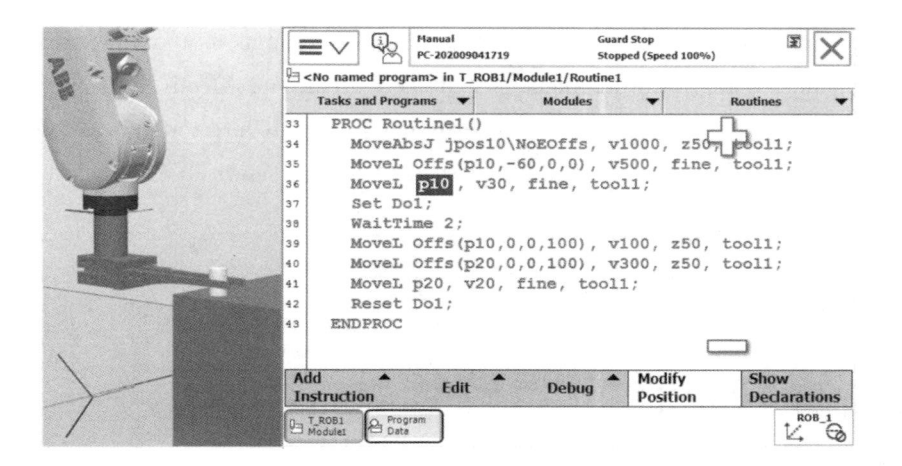

Fig. 8–19　Teaching at p20

pointer to an appropriate position (generally the program header). Control the "Start", "Step forward", "Step backward" and "Stop" buttons on the teach pendant to run the program continuously or step by step, as shown in Fig. 8–20.

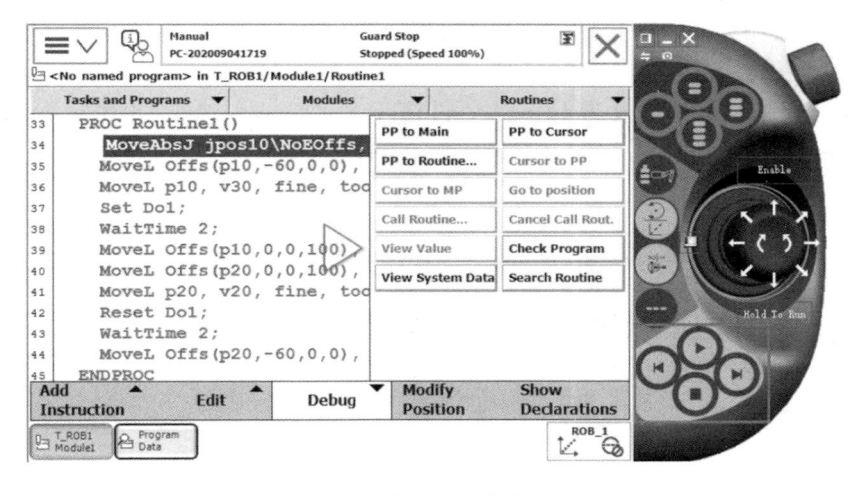

Fig. 8–20　Program Debugging

8.1.4 Writing of stacking program

The motion of stacking robots is similar to handling action. The difference is that handling is generally from one fixed point to another fixed point or multiple

173

Industrial Robot Technology Application

fixed points, while stacking is from one or more fixed points to a group of interrelated points. Using cycle instructions FOR or WHILE can effectively simplify the structure of the stacking program. The specific use of FOR instruction is as follows:

(1)Create Routine2, and add FOR instructions, as shown in Fig. 8-21.

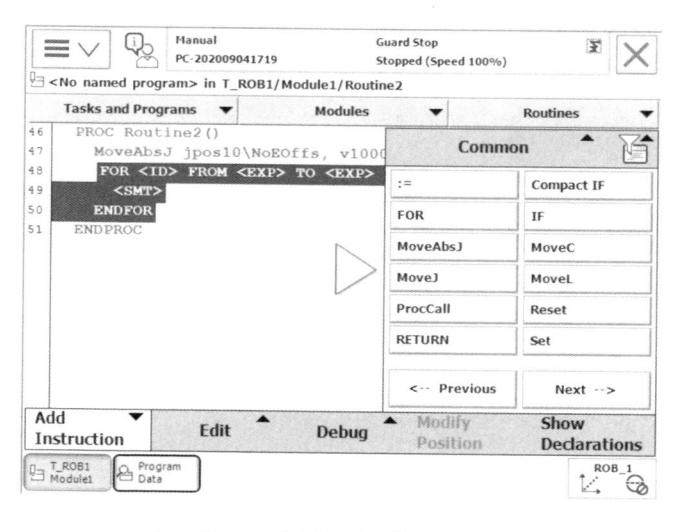

Fig. 8-21　Adding FOR Instruction

(2)Select the FOR field and double-tap it to enter the FOR instruction structure editing interface, as shown in Fig. 8-22.

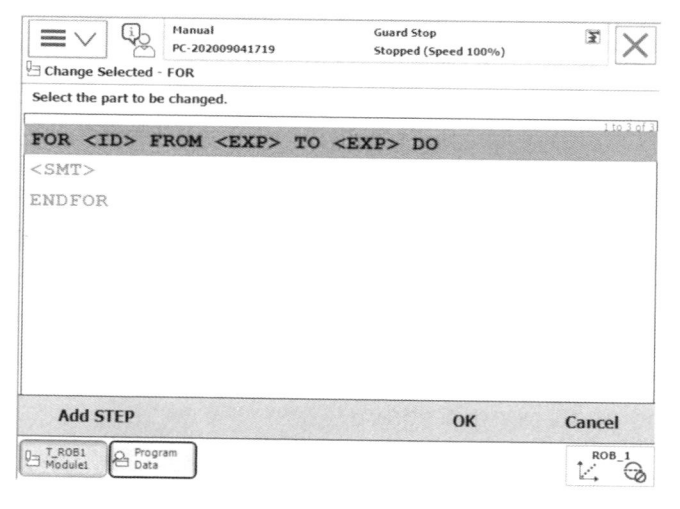

Fig. 8-22　FOR Instruction Editing

174

8. ABB Robot Programming

(3)Tap "Add STEP" button to increase incremental parameters, as shown in Fig. 8–23.

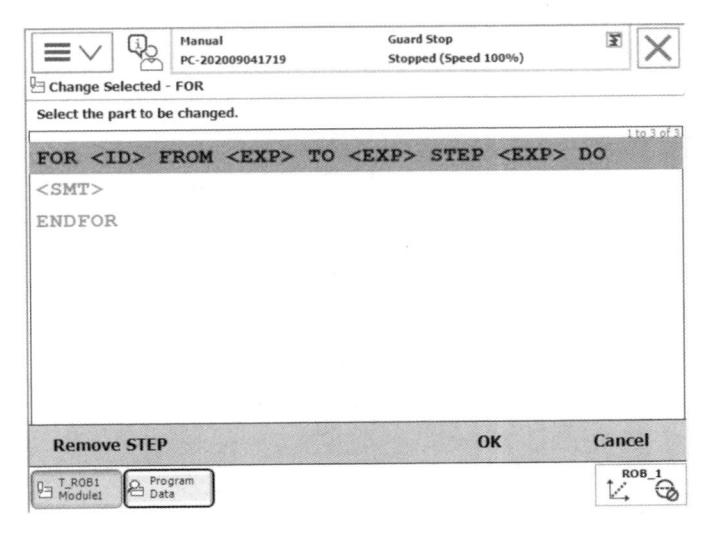

Fig. 8–23   Adding STEP Increment

(4)Double–tap the "<ID>" field to open the input panel and enter the name of counter variable (no need to define it in the program data in advance), as shown in Fig. 8–24.

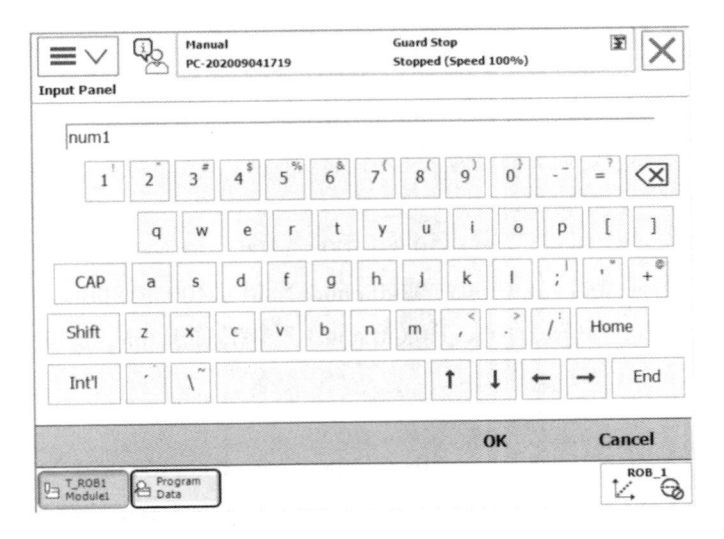

Fig. 8–24   Counter Variable Input Panel

175

Industrial Robot Technology Application

(5)In this way, the expression of counter num1 can be used to express the offset in the placement position Offs, as shown in Fig. 8-25. The offset of Z value is set to num1*50+100.

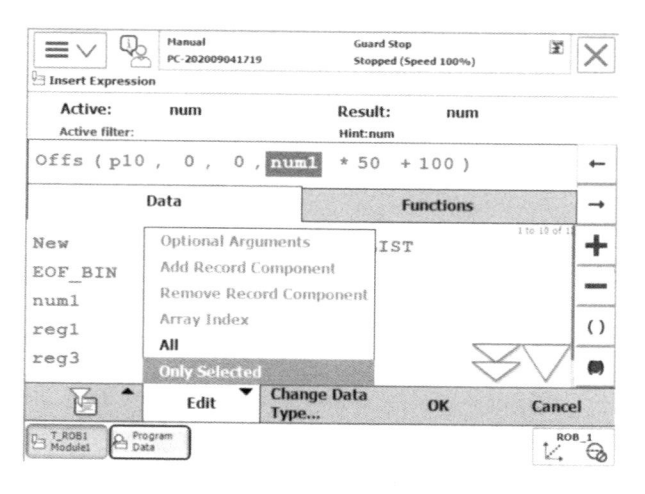

Fig. 8-25   Expression Setting for Offs Parameter

(6)Continue to input instructions in accordance with the above methods, and complete the writing of Routine2( )stacking routine as follows:

```
PROC Routine2( )
    MoveAbsJ jpos10\NoEOffs, v1000, z50, tool0;
    FOR num1 FROM 0 TO 3 STEP 1 DO
        MoveL Offs(p10,-60,0,0), v500, fine, tool0;
        MoveL p10, v30, fine, tool0;
        Set Do1;
        WaitTime 2;
        MoveL Offs(p10,0,0,100), v100, z50, tool0;
        MoveL Offs(p20,0,0,num1 * 50 + 100), v300, z50, tool0;
        MoveL Offs(p20,0,0,num1 * 50), v20, fine, tool0;
        Reset Do1;
        WaitTime 2;
        MoveL Offs(p20,-60,0,num1 * 50), v100, z50, tool0;
        MoveL Offs(p20,-60,0,num1 * 50 + 100), v300, z50, tool0;
    ENDFOR
```

176

ERROR

MoveAbsJ jpos10\NoEOffs, v1000, z50, tool0;

ENDPROC

8.2 Offline programming

Offline programming can achieve a more complicated path. Unlike teaching programming, its accuracy is completely determined by the visual observation of the instructor. It has a high theoretical accuracy. Since there are errors in setting the tool coordinate system and work object coordinate system of the robot, the actual motion track accuracy of offline programming is directly related to the operator's experience. The calibration accuracy of the tool coordinate system and the work object coordinate system can be improved by means of tooling, external high−precision measuring equipment or probes, which can further reduce the influence of human factors on the actual motion track accuracy of offline programming. Offline programming can be divided into 2D offline programming and 3D offline programming.

8.2.1 2D offline programming

As the name implies, 2D offline programming means that the TCP path of the robot is in a plane and Z−axis of the tool remains unchanged. Since Z−axis is fixed, the tool pose changes simply. The path is easy to be controlled, and over−travel and interference are rare. The operation steps are as follows:

(1)Open RobotStudio and create IRB1200 robot workstation, load myTool tool and install it on the flange end face. Take the self−contained model of RobotStudio as an example, click "Basic−Import Model Library−Device−Curve Thing" in turn, as shown in Fig. 8−26.

(2)Determine the placement position of the newly imported model in the robot base coordinate system, and select "Position−Setting Position" in the right−click menu of Curve Thing, as shown in Fig. 8−27.

(3)Set the position and direction parameters in the pop−up dialog box. Generally, it is easier to refer to the world coordinate. The positions before and after motion are shown in Fig. 8−28. Note: The position of the work object in the robot workstation is preferably close to the actual production position. Although this difference can be eliminated by calibrating coordinate system of the work object, too

Industrial Robot Technology Application

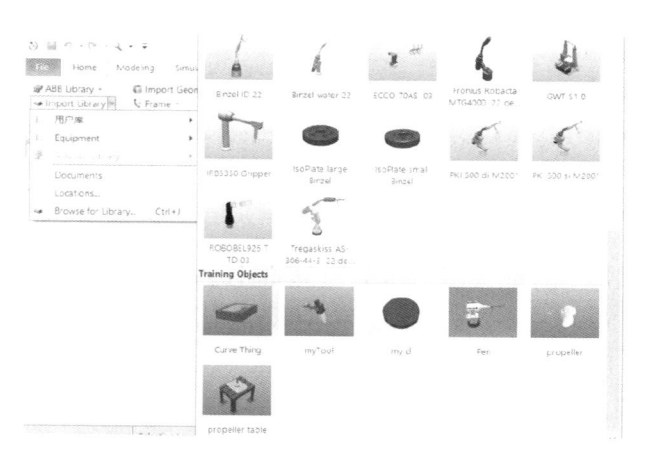

Fig. 8-26　Importing Curve Thing Model

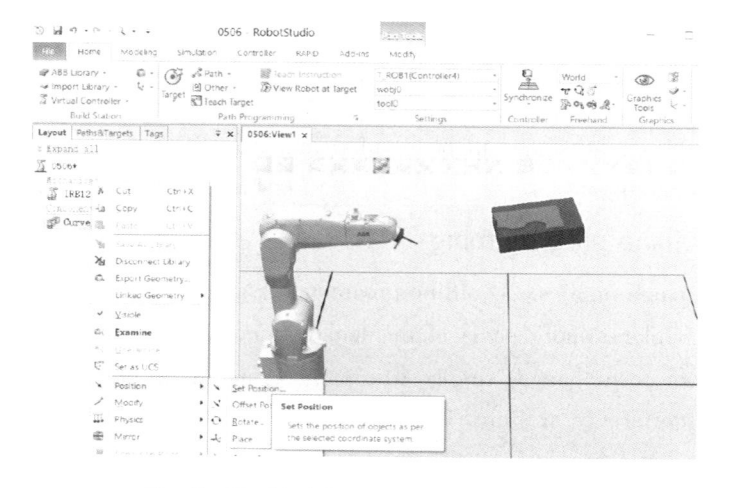

Fig. 8-27　Model Position Setting Option

large position difference may cause robot overreach or interference. In this case, the robot base coordinate system is used as the work object coordinate system.

(4)Click "Basic-Path-Automatic Path" in turn to pop up a dialog box for creating a path by a curve, as shown in Fig. 8-29. When selecting a curve, pay attention to Z-axis direction of the target point. In the figure, the blue axis represents Z direction of the target point, and the red axis represents X direction. In addition, attention should be paid to the setting of options such as motion mode (straight line and circular arc)and tolerance(for straight line approximate curve).

(5)Select each edge of the upper surface in turn, as shown in Fig. 8-30.

178

8. ABB Robot Programming

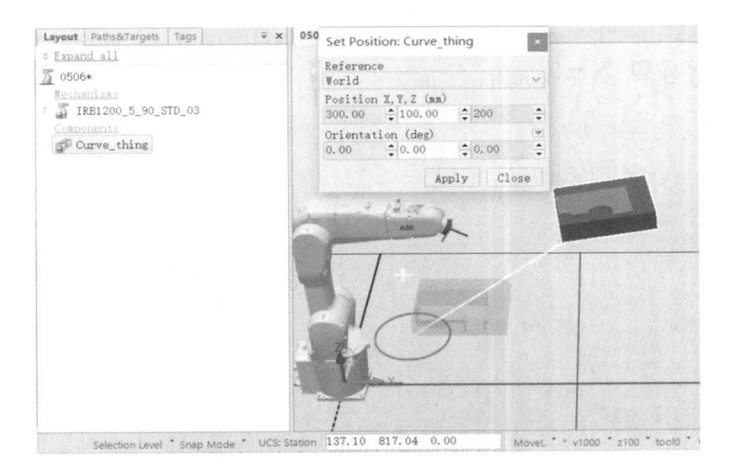

Fig. 8-28　Setting Position Dialog Box

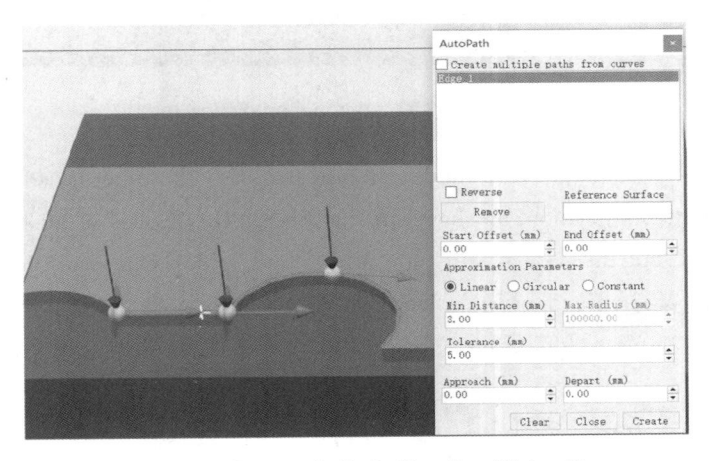

Fig. 8-29　Automatic Path Creation Dialog Box

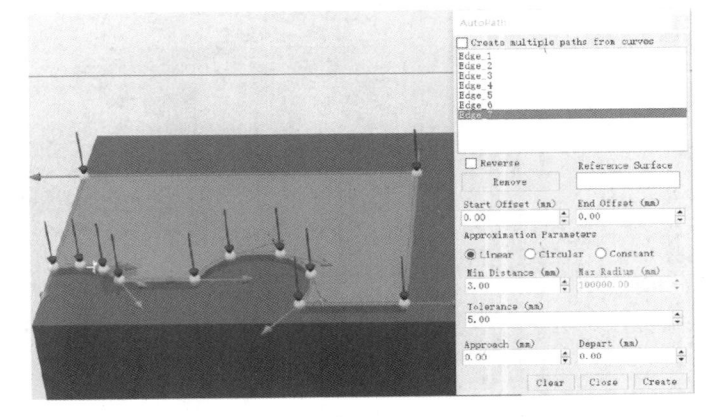

Fig. 8-30　Automatic Path Creation Dialog Box

179

Industrial Robot Technology Application

(6)Click the "Create" button to create Path_10 path, as shown in Fig. 8–31. You can see the newly generated target points and motion instructions in the directory tree.

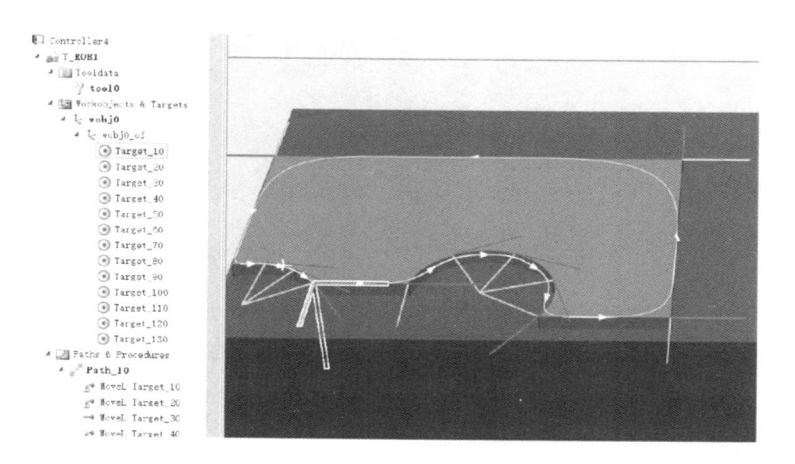

Fig. 8–31　Path_10 Creation Completion Interface

(7)Select the target point and right–click to view the robot target to view the robot pose at this position, as shown in Fig. 8–32. The Z–axis of all target points is downward, but the X–axis direction is different, which can easily lead to self–interference of each axis of the robot. The wiring harness at the end of the tool in the figure is prone to interference.

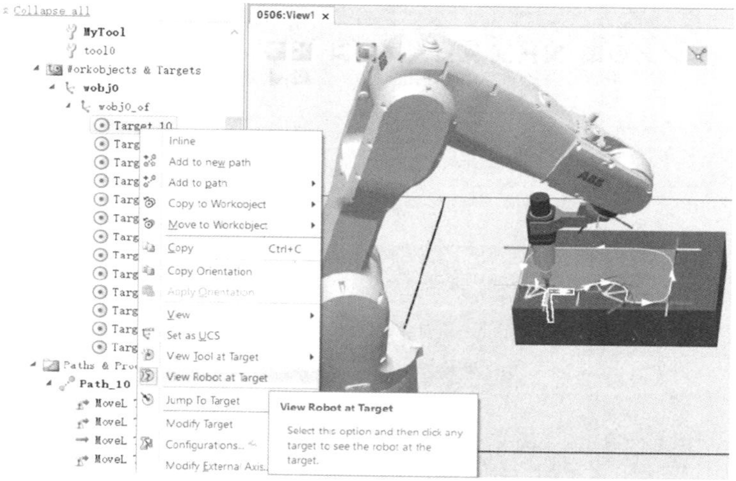

Fig. 8–32　Viewing Robot Target

180

8. ABB Robot Programming

(8)The direction of target points can be modified by using the rotary tool (right-click "Modify Target-Rotate"), as shown in Fig. 8-33. Target_10 target point rotates 180° around Z-axis of its own coordinate system.

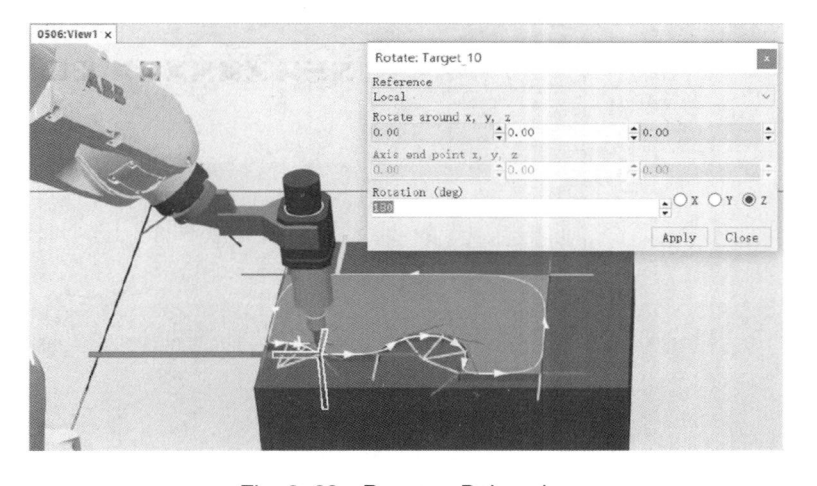

Fig. 8-33   Program Debugging

(9)Copy the Target_10 direction and apply it to other target points in batches to complete batch modification of the direction, as shown in Fig. 8-34.

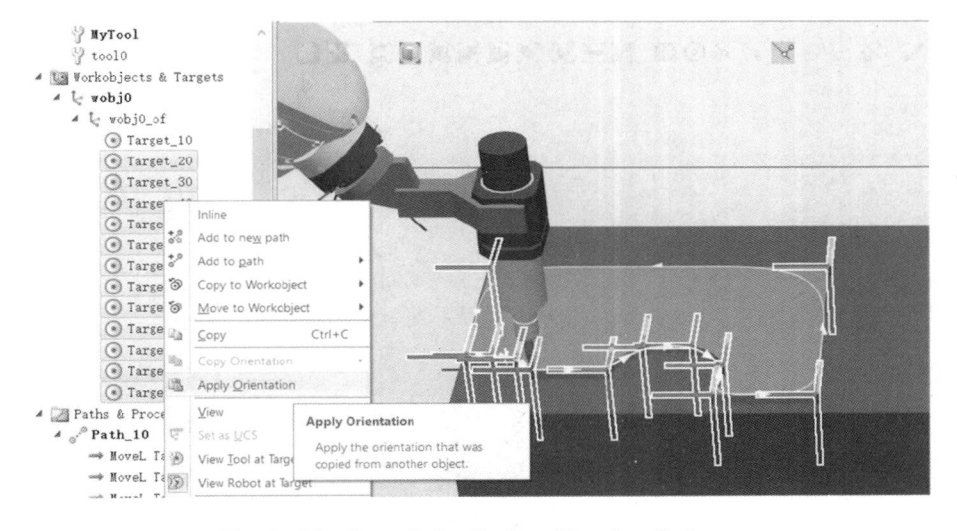

Fig. 8-34   Copy & Application Direction Options

181

Industrial Robot Technology Application

(10)Click "Basic−Target Point−Create Jointtarget" to add the starting point of the program, open the joint data setting dialog box as shown in Fig. 8−35, and click the "Values" button to pop up the input box for modification.

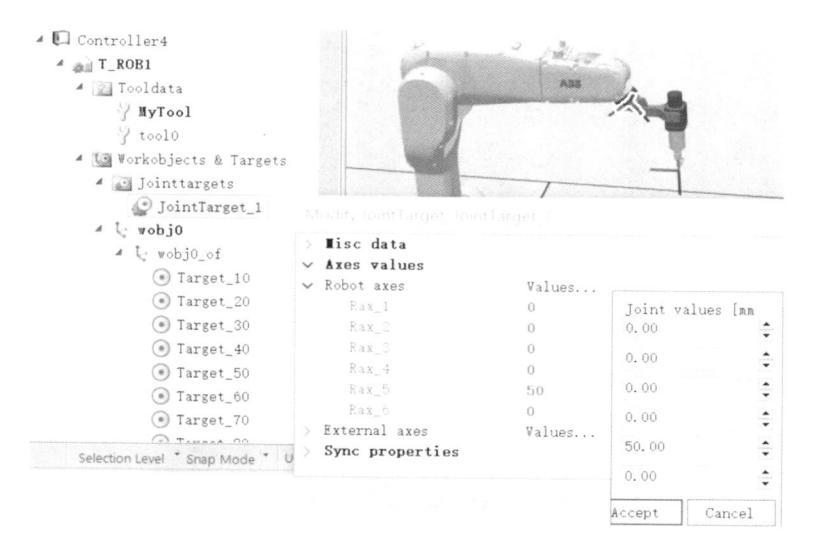

Fig. 8−35   Joint Target Point Creation

(11)Right−click the menu to add the starting point of the program to the beginning and end of the program path, as shown in Fig. 8−36.

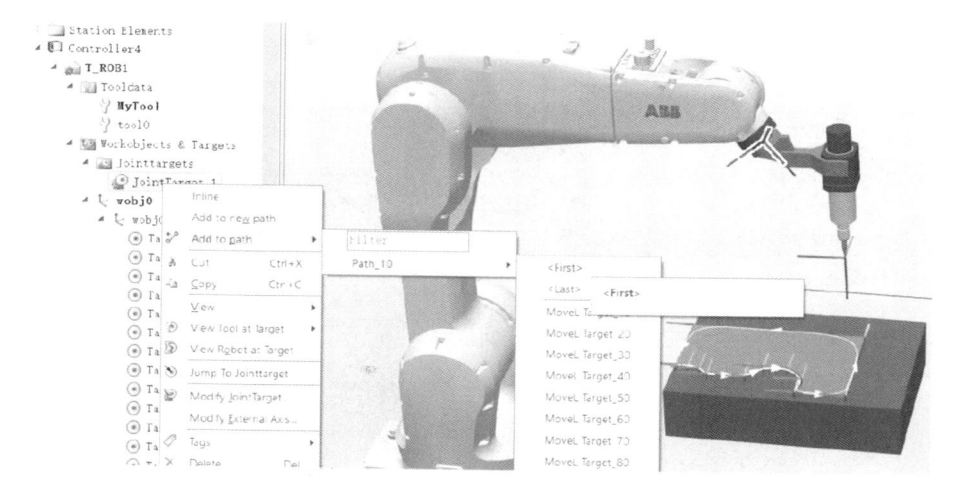

Fig. 8−36   Program Debugging

182

8. ABB Robot Programming

（12）Copy and paste the beginning and end points of the path, which can be used as the path approach points after modification of the Z value, as shown in Fig. 8-37.

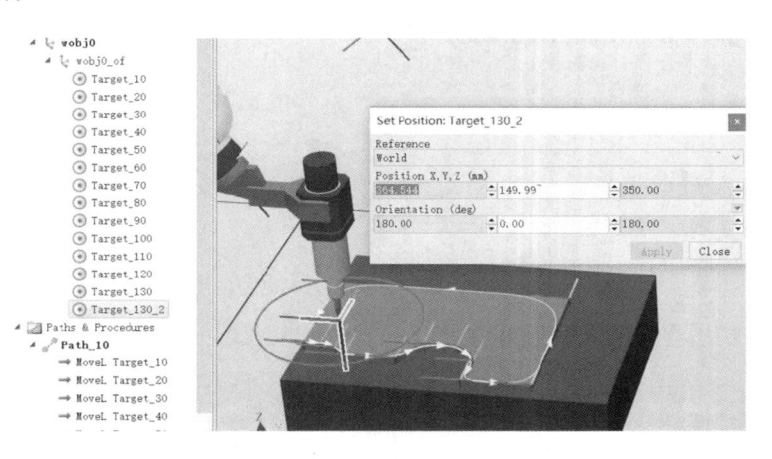

Fig. 8-37　Program Debugging

（13）Select all target points and right-click to select "Modify Instruction - Area - Fine" options in turn to modify area parameters in batches and achieve accurate arrival of all area points, as shown in Fig. 8-38. Parameters such as speed and tools can also be modified in batches.

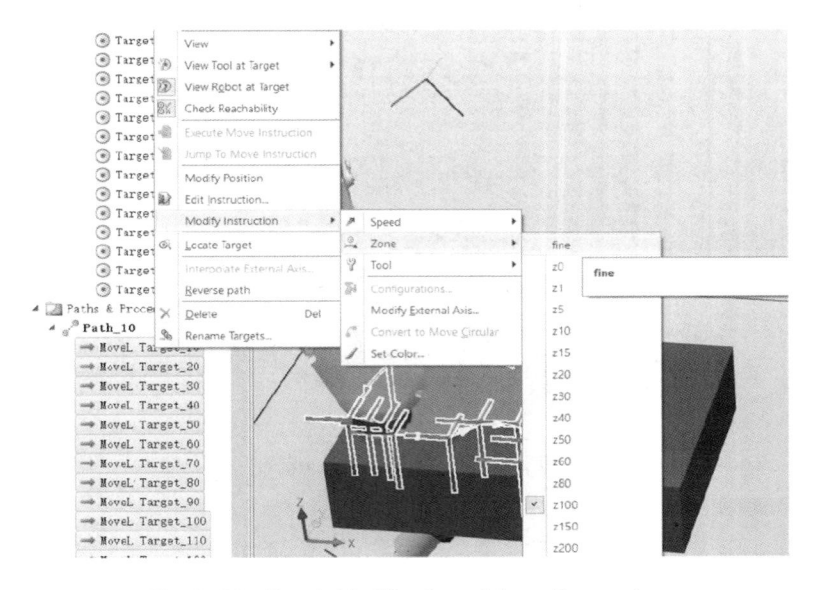

Fig. 8-38　Batch Modification of Area Parameters

183

Industrial Robot Technology Application

(14)After path editing, select Path_10 and right-click the "Synchronize to RAPD" option to pop up the dialog box as shown in Fig. 8–39. Select the path and tool to be synchronized and click "OK" to convert the path into robot instruction and add it to the virtual controller.

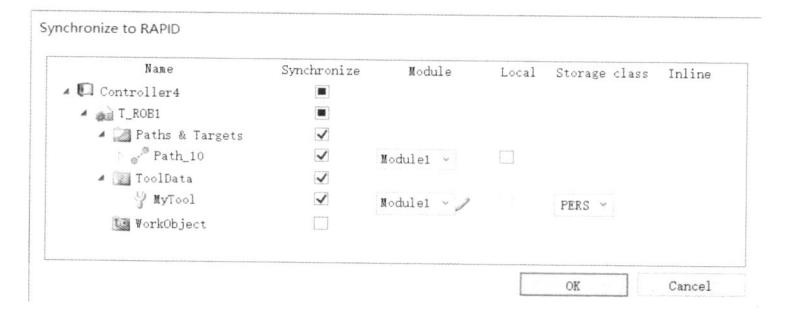

Fig. 8–39　Program Debugging

(15)Open the virtual teach pendant, and you can see Path_10 routine as shown in Fig. 8–40.

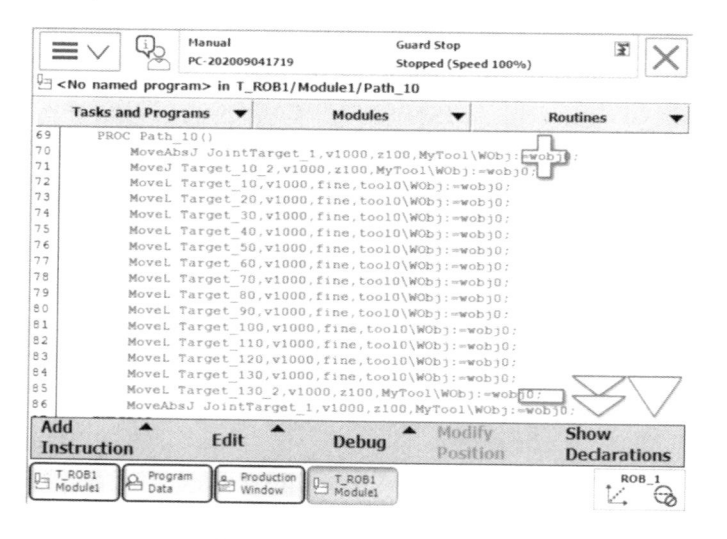

Fig. 8–40　Path_10 Routine

184

8. ABB Robot Programming

(16)Instructions such as I/O signals can be added as required, and programs that have no problems in the simulation test in the virtual workstation can be exported using the save–as module tool, as shown in Fig. 8–41. See the prompt line in the figure for the path of the exported .MOD file. Copy it to the real robot with a USB flash disk and import it into the module to complete the input of offline programs.

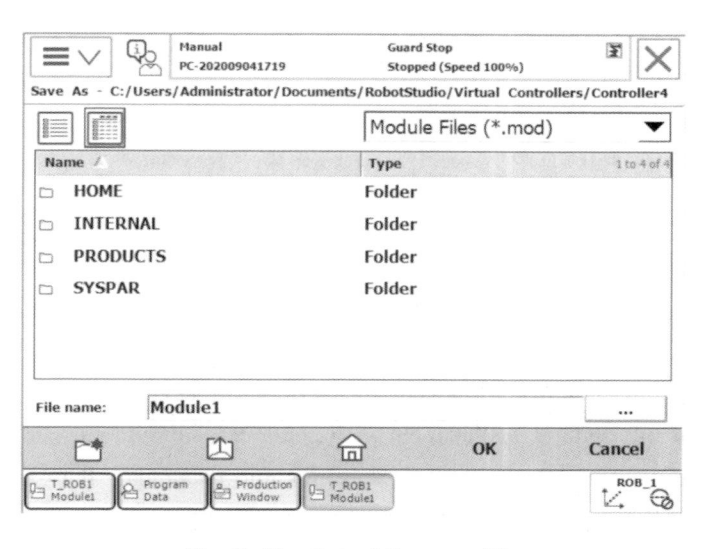

Fig. 8–41   Output Program File

8.2.3 3D offline programming

Generally, the tool coordinate Z–axis of 3D offline programming changes with the shape of the work object. Tools such as laser cutting and laser cladding shall be always perpendicular to the surface of the work object, and tools such as welding and gluing shall maintain a certain angle with the surface of the work object. If the shape of the work object is complicated, the tool pose shall change constantly, and the robot is prone to over–travel and interference. Sometimes, a path may need to be divided into two paths for programming to solve the problem that the robot pose cannot be adjusted continuously. The method of 3D offline programming is basically the same as that of 2D programming, and its main operation steps are as follows:

(1)Open RobotStudio and create IRB1200 robot workstation, load myTool tool

185

Industrial Robot Technology Application

and install it on the flange end face. Import the 3D model for offline programming, and the options are shown in Fig. 8–42.

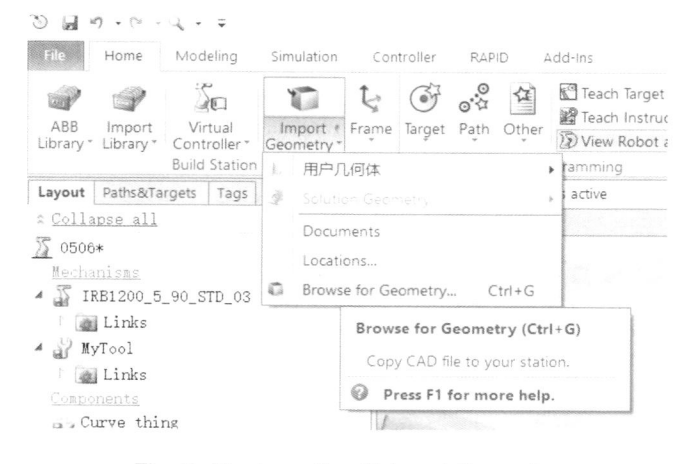

Fig. 8–42　Importing External Geometry

（2）Set the position coordinates of the work object as shown in Fig. 8–43. If the collision and interference of the robot path generated subsequently are difficult to eliminate, try to adjust the work object position and recalculate the path.

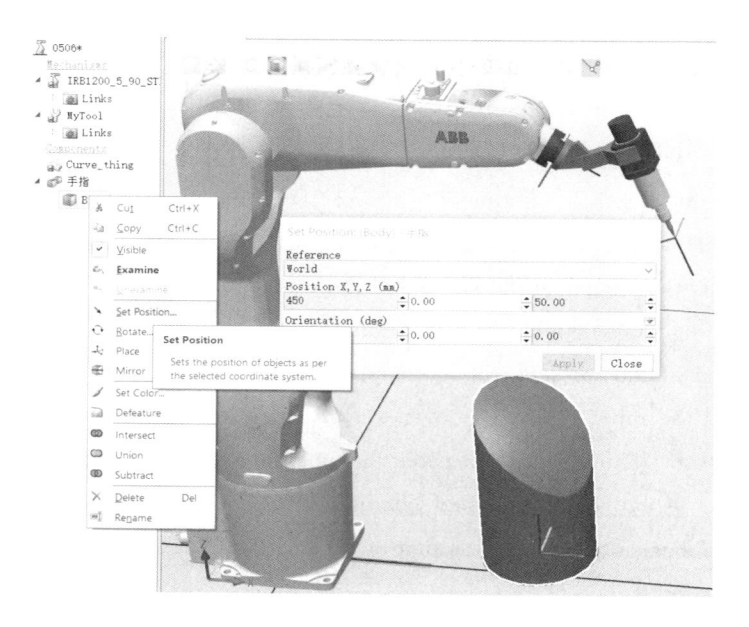

Fig. 8–43　Setting Work Object Position

186

8. ABB Robot Programming

(3)Reasonable setting of the work object coordinate system can simplify its calibration during actual operation. Fig. 8−44 shows the method of creating a co-ordinate system in the software(Menu − Basic − Frame − Create Frame). The origin of the coordinate system in the figure coincides with the center of the bottom surface of the work object.

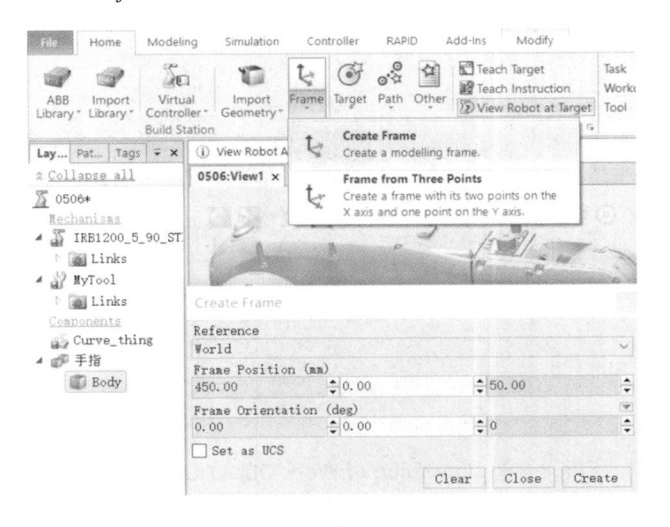

Fig. 8−44　Creating Framework Interface

(4)The frame created by RobotStudio is actually the coordinate system. The newly created frame is displayed in the layout view as shown in Fig. 8−45, and converted into work object coordinates.

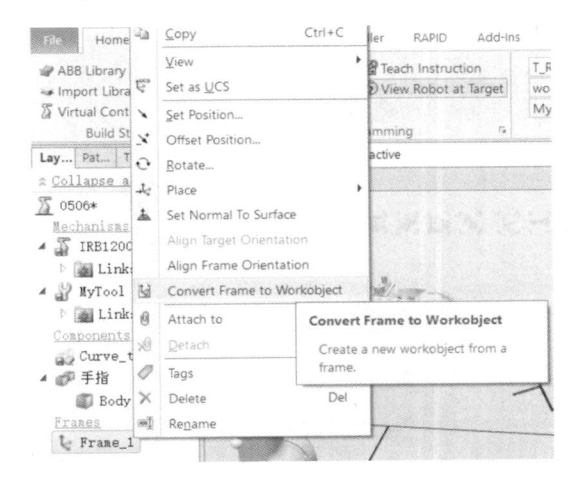

Fig. 8−45　Converting Frame to Work Object Coordinates

187

Industrial Robot Technology Application

(5)The newly converted work object coordinate system is displayed in the "Path and Target Point" view, as shown in Fig. 8-46. Right-click to set it to active mode, and perform the same operation to ensure that T_ROB1 and MyTool are activated to prepare for the next programming.

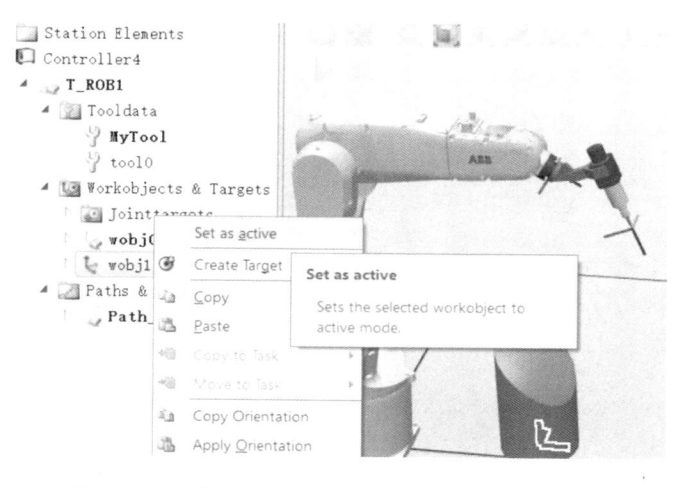

Fig. 8-46　Activation of Work Object Coordinates

(6)Open the "Auto Path" tool to select the upper surface edge of the work object, as shown in Fig. 8-47. Pay attention to setting circular motion and a tolerance of 0.5 mm. Pay attention to the Z-axis direction when selecting, and click the "Create" button after completion.

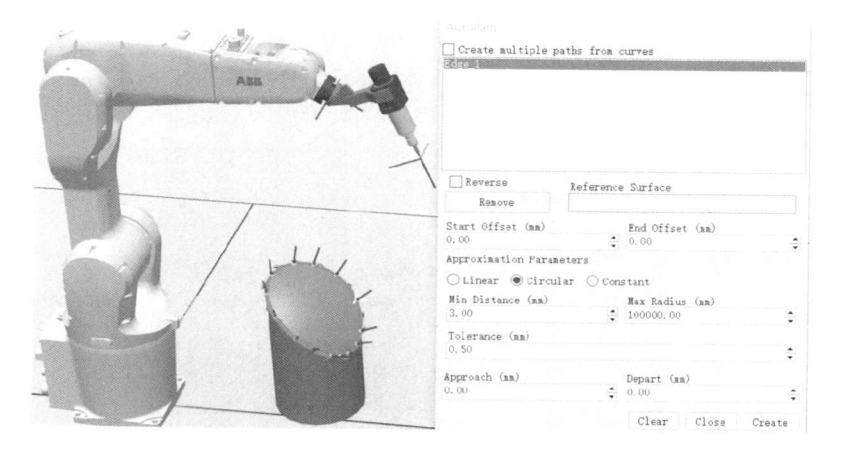

Fig. 8-47　Automatic Creation of Path

8. ABB Robot Programming

(7) The Z direction of the automatically created target point is perpendicular to the curved surface, and the X direction is tangent to the path. You can select the target point and right-click it to view the robot pose, as shown in Fig. 8-48. It can be seen that the tool pose rotates 360° around the Z-axis from the beginning to the end, which can easily cause the robot overreach or interference. The pose shown in the figure is not reasonable.

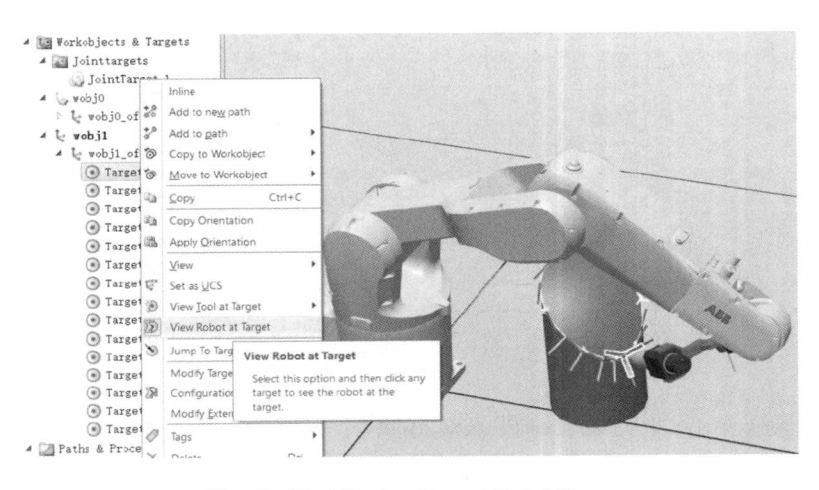

Fig. 8-48　Viewing Target Point Pose

(8) Switch between different target points until a target point with a reasonable robot pose is found, such as Target_60 shown in Fig. 8-49.

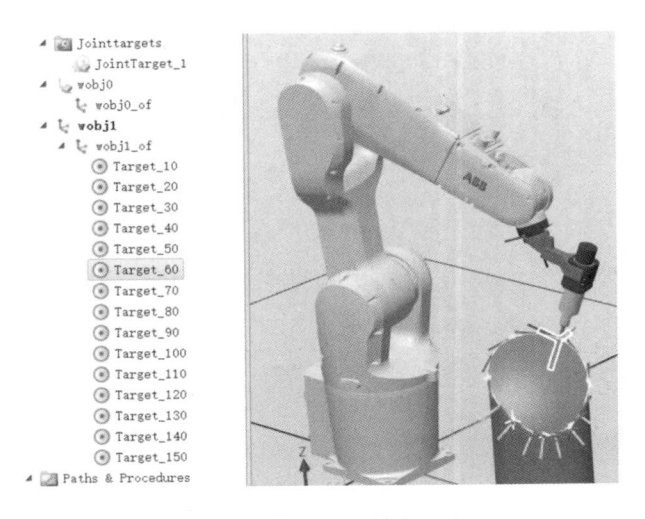

Fig. 8-49　Program Debugging

189

Industrial Robot Technology Application

(9)Select the remaining target points, right-click to modify the target, and se-lect the "Align with target point direction" option, as shown in Fig. 8-50.

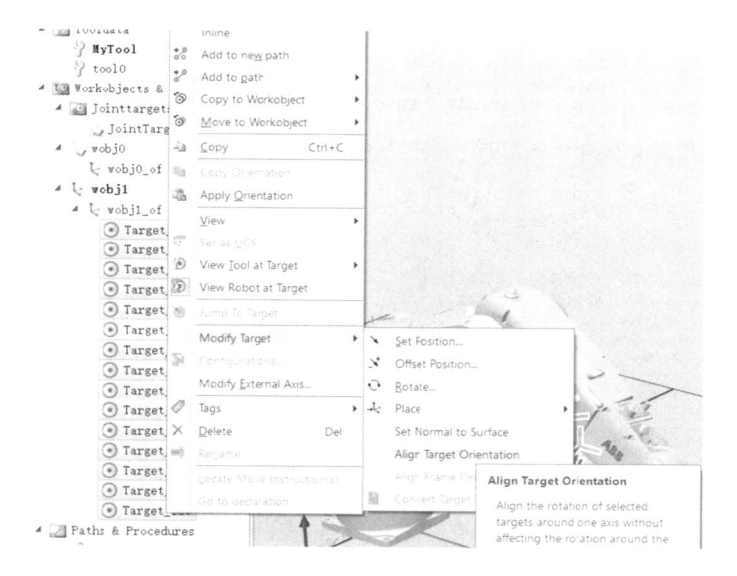

Fig. 8-50   "Align with target point direction" Option

(10)Open the "Align with target point direction" dialog box as shown in Fig. 8-51, select the reference coordinate system as Target_60, select X for the align-ment axis and click the "Apply" button. The alignment effect is shown in the figure.

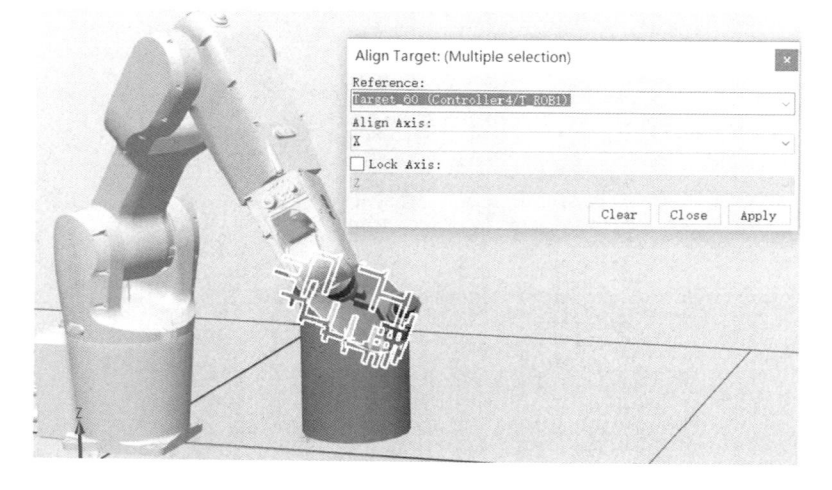

Fig. 8-51   "Align with target point direction" Dialog Box

8. ABB Robot Programming

(11) After aligning the direction, check the original problematic target point, as shown in Fig. 8–52. It can be seen that there is no self–interference in the robot pose, and the current pose can be fine–tuned with a rotating tool alone. Finally, select the overall path in the directory and right–click "Move along the path" for motion simulation.

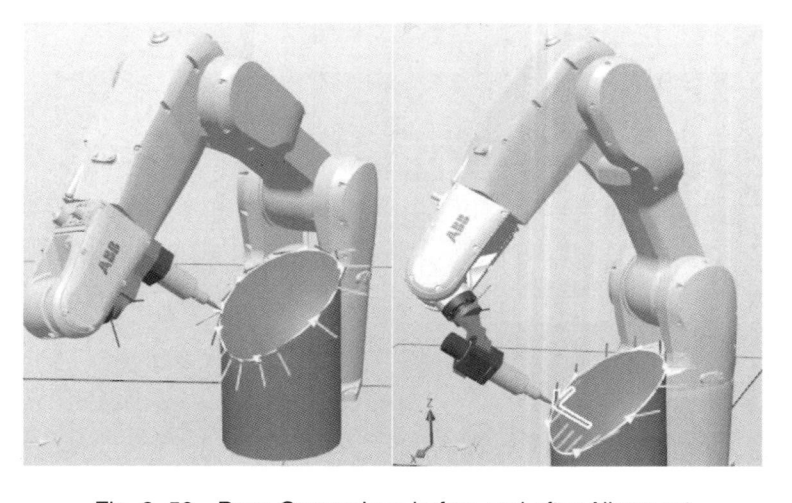

Fig. 8–52   Pose Comparison before and after Alignment

(12) After ensuring that there is no problem with the edited path, you can synchronize the path to the virtual controller, further add peripheral control instructions, perform run simulation in the virtual controller, and export the MOD file after ensuring that there is no problem.

Note: Since there is a certain deviation between the actual work object coordinate system and that set in the software relative to the base coordinate system of the robot, the offline programming program must be tested manually at low speed during the first run on the real robot to ensure that there is no error before automatic operation is allowed.

191

9.

Maintenance and Common Fault Repair for Robot

9.1 ABB robot maintenance

Due to the harsh working conditions, the industrial robot must do routine inspection and preventive maintenance regular. even if there are perfect protection design standards. Proper robot maintenance will prolong the service life of the robot effectively and reduce the probability of failure. It will also greatly help the stable operation of production. Therefore, it is necessary to understand and strictly implement the daily maintenance for industrial robots.

9.1.1 Robot maintenance

(1)General maintenance and repair: Before running, check the brake of each shaft motor, operate each shaft to its maximum load position, turn off the motor to check whether the shaft stops at the original position. If the brake is good, the shaft remains at the original position. Carry out a spot check to confirm the origin position of the robot. Do not change the gear ratio or other motion parameters from the computer or the teach pendant. Otherwise, it will affect the deceleration running function.

(2)Precautions for use during operation: Pay attention to the robot's hydraulic pressure, pneumatic pressure, instrument indication, monitoring signal and safety device, and use the enabler key of the teach pendant correctly. If the key is half-pressed, the system is powered on; if released or pressed fully, the motor is turned off. When the enabler key does not require function programming or de-

193

Industrial Robot Technology Application

bugging, and the robot does not need to move, release the key immediately. When the programmers enter the safe area, they must wear the teach pendant belt to avoid misoperation by others. Enter the work area, and select the manual mode switch for the controller to facilitate the operation of the enabling device to disconnect the computer or remote control operation. Pay attention to the rotation axis of each arm of the manipulator, and other selected components or equipment on the manipulator.

(3)Maintenance and repair cycle: The interval of maintenance cycle can be adjusted properly in accordance with the environment, frequency, strength and temperature of the machine.

(4)Maintenance and repair of body: mainly including cleaning and inspection of the manipulator, inspection of axle brake test and inspection of reducer lubrication.

9.1.2 Robot maintenance standard

9.1.2.1 Maintenance of spare parts

(1)Lubricating grease;

(2)Spare parts kit for maintenance: SMB battery, cooling fan, motor lamp and fuse.

9.1.2.2 Standard body maintenance

(1)Routine inspection

Body cleaning: clean the robot body in accordance with the field work.

Fixing inspection of body and 6−axis tool end: check whether the body and tools are fixed well.

Inspection of the limit stop of each shaft.

Inspection of cable status: check the service and wear conditions of the robot's signal cables, power cables, user cables and body cables.

Inspection of sealing status: check the body gearbox and wrist for oil leakage or seepage.

(2)Functional measurement

Mechanical zero measurement: check whether the current zero position of the robot is consistent with the standard calibration position.

Inspection of motor braking status: check the braking voltage value of the

9. Maintenance and Common Fault Repair for Robot

opened motor, and test the motor braking function of each shaft.

(3)Replacement of maintenance parts

Replacement of body oil: replace oil of robot gearbox, balancing cylinder or connecting rod.

Inspection of robot SMB board and battery replacement: check whether the fixed connection of SMB board is normal, and replace the battery.

9.1.2.3 Standard maintenance of control cabinet

(1)Routine inspection

Cleaning of control cabinet: clean the appearance of robot control cabinet and remove dust inside it.

Soundness inspection of all components of the control cabinet: check the fastening state of all components in the control cabinet.

Cleaning of teach pendant: cleaning and sorting of the teach pendant and cables.

Status of circuit board indicator: check the status indicator of each circuit board in the control cabinet and confirm the board status.

Inspection of cables inside the control cabinet: All cable plugs inside the control cabinet are connected firmly, and the cables are neat.

(2)Control cabinet measurement

Power supply voltage measurement: measure the incoming voltage, drive voltage and power supply module voltage of the robot for overall evaluation.

Safety circuit detection: check whether the operation status of safety circuit (AS, GS, ES)is normal.

Teach pendant function detection: check the effectiveness of all keys, whether the emergency stop circuit is normal, and test the functions of the touch screen and display screen.

System calibration compensation value detection: check whether calibration compensation value parameters of the robot is consistent with the factory configuration values.

System backup and import detection: check whether the robot can perform the program backup and re-import functions normally.

Hard disk space detection: optimize the hard disk space of the robot control cabinet to ensure normal operation space.

Industrial Robot Technology Application

(3)Replacement of maintenance parts

Replacement of drive unit cooling fan;

Replacement of control cabinet fuse;

Replacement of power-on button indicator of the motor on the operation panel of control cabinet;

9.1.2.4 Summary of standard maintenance report

(1)Suggestions after standard robot maintenance;

(2)List of recommended spare parts after standard robot maintenance.

9.1.2.5 Standard maintenance cycle: one year or 3,000-4,000 hours of operation.

9.1.3 Replacement of SMB battery

After the main power supply is turned off, the robot will rely on SMB battery to save the data of 6 axes. Battery run-out will cause zero information loss of the robot, thus affecting the accuracy and normal operation of the robot. When the remaining backup power of the battery(robot power-off)lasts for less than 2 months, the ABB robot system will display the low battery alarm(38213 Low battery).

SMB and battery are installed in the robot body, and their installation positions vary according to the robot model. For the specific positions, you can query the product manual of the corresponding robot model. The battery of IRB1200 robot is shown in Fig. 9-1. Replacing the SMB battery before run-out can save the trouble of manually finding the zero position. Therefore, the operation specification for battery replacement is particularly important. The specific steps are as follows:

(1)Return all joints of the robot to zero. The purpose of this step is to update the revolution counter after battery replacement. Each joint is marked with zero.

(2)Turn off all power, air, and hydraulic sources connected to the robot.

(3)First conduct electrostatic dis-

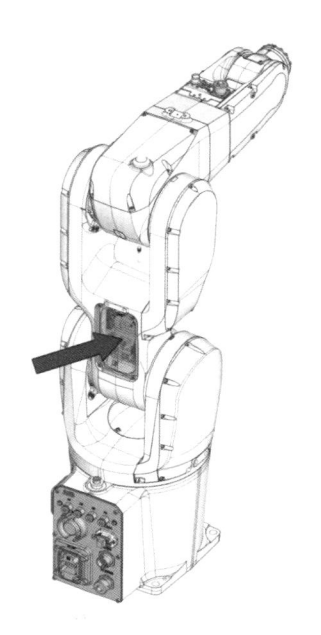

Fig. 9-1　Battery Position

196

9. Maintenance and Common Fault Repair for Robot

charge (ESD), then remove connecting screws of the lower arm connector cover and open the cover carefully (The cover needs to be cleaned before opening, because conductive impurities may cause short−circuit fault of the robot), as shown in Fig. 9−2. (Note: ⚠ ESD is the electrostatic conduction between two objects with different potentials. It can be conducted by direct contact or induced electric field. Ungrounded personnel may transmit large amounts of electrostatic charge when handling components or component containers. This discharge process may damage sensitive electronic equipment. Therefore, protection against electrostatic discharge shall be provided when this symbol is displayed.)

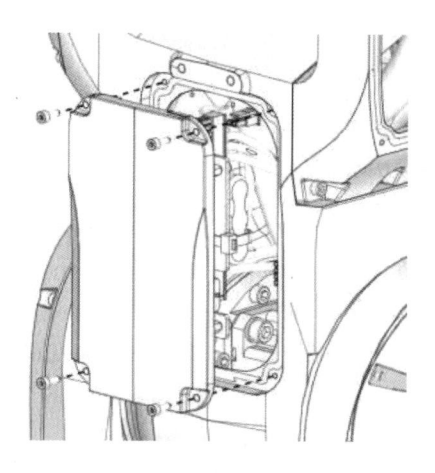

Fig. 9−2 Opening Connector Cover

(4)Disconnect battery cables from the encoder interface circuit board(Note: the clip needs to be pressed, and do not unplug the plug forcibly). Cut off the tie with scissors and remove the old battery, as shown in Fig. 9−3.

(5)Reinstall the new battery, fix it with a tie, and cut off the excessive tie. Connect battery cables to the encoder interface circuit board, and reinstall the connector cover to the robot body with its connecting screws.

(6)Update the revolution counter as explained in Section 4.4.2, and replace the battery.

(Note: Some robots can be powered on during battery replacement. In this case, the revolution counter does not need to be updated after battery replacement, and the robots can continue to be used. Refer to the User's Manual of the

197

Industrial Robot Technology Application

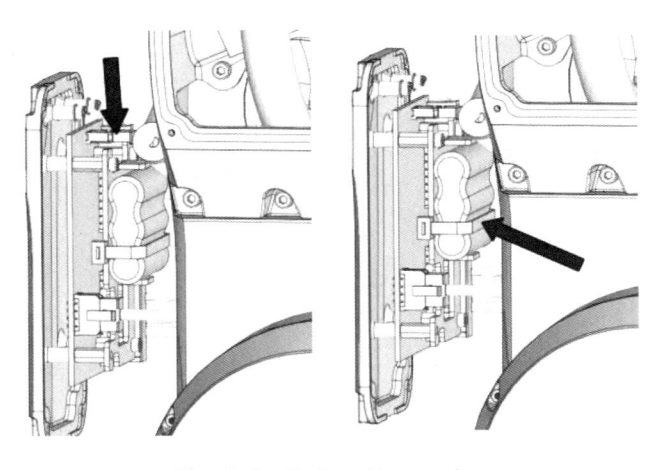

Fig. 9-3　Battery Removal

corresponding robot for specific requirements.)

9.1.4 Cleaning

Regular cleaning is very important for long-term normal operation of the robot. The cleaning frequency depends on the operating environment of the robot. Different cleaning methods can be adopted for different protection types of robots. Be sure to confirm the protection type of the robot and cleaning precautions specified in the instructions before cleaning. The specific requirements are as follows:

(1)Always use the required cleaning equipment. Otherwise, any other cleaning equipment may shorten the service life of the robot.

(2)Be sure to check whether all protective covers have been installed on the robot before cleaning.

(3)Do not spray water on connectors, joints, seals or gaskets.

(4)Do not clean the robot with compressed air.

(5)Never use solvents not approved by ABB to clean the robot.

(6)Do not spray water within 0.4m.

(7)Do not remove any covers or other protective devices before cleaning the robot.

Note: Robots with standard IP grade shall be cleaned by the vacuum cleaner or cleaning cloth. Do not to flush them with water. Robots with IP67 grade can be flushed with water, but it is strongly recommended to add anti-rusting agents to the water and dry the manipulator after cleaning.

198

9. Maintenance and Common Fault Repair for Robot

9.2 Common fault repair for robot

9.2.1 Manual operation error of robot

It shows that the robot cannot be operated manually with a teach pendant. The troubleshooting method is as follows:

(1)Firstly, make sure that FlexPendant is connected to the Control Module correctly.

(2)The possible reason is that the joystick on the teach pendant was touched accidentally during the robot startup(At this moment, the error message will show that the joystick cannot be used, as shown in Fig. 9–4). Tap the "Restart" option in the ABB menu to restart the robot.

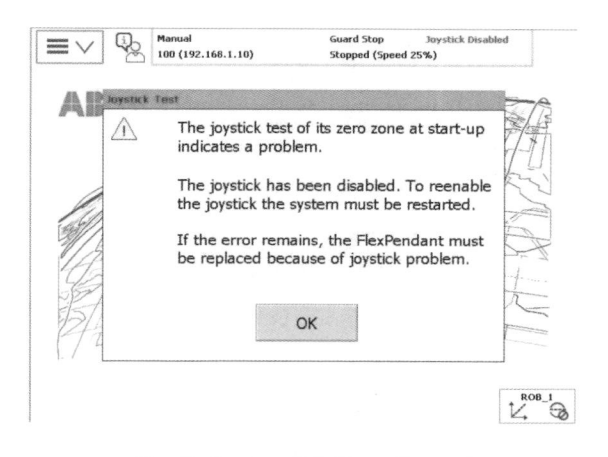

Fig. 9–4   Joystick Error Prompt

(3)The robot system can be started normally, but the joystick on FlexPen-dant cannot work. The possible reason is that the joystick is faulty or deflected. It is necessary to ensure to press the reset button on the back of FlexPendant with the controller in manual mode. Pressing the reset button will reset FlexPendant, but not reset the system on the controller.

9.2.2 Power–on Connection Error

If the robot is powered on, but the teach pendant keeps displaying "Connect-ing to the robot controller" and it fails to enter the system, as shown in Fig. 9–5,

199

Industrial Robot Technology Application

Fig. 9–5　System Connection Error

the reason is that there is no communication connection between the teach pendant and the robot master controller. The reasons for not establishing the connection include:

(1)The robot master controller is faulty;

(2)The built–in CF card(SD card)of the robot master controller is faulty;

(3)The network cable between the teach pendant and the master controller is loose.

Handling methods are as follows:

(1)Check whether the master controller and the SD card inside it are normal?

(2)Check whether the network cable between the teach pendant and the master controller is connected normally? The network ports from the teach pendant to the master controller are all green tag ports of the master controller, as shown in Fig. 9–6.

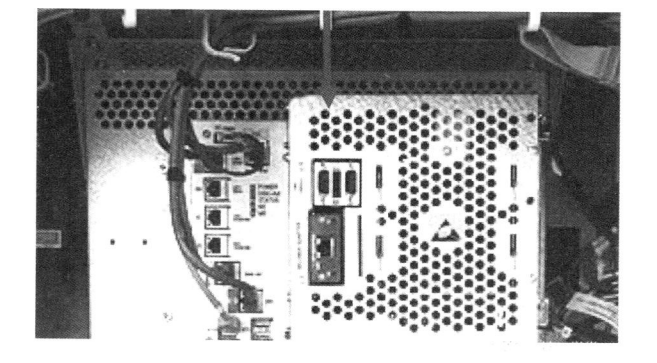

Fig. 9–6　Network Cable Connection

200

9. Maintenance and Common Fault Repair for Robot

9.2.3 SMB memory data difference(50296)

If the robot gives a prompt of 50296 alarm(SMB memory difference)as shown in Fig. 9–7, the fault reason is due to inconsistent data in SMB and control cabinet after SMB is replaced.

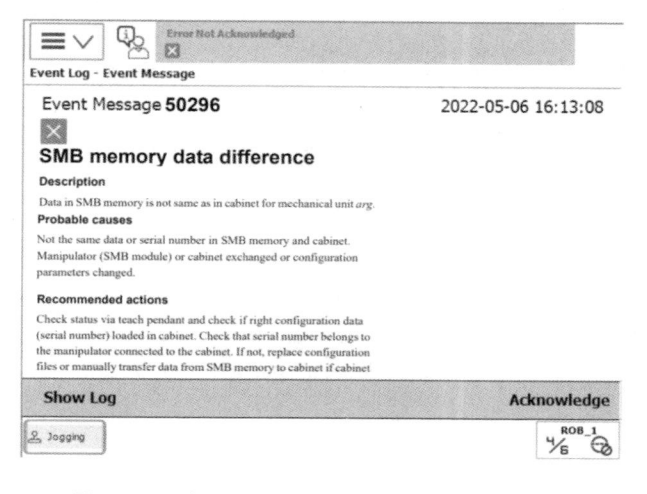

Fig. 9–7   SMB Memory Data Difference Alarm

Handling methods are as follows:

(1)Tap "ABB"–"Calibration", and then "SMB memory", as shown in Fig. 9–8.

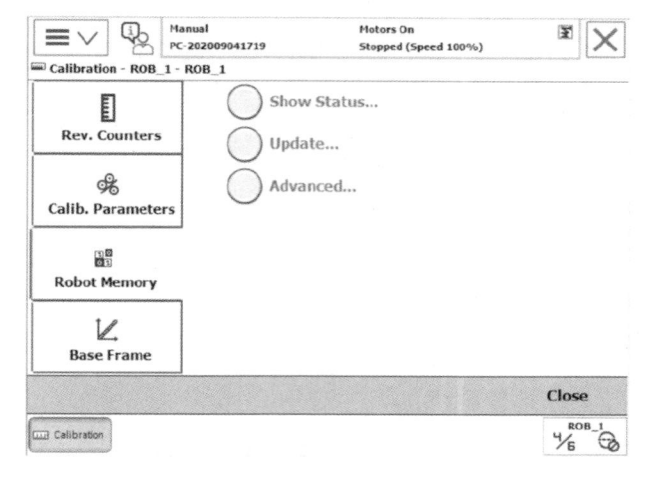

Fig. 9–8   SMB Memory Options

201

Industrial Robot Technology Application

(2)Select "Advanced" and then "Clear SMB" (Since SMB is replaced, select "Clear control cabinet memory" if the controller card is replaced), as shown in Fig. 9–9.

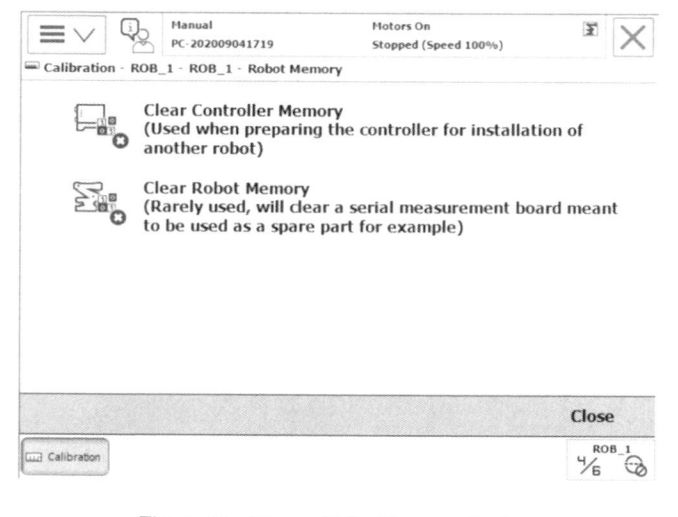

Fig. 9–9   Clean SMB Memory Option

(3)Tap "Close" and select "Update", as shown in Fig. 9–10.

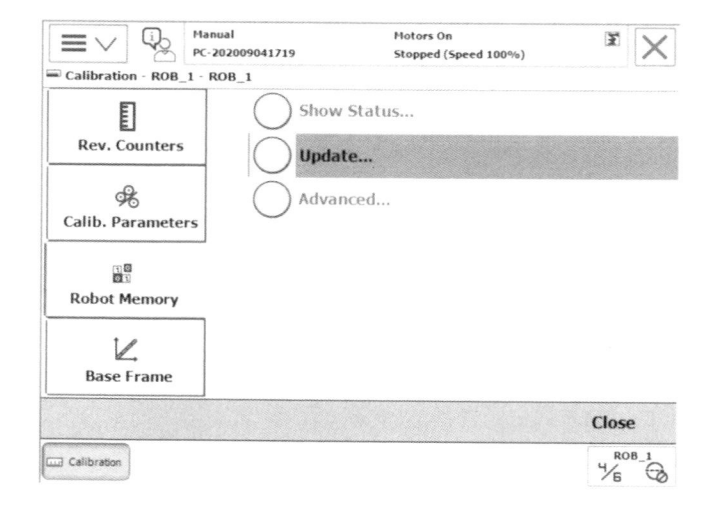

Fig. 9–10   Clear SMB Memory Option

9. Maintenance and Common Fault Repair for Robot

(4)Select "Replace the SMB circuit board." option, and finally re—update the revolution counter(see Section 4.4.2 for details), as shown in Fig. 9—11.

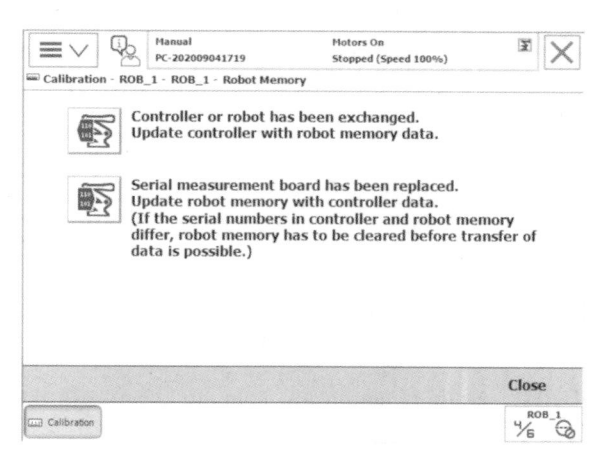

Fig. 9–11   Replace the SMB Circuit Board Option

9.2.4 Fault of startup system

The robot enters the system fault state upon startup, and the handling methods are as follows:

(1)Restart the robot.

(2)If it doesn't work, check whether there are more detailed alarm prompts on the teach pendant, handle these prompts, and then restart the robot.

(3)If the alarm still cannot be cleared, please try B startup. After restarting the controller, it will return to the last auto—saved state.

(4)If it still doesn't work, please try P startup. After restarting the controller, RAPID will be reset, and the program will be deleted, but the configuration data will be kept.

(5)If it still doesn't work, please try I startup, which will restore the robot to the factory setting status and delete all programs and configuration data. It is necessary to operate with caution.

9.2.5 Poor controller performance

If the controller is completely unresponsive, and the FlexPendant cannot operate the system, it may be that the controller is not connected to the main power

Industrial Robot Technology Application

supply, or the main fuse(Q1)may have been disconnected. Check and replace the controller.

If the controller performance is poor, and the program is executed slowly, cannot be executed properly or sometimes stops, it may be caused by heavy load of computer system, specifically by one or more of the following reasons:

(1)The program only contains logic instructions with too high level, resulting in too fast program cycle and the processor overload.

(2)The I/O update interval is set to a low value, resulting in frequent updates and excessive I/O loads.

(3)Internal system cross-connection and logical functions are used too frequently.

(4)External PLC or other monitoring computers address the system too frequently, resulting in system overload.

Try the following solutions:

(1)Check whether the program contains logic instructions(or other instructions that "take no time" to execute), because such programs will cause execution cycles when conditions are not met. The test can be conducted by adding one or more WAIT instructions, only using shorter WAIT instructions to avoid slowing down the program unnecessarily. WAIT instructions can be added properly in the main routine, better to approach the end; in the WHILE/FOR/GOTO cycle, better to approach the parts of ENDWHILE/ENDFOR instructions at the end.

(2)Ensure that the I/O update interval value of each I/O board is not too low, so that I/O units that are not frequently read can perform "state change" operations.

(3)Check whether there is a lot of cross-connection and I/O communication between the PLC and the robot system, and disconnect unnecessary connection and communication.

(4)Try to edit the PLC program with event-driven instructions instead of using cyclic instructions. The robot system has many fixed inputs and outputs to achieve this. Try to optimize the program.

9.2.6 Inconsistent path accuracy

It is manifested by inconsistent path of the robot TCP, sometimes accompa-

nied by noise from bearings, gearboxes or other locations, and frequent changes in accuracy that affect normal production. Possible causes and solutions are as follows:

(1)If the robot is not calibrated properly, check the position of the revolution counter, and recalibrate the axes of the robot if necessary.

(2)If the robot TCP is not defined correctly, check whether the robot tool is defined correctly by reorienting movement mode or auxiliary tools.

(3)Damaged mechanical joints between the motor and gear often make the faulty motor noisy. Identify the noise source and check the corresponding motor.

(4)If the bearing is damaged or broken, especially if the coupling paths are inconsistent, and there is a tap or friction noise in one or more bearings, identify the noise source and check the corresponding bearing.

(5)The brake is not released properly.

9.3 Safety operation specification for robot

Since there are safety risks in the transportation, installation, commissioning and operation of industrial robots, it is necessary to understand these risks, master the general safety specifications for industrial robots, and establish correct safety operation awareness. This section focuses on the general safety specifications for operation of industrial robots.

(1)Operators must wear the personal protective equipment properly;

(2)The robot operators must have received professional training, be familiar with various safety warning signs on the robot body and control cabinet, and can control the robot action by manual or automatic programming according to key points. After power−on, the personnel not receiving training shall not touch the robot and teach pendant to avoid personal injury or damage to the equipment.

(3)Safety separation belts must be set up around the robot equipment, which must be clean and free of oil, water and sundries.

(4)Before startup, check whether the electricity and gas supply is normal.

(5)During commissioning and operation of the robot, some unexpected or non−standard movements may be performed, and all movements will generate great force, which will cause serious personal injury or damage to any equipment

within the working range of the robot. Therefore, always keep a sufficient safe distance from the robot.

(6)When entering the robot operation area, if there are workers in the robot operation area, be sure to operate the robot system manually. Prepare the teach pendant to control the robot at any time. Before approaching the robot operation area, make sure that the robot and moving tools have stopped.

(7)Pay attention to the fixture and ensure to clamp the work object properly. If the fixture is opened, the work object will fall off, resulting in personal injury or damage to the equipment. Some fixtures are very powerful, and may cause personal injury if not operated in the correct way.

(8)Pay attention to hydraulic and pneumatic systems and live parts. The residual power on these circuits is dangerous even in case of power failure.

(9)Do not operate the robot teach pendant while wearing gloves. If it is necessary to control the robot manually, ensure that there are no personnel and obstacles within the motion range of the robot, and adjust the speed from slow to fast gradually to avoid personnel injury or damage to the equipment caused by sudden speed change.

(10)Handle the FlexPendant with care. Do not drop, throw or hit it to avoid damage or malfunction. During programming and testing, the enabler must be released as soon as the robot does not need to move. When not in use, the FlexPendant shall be stored on a wall bracket to ensure that it will not fall to the ground accidentally or someone doesn't trip over its cables.

(11)Before automatic execution of the program, make sure that there are no irrelevant personnel, tools or articles in the robot operation area, and that the work objects are clamped reliably and confirmed.

(12)Be sure to know the position and status of all switches, sensors and control signals that will affect the robot motion.

(13)Be sure to know the position of emergency stop buttons on the robot controller and peripheral control equipment, and get ready to use these buttons in case of emergency.

10.

Practical Applications of Industrial Robot

10.1 Basic operations of ABB robot

By operating the ABB IRB1200 industrial robot, all students are required to get familiar with its basic structure, the structure and basic operation methods of its teach pendant, and master the single−axis motion, linear motion and reorientation motion operations of the ABB robot.

The practical training is detailed as follows:

(1)Understand main technical parameters of the training robot;

Table 10−1 Main Parameters of IRB1200 Robot

Model	IRB 1200−5
Operating range	901 mm
Load	5 kg
Number of axes	6
Accuracy	0.02 mm
Installation mode	At any angle
IP rating	IP40, IP67
Controller	IRC5 compact type
Range of motion of each axis	
Axis 1 − rotating	+170°~ −170°
Axis 2 − arm motion	+130°~ −100°
Axis 3 − arm motion	+70° ~ −200°
Axis 4 − wrist motion	+270°~ −270°
Axis 5 − bending	+130°~ −130°
Axis 6 − tilting	+400°~−400°

Industrial Robot Technology Application

(2)Understand the components and working principles of the robot system;

(3)Master the operation methods for startup and shutdown of the ABB robot and know about safety precautions;

(4)Get familiar with the teach pendant(i.e., FlexPendant), master its holding method, and master its components and operation methods, especially the following:

- Status key
- Enabler
- Joystick
- Stylus pen

(5)Master the name and meaning of each of the following keys on the teach pendant:

- Preset keys(4 in total)
- Select mechanical unit
- Switch motion mode(reorient or linear)
- Switch motion mode(axes 1–3 or axes 4–6)
- Switch to incremental mode
- Step BACKWARD button
- START button
- L Step FORWARD button
- M STOP button

(6)Master the single–axis motion operation of the ABB industrial robot;

(7)Master the linear motion operation of the ABB industrial robot;

(8)Master the reorientation motion operation of the ABB industrial robot.

10.2 Updating revolution counter for robot zero point

All students are required to understand the concept of robot zero point, and master the operation method and skills for manual return to the robot zero point, as well as the operation method and precautions for updating the revolution counter for ABB robot zero point.

The practical training is detailed as follows:

(1)Understand the concept of robot zero point and know about the working

10. Practical Applications of Industrial Robot

conditions under which the robot revolution counter must be updated;

(2)Be able to manually operate the robot to quickly return to its zero point;

(3)Be able to find the synchronization marks of axes 1–6. As shown in Fig. 10–1, A~F are the mechanical zero points of axes 1–6 respectively;

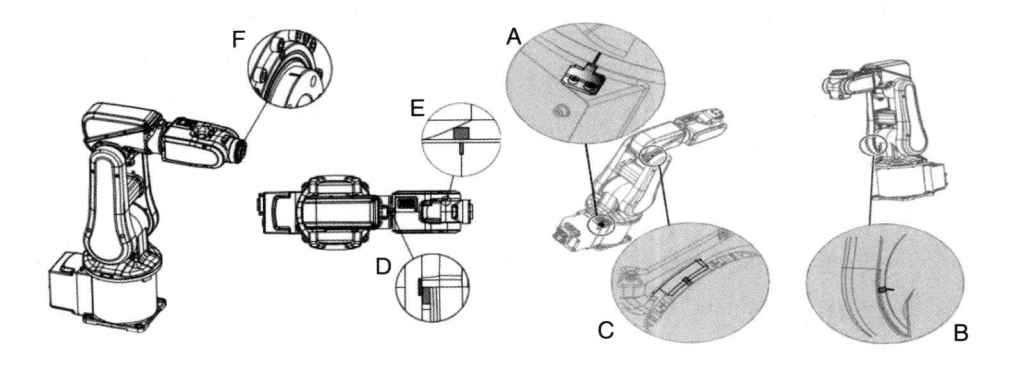

Fig. 10–1　Zero Point of Each Axis

(4)Be able to manually operate the robot to make axes 1–6 correctly return to their zero points in sequence(4→5→6→3→2→1);

(5)Know about the battery position of ABB IRB1200 robot and get familiar with the battery replacement method;

(6)Master the operation method for updating the speed counter;

(7)Master the main points for backup and restoration of the robot control system.

10.3 Configuration and monitoring of I/O signals and definitions of quickset keys

After receiving the training on this topic, all students are required to master the I/O communication types of ABB robot, the bus connection of commonly used I/O boards, and the configuration and testing methods of I/O signals.

The specific requirements are as follows:

(1)Define and correctly configure an DSQC651 I/O board in the teach pendant;

(2)Define Di1, Di2, Do1, Do2, Gi1, Go1 and Ao1 signals on the DSQC651 board;

209

Industrial Robot Technology Application

(3)Correctly connect the Do1 signal terminal to the relay shown in Fig. 10-2, and make sure that the relay is connected when Do1 outputs 1 and is disconnected when Do1 outputs 0;

(4)Correctly connect the relay to the two -position five -way solenoid valve to realize control of the parallel cylinder by Do1 signal;

(5)Correctly connect the magnetic switch sensor on the parallel cylinder to

Fig. 10-2　Relay Wiring Diagram

the Di1 signal terminal, and confirm that the cylinder is opened in position when Di1 is 1;

(6)Correctly configure shortcut keys 1 and 2 of the teach pendant to correspond to Do1 and Di1, and take testing on them.

(7)Correctly configure the Go1 signal to realize control of the pneumatic sucker;

(8)Correctly configure the connection between Ao1 terminal and 3010 DC fan(24V)to realize control of the fan voltage by Ao1 signal and achieve speed regulation;

10.4 Setting of tool coordinate system and simple teach programming

After receiving the training on this topic, all students are required to master the creation method, operation and precautions for use of tooldata for the ABB robot, master the tool center point (TCP)and coordinate axis directions of the default tool (tool0)in the system, understand the significance and setting methods of TCP, mass, center of gravity and other parameters of the tool coordinate tooldata, and master the basic operation methods for ABB teach programming.

Principles for setting the TCP are as follows:

(1)Find a very accurate fixed tip in the operating range of the robot as the fixed reference point.

10. Practical Applications of Industrial Robot

(2)Find a tip on the tool(preferably at the TCP)as the reference point.

(3)Control the robot to make the reference point on the tool move as close as possible to the fixed reference point with at least four different robot poses. Generally, the calibration accuracy can be effectively improved by selecting calibration points 1, 2, 3 and 4 in such a way that the angle between the lines passing through two adjacent calibration points is 90° and all of these calibration points are not on the same plane, as shown in Fig. 10–3.

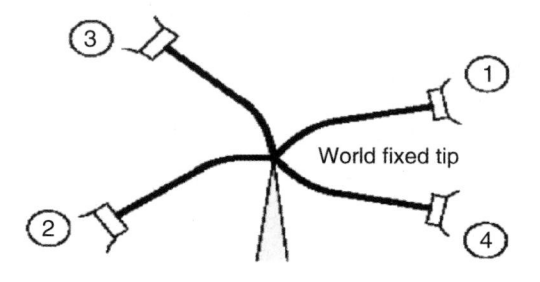

Fig. 10–3   4–Point Calibration Method

(4)Perform calculation by four sets of solutions to get the coordinates of TCP relative to the robot's flange coordinate system.

The specific requirements are as follows:

(1)Create a new tool(tool1);

(2)Select the suitable tooling to hold a 0.5 mm marker pen;

(3)Select a proper position to place the tip reference for calibration;

(4)Calibrate tool1 using the 4–point calibration method, with the error controlled to be not more than 0.5 mm;

(5)Upon completion of calibration, manually operate the robot to make it perform reorientation, select tool1, and observe position changes of the pen tip when the robot rotates around the TCP of tool1;

(6)Create a routine, select different practice templates(triangle, quadrangle, pentagon and circle)and place on the workbench, reasonably select MoveL, MoveJ and other instructions for programming, and teach each apex in sequence using the pen tip;

(7)Add the Home point for the program(with the pen pointing vertically

Industrial Robot Technology Application

downwards). Make sure that the program starts from the Home point and returns to this point again after running;

(8)Add approach and departure points. Make sure that the approach point is 30 mm above the first point taught and the departure point is 30 mm above the last point taught;

(9)Move the program pointer(PP)to the beginning of the program, and manually control the robot to complete graph drawing.

10.5 Setting of work object coordinate system and 2D offline programming

After receiving the training on this topic, all students are required to master the creation of work object coordinate data (wobjdata)and payload data of ABB robot, operate the RobotStudio software to perform 2D offline programming for motion tracks, and master the export of RobotStudio software programs and the import of ABB robot programs.

A robot can possess several work object coordinate systems, which either represent different work object positions or represent several copies of the same work object at different positions. Programming a robot is to create the robot's target point and path in the work object coordinate system.

The specific requirements are as follows:

(1)Select the calibrated tool1 described under topic 4 as the tool;

(2)Create a work object coordinate(wobj1);

(3)Use the 3-point calibration method and calibrate the rectangular workbench with the pen tip;

(4)Verification of work object coordinate accuracy: select the linear motion mode and the work object coordinate(wobj1), move each coordinate axis linearly, and check whether the coordinate change shown in the upper right corner of the teach pendant's display screen and the robot motion are consistent with expected results;

(5)Start up the RobotStudio software and define the work object coordinate system. Make sure that relevant points are close to the work object coordinate system points calibrated in Step 3, and that the included angle between X direc-

10. Practical Applications of Industrial Robot

tions of the work object coordinate system and the base coordinate system is 45°. Import the 2D graph(with hollow words).

(6)Define the tool Pen and make sure that its TCP is consistent with that of the robot tool(tool1);

(7)Perform 2D offline programming for motion tracks, and add approach and departure points.(Note that multiple groups of approach and departure points need to be added for the graphs with multiple non-connecting tracks.)

(8)Add the Home point for the program(with the pen pointing vertically downwards). Make sure that the program starts from the Home point and returns to this point again after running;

(9)Perform simulation in the RobotStudio software, and export the program after confirmation for no error;

(10)Use a USB flash disk to import the 2D graph program into the robot controller;

(11)Move the program pointer(PP)to the beginning of the program, and manually control the robot to complete graph drawing.

10.6 Design and teaching of stacking program

After receiving the training on this topic, all students are required to master the methods for using cycle instructions and coordinate offset instructions and the operation skills of point teaching for ABB robot, and solve various stacking problems encountered in actual engineering applications.

Fig. 10-4 shows an example of single-row stacking, with the material size of Φ90*8 mm. It is required that the industrial robot should use its pneumatic sucker to move the three straightly-arranged materials at position P10 to position P20 in sequence and stack them into 3 layers. The specific motion requirements are as follows:

(1)Control the robot to make its TCP move from the Home Point(jpos10)in joint motion mode to 20 mm above the middle point(P10);

(2)Wait until the robot moves to P10 in low-speed linear motion mode, and suck the material with the pneumatic sucker;

(3)Control the robot to return to 20 mm above P10 in low-speed linear mo-

213

Industrial Robot Technology Application

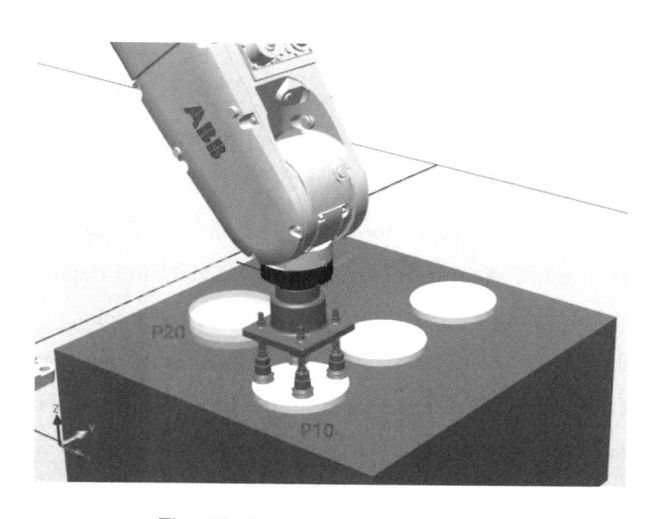

Fig. 10-4   Single-Row Stacking

tion mode, then move it in medium-speed joint motion mode to 100 mm above the target point(P20);

(4)Lower the robot to P20 in low-speed linear motion mode, release the pneumatic sucker to place the material, then quickly lift the robot by 100 mm in the Z direction to complete the handling of material on layer 1;

(5)Control the robot to quickly move to 20 mm above the second material (with a material spacing of 120 mm in the Y direction), and repeat the steps given above to complete the stacking of materials on the remaining two layers in turn;

(6)Check that the robot returns to the Home point(jpos10)after the completing the stacking operation.

It is required that the program structure of routine called out by main should be used. An example of the program structure is as follows:

    PROC main( )   //主程序
      RInit;   //初始化子程序
      SuckerTool；   //工具拾取子程序,拾取吸盘工具。
      Stack；   //工件码垛子程序
      ReleaseTool；   //工具释放子程序,释放吸盘工具。
    ENDPROC

214

10. Practical Applications of Industrial Robot

```
PROC Stack()
    MoveAbsJ jops10\NoEoffs,V500,z50,tool1;
    FOR I FROM 0 TO 2 DO
        MoveJ Offs(p10,0,I*100,20),V500,z50,tool1;          //接近点
        MoveL Offs(p10,0,I*100,0),V100,fine,tool1;          //抓取点
        WaitTime 0.5;
        Set Do1;
        WaitTime 0.5;
        MoveL Offs(p10,0,I*120,20),V100,z50,tool1;          //低速抬升
        MoveJ Offs(p10,0,I*120,100),V200,z50,tool1;         //中速关节运动
        MoveJ Offs(p20,0,0,8*I+20),V200,z50,tool1;          //中速关节运动
        MoveL Offs(p20,0,0,8*I),V100,z50,tool1;             //低速运动至放置点
        WaitTime 0.5;
        Reset Do1;
        WaitTime 0.5;
        MoveL Offs(p20,0,0,8*I+100),V500,z50,tool1;         //快速抬升
    ENDFOR
    MoveAbsJ jops10\NoEoffs,V500,z50,tool1;初始原点
ENDPROC
```

215